Open innovation in the food and beverage industry

Related titles:

Consumer-driven innovation in food and personal care products
(ISBN 978-1-84569-567-5)

Delivering performance in food supply chains (ISBN 978-1-84569-471-5)

Case studies in food product development (ISBN 978-1-84569-260-5)

Details of these books and a complete list of titles from Woodhead Publishing can be obtained by:

- visiting our web site at www.woodheadpublishing.com
- contacting Customer Services (e-mail: sales@woodheadpublishing.com; fax: +44 (0) 1223 832819; tel.: +44 (0) 1223 499140 ext. 130; address: Woodhead Publishing Limited, 80 High Street, Sawston, Cambridge CB22 3HJ, UK)
- in North America, contacting our US office (e-mail: usmarketing@woodheadpublishing.com; tel.: (215) 928 9112; address: Woodhead Publishing, 1518 Walnut Street, Suite 1100, Philadelphia, PA 19102-3406, USA)

If you would like e-versions of our content, please visit our online platform: www.woodheadpublishingonline.com. Please recommend it to your librarian so that everyone in your institution can benefit from the wealth of content on the site.

We are always happy to receive suggestions for new books from potential editors. To enquire about contributing to our Food Science, Technology and Nutrition series, please send your name, contact address and details of the topic/s you are interested in to sarah.hughes@woodheadpublishing.com. We look forward to hearing from you.

The Woodhead team responsible for publishing this book:
Commissioning Editor: Sarah Hughes
Publications Coordinator: Steven Mathews
Project Editor: Cathryn Freear
Editorial and Production Manager: Mary Campbell
Production Editor: Adam Hooper
Copyeditor: Janice Gordon
Proofreader: Trevor Birch
Cover Designer: Terry Callanan

Woodhead Publishing Series in Food Science, Technology and Nutrition:
Number 243

Open innovation in the food and beverage industry

Edited by
Marian Garcia Martinez

WOODHEAD PUBLISHING

Oxford Cambridge Philadelphia New Delhi

Published by Woodhead Publishing Limited,
80 High Street, Sawston, Cambridge CB22 3HJ, UK
www.woodheadpublishing.com
www.woodheadpublishingonline.com

Woodhead Publishing, 1518 Walnut Street, Suite 1100, Philadelphia,
PA 19102-3406, USA

Woodhead Publishing India Private Limited, G-2, Vardaan House,
7/28 Ansari Road, Daryaganj, New Delhi – 110002, India
www.woodheadpublishingindia.com

First published 2013, Woodhead Publishing Limited

British Library Cataloguing in Publication Data
A catalogue record for this book is available from the British Library.

Library of Congress Control Number: 2012952907

ISBN 978-0-85709-595-4 (print)
ISBN 978-0-85709-724-8 (online)
ISSN 2042-8049 Woodhead Publishing Series in Food Science, Technology and Nutrition (print)
ISSN 2042-8057 Woodhead Publishing Series in Food Science, Technology and Nutrition (online)

The publisher's policy is to use permanent paper from mills that operate a sustainable forestry policy, and which has been manufactured from pulp which is processed using acid-free and elemental chlorine-free practices. Furthermore, the publisher ensures that the text paper and cover board used have met acceptable environmental accreditation standards.

Typeset by Ann Buchan (Typesetters), Middlesex, UK
Printed and bound in the UK by the MPG Books Group

Contents

Contributor contact details . . . . *xiii*
Woodhead Publishing Series in Food Science, Technology and Nutrition . . . *xix*
Foreword by J. Hyman . . . . *xxvii*
Foreword by W. H. Noordman and E. M. Meijer . . . . *xxxiii*

Part I The changing nature of innovation in the food and drink industry

1 Trends in the acquisition of external knowledge for innovation in the food industry . . . 3
M. Acosta, D. Coronado and E. Ferrándiz, University of Cadiz, Spain
1.1 Introduction . . . 3
1.2 Reasons for open innovation in the food industry . . . 4
1.3 Measuring open innovation in the food industry . . . 6
1.4 Sources and types of data . . . 8
1.5 Results of the open-innovation study . . . 9
1.6 Conclusions . . . 19
1.7 Acknowledgements . . . 20
1.8 References . . . 20
1.9 Appendix: concordance between agri-food technological sectors and International Patent Classification (IPC) . . . 23

2 The tension between traditional innovation strategies and openness: Lindt's controlled open innovation approach . . . 25
V. Lazzarotti and R. Manzini, Cattaneo University – LIUC, Italy
2.1 Introduction . . . 25
2.2 Literature review . . . 26
2.3 Research method for Lindt case study . . . 28
2.4 Open and closed innovation at Lindt . . . 29

2.5 Lindt's open-innovation approach in practice: the innovation project Noccior 32
2.6 Results of controlled open innovation in the Lindt case 34
2.7 Conclusions 36
2.8 References 36

3 The role of open innovation in the industry convergence between foods and pharmaceuticals 39
S. Bröring, University of Applied Sciences Osnabrück, Germany and Wageningen University and Research Centre, The Netherlands
3.1 Introduction 39
3.2 A brief literature review on industry convergence 40
3.3 Convergence-related challenges and the role of open innovation 43
3.4 Evidence for industry convergence between foods and pharmaceuticals 47
3.5 Open innovation in order to cope with convergence in the neutraceuticals and functional foods (NFF) sector 51
3.6 Conclusion 56
3.7 Future trends 58
3.8 References 59

4 Accelerating the innovation cycle through intermediation: the case of Kraft's melt-proof chocolate bars 63
R. Wielens, NineSigma, Belgium
4.1 Introduction 63
4.2 From research to search in company innovation 64
4.3 Key capabilities in open innovation 65
4.4 From idea-driven innovation to need-driven innovation 68
4.5 Case study: melt-proof chocolate bars from Kraft 68
4.6 Conclusions 70
4.7 Future trends 71
4.8 References 72

5 The impact of open innovation on innovation performance: the case of Spanish agri-food firms 74
C. Bayona-Sáez, T. García-Marco and M. Sanchez-García, Universidad Pública de Navarra, Spain and C. Cruz-Càzares, University of Barcelona, Spain
5.1 Introduction: the agri-food sector and innovation 74
5.2 How innovative are Spanish agri-food firms? 77
5.3 Measuring open innovation in Spanish agri-food firms 80
5.4 The effect of openness on the innovative performance of firms 85
5.5 Conclusions 91
5.6 References 93

Part II Partners and networks for open innovation

6 Partnering with public research centres and private technical and scientific service providers for innovation: the case of Italian rice company, Riso Scotti 97
V. Lazzarotti and R. Manzini, Cattaneo University – LIUC, Italy
6.1 Introduction 97
6.2 The role of private technical and scientific service (TSS) providers: advantages and limitations 99
6.3 The role of universities and public research centres: advantages and limitations 100
6.4 Riso Scotti case study 101
6.5 Conclusions and managerial implications 105
6.6 References 107

7 Consumers as part of food and beverage industry innovation 109
S. E. Kemp, Consultant, UK
7.1 Introduction 109
7.2 Understanding food and beverage consumers and their world 111
7.3 Consumer-centric company culture for innovation 115
7.4 Consumer-driven innovation process 117
7.5 Consumers as co-creators 127
7.6 Conclusion 132
7.7 Future trends 133
7.8 Sources of further information and advice 134
7.9 References 135

8 Co-creation of value with consumers as an innovation strategy in the food and beverage industry: the case of Molson Coors' 'talking can' 139
M. Garcia Martinez, Kent Business School, University of Kent, UK
8.1 Introduction 139
8.2 Co-creation of value with consumers 141
8.3 Research design 142
8.4 Molson-Coors Brewing Company UK (MCBC-UK): the need for consumer-driven innovation 142
8.5 The discover style of open innovation 145
8.6 The 'talking can': co-creating value with consumers 146
8.7 Discussion 149
8.8 Conclusion 151
8.9 References 151

9 Collaborative product innovation in the food service industry. Do too many cooks really spoil the broth? **154**
A. I. A. Costa, The Catholic University of Portugal, Portugal
9.1 Introduction 154
9.2 A review of open-innovation practices in the food industry 156
9.3 Collaborative product innovation (CPI) in the food-service industry: the path of diffusion of sous vide cooking in the US 160
9.4 Conclusions and future trends 170
9.5 References 171

10 Effectiveness of cluster organizations in facilitating open innovation in regional innovation systems: the case of Food Valley in the Netherlands **174**
S. W. F. Omta, Wageningen University and Research Centre, The Netherlands and F. T. J. M. Fortuin, Food Valley Organization, The Netherlands and Wageningen University and Research Centre, The Netherlands
10.1 Introduction 174
10.2 Theoretical background 175
10.3 The Dutch agri-food sector and Food Valley Organization 177
10.4 Conclusions 183
10.5 Future trends 184
10.6 Sources of further information and advice 184
10.7 References 185
10.8 Appendix 1: the four main functions of Food Valley, including fifteen services, activities and information sources 186
10.9 Appendix 2: assessment of small to medium-sized enterprises and large companies of the importance of Food Valley functions 187
10.10 Appendix 3: member company assessment of the importance of Food Valley functions by company type 188

11 The importance of networks for knowledge exchange and innovation in the food industry **189**
B. Kühne, V. Lefebvre, C. Cochez and X. Gellynck, Ghent University, Belgium
11.1 Introduction 189
11.2 Knowledge exchange and innovation and the importance of networks 190
11.3 Network methodology: a case study approach 195
11.4 Results of the three Flemish case studies 198
11.5 Conclusions and future trends 207
11.6 References 208

Part III Establishing and managing open-innovation partnerships and networks

12 Managing open-innovation communities: the development of an open-innovation community scorecard 215
I. Blohm, University of St. Gallen, Switzerland, J. M. Leimeister, Kassel University, Germany and H. Krcmar, Technische Universität München, Germany
12.1 Introduction 215
12.2 Introduction to open-innovation communities and their management 217
12.3 Development of an open-innovation community scorecard 219
12.4 Implementation of the open-innovation scorecard 228
12.5 Conclusion and future trends 229
12.6 References 230
12.7 Appendix 1: data sources of success measures 232
12.8 Appendix 2: member survey 234

13 The evolution of partnering in open innovation: from transactions to communities 235
K. G. McFarthing, Innovation Fixer Ltd, UK
13.1 Introduction 235
13.2 Identifying and securing partners 236
13.3 Building and structuring relationships 239
13.4 Ecosystems 243
13.5 Human factors 246
13.6 Building a community 249
13.7 Conclusion 251
13.8 Acknowledgements 252
13.9 References 252

14 Managing co-innovation partnerships: the case of Unilever and its preferred flavour suppliers 254
M. Tepic and S. W. F. Omta, Wageningen University and Research Centre, The Netherlands, F. T. J. M. Fortuin, Food Valley Organization, The Netherlands and Wageningen University and Research Centre, The Netherlands and A. Saris, Wageningen University and Research Centre, The Netherlands
14.1 Introduction 254
14.2 Co-innovation 256
14.3 The co-innovation partnership between Unilever and flavour suppliers 260
14.4 Implementation and development of the Flavour Operating Framework partnership 262
14.5 Conclusion 268
14.6 Future trends 272
14.7 References 272

15 Managing asymmetric relationships in open innovation: lessons from multinational companies and SMEs **276**
D. Oughton, University of Cambridge, UK and L. Mortara and T. Minshall, IfM Centre for Technology Management, UK
15.1 Introduction: the importance of large and small company partnerships in the food industry 276
15.2 The difficulties of open innovation 278
15.3 Culture, complexity and communication problems 281
15.4 The importance for companies of focusing on risk, reward and balance 283
15.5 Overcoming obstacles to achieve successful company partnerships 284
15.6 Collaborations between companies: case studies 287
15.7 Conclusion 289
15.8 Acknowledgements 291
15.9 References 292

16 Challenges faced by multinational food and beverage corporations when forming strategic external networks for open innovation ... **294**
L. Tolhurst and J. Brown, Oakland Innovation, UK
16.1 Introduction 294
16.2 Strategic external networks for open innovation 295
16.3 Research methodology 298
16.4 Findings 299
16.5 Discussion 304
16.6 Future trends 305
16.7 Conclusions and recommendations 305
16.8 References 310

Part IV Open innovation tools, process and managerial frameworks

17 The 'want find get manage' (WFGM) framework for open-innovation management and its use by Mars, Incorporated **315**
M. Garcia Martinez, Kent Business School, University of Kent, UK
17.1 Introduction 315
17.2 History of open innovation at Mars, Incorporated 317
17.3 Mars' open-innovation model 317
17.4 The open-innovation framework: 'want find get manage' 322
17.5 Conclusions 329
17.6 References 330

18 Crowdsourcing: the potential of online communities as a tool for data analysis **332**
M. Garcia Martinez and B. Walton, Kent Business School, University of Kent, UK
18.1 Introduction 332
18.2 Predictive modelling competitions 333

18.3 Design and management of predictive modelling competitions . 335
18.4 Case study: Kaggle 338
18.5 Conclusions 340
18.6 References 340

19 The role of information systems in innovative food and beverage organizations 343
J. G. Caudill, University of Tennessee, USA
19.1 Introduction 343
19.2 The role of technology in innovation 344
19.3 Innovative technologies in agriculture and food production . . . 345
19.4 Technology's support of innovation 347
19.5 Free tools for innovation 352
19.6 Future trends 353
19.7 Conclusion 353
19.8 Sources of further information and advice 354
19.9 References 354

20 Effective organizational and managerial frameworks to support open innovation: overview and the case of Heinz 356
V. Lazzarotti and R. Manzini, Cattaneo University – LIUC, Italy
20.1 Introduction 356
20.2 The need for organizational and management tools to support open innovation 357
20.3 Case study: Heinz's strategy, business and organization 360
20.4 Conclusions and managerial implications 366
20.5 References 367

21 Innovating with brains: the psychology of open innovation 369
B. A. Sabel, University of Magdeburg Medical School, Germany and Institute of Automation, China, K. Sabel, Consulting-Berlin, Germany and J. T. Preston, TEM Capital and Continuum Energy Technologies, USA
21.1 Introduction 369
21.2 Innovation is all about psychology 370
21.3 Phases of innovation 371
21.4 The influence of soft factors on the success of innovation 373
21.5 The psychology of the innovation team 380
21.6 The innovative environment of academia 382
21.7 Start-ups and small to medium-sized enterprises (SMEs): open innovation by default 385
21.8 Predicting innovation success: the 'Preston' equation 385
21.9 Future trends 386
21.10 References 387

Index 389

Contributor contact details

(* = main contact)

Editor and chapters 8 and 17

Dr Marian Garcia Martinez
Kent Business School
University of Kent
Canterbury
CT2 7PE
UK

E-mail: m.garcia@kent.ac.uk

Chapter 1

Manuel Acosta*, Daniel Coronado and Esther Ferrándiz
Departamento de Economía General
University of Cadiz. Campus de Excelencia Internacional Agroalimentario, ceiA3
Glorieta Carlos Cano s/n
11002 Cádiz
Spain

E-mail: manuel.acosta@uca.es

Chapters 2, 6 and 20

Assistant Professor Valentina Lazzarotti* and Raffaella Manzini
Cattaneo University - LIUC
Uni.TIS, Management Engineering
Corso Matteotti 22, 21052
Castellanza
Varese
Italy

E-mail: vlazzarotti@liuc.it; rmanzini@liuc.it

Chapter 3

Professor Stefanie Bröring
University of Applied Sciences Osnabrück
Oldenburger Landstr. 62
49090 Osnabrück
Germany

E-mail: s.broering@hs-osnabrueck.de

and

Management Studies/Business Administration
Wageningen University and Research Centre
P.O. Box 8130
NL-6700 EW Wageningen
The Netherlands

Chapter 4

Rick Wielens
NineSigma
Koning Leopold 1 Straat 3
3000 Leuven
Belgium

E-mail: wielens@ninesigma.com

Chapter 5

Cristina Bayona-Sáez, Teresa García-Marco* and Mercedes Sanchez-García
Universidad Pública de Navarra
Departamento de Gestión de Empresas
Campus de Arrosadía s/n
31006 Pamplona
Spain
E-mail: bayona@unavarra.es; tgmarco@unavarra.es; mersan@unavarra.es

Claudio Cruz-Cázares
University of Barcelona
Diagonal Avenue 690
Tower 2, 3rd floor, 08034
Barcelona
Spain
E-mail: claudio.cruz@ub.es

Chapter 7

Dr Sarah E. Kemp
Consultant
31 Crow Hill
Broadstairs
Kent
CT10 1HT
UK
E-mail: sarahekemp@btinternet.com

Chapter 9

Ana I. A. Costa
Católica-Lisbon School of Business and Economics
Universidade Católica Portuguesa
Palma de Cima
1649-023 Lisboa
Portugal
E-mail: anacosta@clsbe.lisboa.ucp.pt

Chapter 10

Professor Dr S.W.F. (Onno) Omta*
Management Studies/Business Administration
Wageningen University and Research Centre
P.O. Box 8130
NL-6700 EW Wageningen
The Netherlands
E-mail: onno.omta@wur.nl

Dr Frances T. J. M. Fortuin
Food Valley Organization
Nieuwe Kanaal 9D-3
P.O. Box 294
6700 AG Wageningen
The Netherlands
E-mail: frances.fortuin@foodvalley.nl

Management Studies/Business Administration
Wageningen University and Research Centre
P.O. Box 8130
NL-6700 EW Wageningen
The Netherlands

Chapter 11

Dr Bianka Kühne*, Virginie Lefebvre, Carl Cochez and Xavier Gellynck
Ghent University
Department of Agricultural Economics
Division Agri-Food Marketing and Chain Management
Coupure Links 653
B-9000 Gent
Belgium
E-mail: Bianka.Kuhne@UGent.be

Chapter 12

Ivo Blohm*
University of St. Gallen
Institute of Information Management
Müller-Friedberg-Strasse 8
9000 St. Gallen
Switzerland

E-mail: ivo.blohm@unisg.ch

Professor Dr Jan Marco Leimeister
Kassel University
Chair of Information Systems
Nora-Platiel-Str. 4
34137 Kassel
Germany

E-mail: leimeister@uni-kassel.de

Professor Dr Helmut Krcmar
Technische Universität München
Chair of Information Systems (I17)
Boltzmannstr. 3
85748 Garching b. München
Germany

E-mail: krcmar@in.tum.de

Chapter 13

Dr Kevin G. McFarthing
Innovation Fixer Ltd
62 Lark Hill
Oxford
OX2 7DR
UK

E-mail: kevin.mcfarthing@innovationfixer.com

Chapter 14

Dr Mersiha Tepic
Management Studies/Business Administration
Wageningen University and Research Centre
P.O. Box 8130
NL-6700 EW Wageningen
The Netherlands

E-mail: Mersiha.tepic@wur.nl

Professor Dr S.W.F. (Onno) Omta*
Management Studies/Business Administration
Wageningen University and Research Centre
P.O. Box 8130
NL-6700 EW Wageningen
The Netherlands

E-mail: onno.omta@wur.nl

Dr Frances T. J. M. Fortuin
Food Valley Organization
Nieuwe Kanaal 9D-3
P.O. Box 294
6700 AG Wageningen
The Netherlands

E-mail: frances.fortuin@foodvalley.nl

and

Management Studies/Business Administration
Wageningen University and Research Centre
P.O. Box 8130
NL-6700 EW Wageningen
The Netherlands

Anne Saris
Management Studies/Business Administration
Wageningen University and Research Centre
P.O. Box 8130
NL-6700 EW Wageningen
The Netherlands

Chapter 15

Dominic Oughton*
Principal Industrial Fellow
IfM Education and Consultancy Services
Institute for Manufacturing
University of Cambridge
Charles Babbage Road
Cambridge
CB3 0FS
UK

E-mail: do251@cam.ac.uk

Dr Letizia Mortara
Research Associate
IfM Centre for Technology Management
Cambridge
UK

E-mail: lm367@cam.ac.uk

Dr Tim Minshall
Senior Lecturer in Technology Management
IfM Centre for Technology Management
Cambridge
UK

E-mail: thwm100@cam.ac.uk

Chapter 16

Jennifer Brown
Oakland Innovation Ltd
328/329 Cambridge Science Park
Milton Road
Cambridge
CB4 0WG
UK

E-mail: jb@oakland.co.uk

Lukie Tolhurst*
Oakland Innovation Ltd
328/329 Cambridge Science Park
Milton Road
Cambridge
CB4 0WG
UK

E-mail: lucinda.tolhurst@oakland.co.uk

Chapter 18

Marian Garcia Martinez* and Bryn Walton
Kent Business School
University of Kent
Canterbury
Kent
CT2 7PE
UK

E-mail: m.garcia@kent.ac.uk; bcmw@kent.ac.uk

Chapter 19

Dr J. G. Caudill
525 Jane and David Bailey Educational Complex
University of Tennessee
1122 Volunteer Boulevard
Knoxville
TN 37996
USA

E-mail: jason.caudill@gmail.com

Chapter 21

Dr Bernhard A. Sabel*
Institute of Medical Psychology
University of Magdeburg Medical School
39120 Magdeburg
Germany

E-mail: Bernhard.Sabel@med.ovgu.de

Neural Imaging Computation and Analysis Group (NICA)
State Key Laboratory of Management and Control for Complex Systems
Institute of Automation
Chinese Academy of Science
ZhongGuanCun, East Rd. 95#
Beijing, 100190
China

Kornelia Sabel
Consulting-Berlin
Fürstenstrasse 5
14163 Berlin
Germany

E-mail: Kornelia.sabel@consulting-berlin.de

John T. Preston
Transformative Energy and Materials Capital, LLC (TEM Capital)
Boston, MA
USA

E-mail: Preston@mit.edu

Woodhead Publishing Series in Food Science, Technology and Nutrition

1 **Chilled foods: a comprehensive guide** *Edited by C. Dennis and M. Stringer*
2 **Yoghurt: science and technology** *A. Y. Tamime and R. K. Robinson*
3 **Food processing technology: principles and practice** *P. J. Fellows*
4 **Bender's dictionary of nutrition and food technology Sixth edition** *D. A. Bender*
5 **Determination of veterinary residues in food** *Edited by N. T. Crosby*
6 **Food contaminants: sources and surveillance** *Edited by C. Creaser and R. Purchase*
7 **Nitrates and nitrites in food and water** *Edited by M. J. Hill*
8 **Pesticide chemistry and bioscience: the food-environment challenge** *Edited by G. T. Brooks and T. Roberts*
9 **Pesticides: developments, impacts and controls** *Edited by G. A. Best and A. D. Ruthven*
10 **Dietary fibre: chemical and biological aspects** *Edited by D. A. T. Southgate, K. W. Waldron, I. T. Johnson and G. R. Fenwick*
11 **Vitamins and minerals in health and nutrition** *M. Tolonen*
12 **Technology of biscuits, crackers and cookies Second edition** *D. Manley*
13 **Instrumentation and sensors for the food industry** *Edited by E. Kress-Rogers*
14 **Food and cancer prevention: chemical and biological aspects** *Edited by K. W. Waldron, I. T. Johnson and G. R. Fenwick*
15 **Food colloids: proteins, lipids and polysaccharides** *Edited by E. Dickinson and B. Bergenstahl*
16 **Food emulsions and foams** *Edited by E. Dickinson*
17 **Maillard reactions in chemistry, food and health** *Edited by T. P. Labuza, V. Monnier, J. Baynes and J. O'Brien*
18 **The Maillard reaction in foods and medicine** *Edited by J. O'Brien, H. E. Nursten, M. J. Crabbe and J. M. Ames*
19 **Encapsulation and controlled release** *Edited by D. R. Karsa and R. A. Stephenson*
20 **Flavours and fragrances** *Edited by A. D. Swift*
21 **Feta and related cheeses** *Edited by A. Y. Tamime and R. K. Robinson*
22 **Biochemistry of milk products** *Edited by A. T. Andrews and J. R. Varley*
23 **Physical properties of foods and food processing systems** *M. J. Lewis*

24 **Food irradiation: a reference guide** *V. M. Wilkinson and G. Gould*
25 **Kent's technology of cereals: an introduction for students of food science and agriculture Fourth edition** *N. L. Kent and A. D. Evers*
26 **Biosensors for food analysis** *Edited by A. O. Scott*
27 **Separation processes in the food and biotechnology industries: principles and applications** *Edited by A. S. Grandison and M. J. Lewis*
28 **Handbook of indices of food quality and authenticity** *R. S. Singhal, P. K. Kulkarni and D. V. Rege*
29 **Principles and practices for the safe processing of foods** *D. A. Shapton and N. F. Shapton*
30 **Biscuit, cookie and cracker manufacturing manuals Volume 1: ingredients** *D. Manley*
31 **Biscuit, cookie and cracker manufacturing manuals Volume 2: biscuit doughs** *D. Manley*
32 **Biscuit, cookie and cracker manufacturing manuals Volume 3: biscuit dough piece forming** *D. Manley*
33 **Biscuit, cookie and cracker manufacturing manuals Volume 4: baking and cooling of biscuits** *D. Manley*
34 **Biscuit, cookie and cracker manufacturing manuals Volume 5: secondary processing in biscuit manufacturing** *D. Manley*
35 **Biscuit, cookie and cracker manufacturing manuals Volume 6: biscuit packaging and storage** *D. Manley*
36 **Practical dehydration Second edition** *M. Greensmith*
37 **Lawrie's meat science Sixth edition** *R. A. Lawrie*
38 **Yoghurt: science and technology Second edition** *A. Y. Tamime and R. K. Robinson*
39 **New ingredients in food processing: biochemistry and agriculture** *G. Linden and D. Lorient*
40 **Benders' dictionary of nutrition and food technology Seventh edition** *D. A. Bender and A. E. Bender*
41 **Technology of biscuits, crackers and cookies Third edition** *D. Manley*
42 **Food processing technology: principles and practice Second edition** *P. J. Fellows*
43 **Managing frozen foods** *Edited by C. J. Kennedy*
44 **Handbook of hydrocolloids** *Edited by G. O. Phillips and P. A. Williams*
45 **Food labelling** *Edited by J. R. Blanchfield*
46 **Cereal biotechnology** *Edited by P. C. Morris and J. H. Bryce*
47 **Food intolerance and the food industry** *Edited by T. Dean*
48 **The stability and shelf-life of food** *Edited by D. Kilcast and P. Subramaniam*
49 **Functional foods: concept to product** *Edited by G. R. Gibson and C. M. Williams*
50 **Chilled foods: a comprehensive guide Second edition** *Edited by M. Stringer and C. Dennis*
51 **HACCP in the meat industry** *Edited by M. Brown*
52 **Biscuit, cracker and cookie recipes for the food industry** *D. Manley*
53 **Cereals processing technology** *Edited by G. Owens*
54 **Baking problems solved** *S. P. Cauvain and L. S. Young*
55 **Thermal technologies in food processing** *Edited by P. Richardson*
56 **Frying: improving quality** *Edited by J. B. Rossell*
57 **Food chemical safety Volume 1: contaminants** *Edited by D. Watson*
58 **Making the most of HACCP: learning from others' experience** *Edited by T. Mayes and S. Mortimore*
59 **Food process modelling** *Edited by L. M. M. Tijskens, M. L. A. T. M. Hertog and B. M. Nicolaï*
60 **EU food law: a practical guide** *Edited by K. Goodburn*

61 **Extrusion cooking: technologies and applications** *Edited by R. Guy*
62 **Auditing in the food industry: from safety and quality to environmental and other audits** *Edited by M. Dillon and C. Griffith*
63 **Handbook of herbs and spices Volume 1** *Edited by K. V. Peter*
64 **Food product development: maximising success** *M. Earle, R. Earle and A. Anderson*
65 **Instrumentation and sensors for the food industry Second edition** *Edited by E. Kress-Rogers and C. J. B. Brimelow*
66 **Food chemical safety Volume 2: additives** *Edited by D. Watson*
67 **Fruit and vegetable biotechnology** *Edited by V. Valpuesta*
68 **Foodborne pathogens: hazards, risk analysis and control** *Edited by C. de W. Blackburn and P. J. McClure*
69 **Meat refrigeration** *S. J. James and C. James*
70 **Lockhart and Wiseman's crop husbandry Eighth edition** *H. J. S. Finch, A. M. Samuel and G. P. F. Lane*
71 **Safety and quality issues in fish processing** *Edited by H. A. Bremner*
72 **Minimal processing technologies in the food industries** *Edited by T. Ohlsson and N. Bengtsson*
73 **Fruit and vegetable processing: improving quality** *Edited by W. Jongen*
74 **The nutrition handbook for food processors** *Edited by C. J. K. Henry and C. Chapman*
75 **Colour in food: improving quality** *Edited by D. MacDougall*
76 **Meat processing: improving quality** *Edited by J. P. Kerry, J. F. Kerry and D. A. Ledward*
77 **Microbiological risk assessment in food processing** *Edited by M. Brown and M. Stringer*
78 **Performance functional foods** *Edited by D. Watson*
79 **Functional dairy products Volume 1** *Edited by T. Mattila-Sandholm and M. Saarela*
80 **Taints and off-flavours in foods** *Edited by B. Baigrie*
81 **Yeasts in food** *Edited by T. Boekhout and V. Robert*
82 **Phytochemical functional foods** *Edited by I. T. Johnson and G. Williamson*
83 **Novel food packaging techniques** *Edited by R. Ahvenainen*
84 **Detecting pathogens in food** *Edited by T. A. McMeekin*
85 **Natural antimicrobials for the minimal processing of foods** *Edited by S. Roller*
86 **Texture in food Volume 1: semi-solid foods** *Edited by B. M. McKenna*
87 **Dairy processing: improving quality** *Edited by G. Smit*
88 **Hygiene in food processing: principles and practice** *Edited by H. L. M. Lelieveld, M. A. Mostert, B. White and J. Holah*
89 **Rapid and on-line instrumentation for food quality assurance** *Edited by I. Tothill*
90 **Sausage manufacture: principles and practice** *E. Essien*
91 **Environmentally-friendly food processing** *Edited by B. Mattsson and U. Sonesson*
92 **Bread making: improving quality** *Edited by S. P. Cauvain*
93 **Food preservation techniques** *Edited by P. Zeuthen and L. Bøgh-Sørensen*
94 **Food authenticity and traceability** *Edited by M. Lees*
95 **Analytical methods for food additives** *R. Wood, L. Foster, A. Damant and P. Key*
96 **Handbook of herbs and spices Volume 2** *Edited by K. V. Peter*
97 **Texture in food Volume 2: solid foods** *Edited by D. Kilcast*
98 **Proteins in food processing** *Edited by R. Yada*
99 **Detecting foreign bodies in food** *Edited by M. Edwards*

100 **Understanding and measuring the shelf-life of food** *Edited by R. Steele*
101 **Poultry meat processing and quality** *Edited by G. Mead*
102 **Functional foods, ageing and degenerative disease** *Edited by C. Remacle and B. Reusens*
103 **Mycotoxins in food: detection and control** *Edited by N. Magan and M. Olsen*
104 **Improving the thermal processing of foods** *Edited by P. Richardson*
105 **Pesticide, veterinary and other residues in food** *Edited by D. Watson*
106 **Starch in food: structure, functions and applications** *Edited by A.-C. Eliasson*
107 **Functional foods, cardiovascular disease and diabetes** *Edited by A. Arnoldi*
108 **Brewing: science and practice** *D. E. Briggs, P. A. Brookes, R. Stevens and C. A. Boulton*
109 **Using cereal science and technology for the benefit of consumers: proceedings of the 12th International ICC Cereal and Bread Congress, 24 – 26th May, 2004, Harrogate, UK** *Edited by S. P. Cauvain, L. S. Young and S. Salmon*
110 **Improving the safety of fresh meat** *Edited by J. Sofos*
111 **Understanding pathogen behaviour: virulence, stress response and resistance** *Edited by M. Griffiths*
112 **The microwave processing of foods** *Edited by H. Schubert and M. Regier*
113 **Food safety control in the poultry industry** *Edited by G. Mead*
114 **Improving the safety of fresh fruit and vegetables** *Edited by W. Jongen*
115 **Food, diet and obesity** *Edited by D. Mela*
116 **Handbook of hygiene control in the food industry** *Edited by H. L. M. Lelieveld, M. A. Mostert and J. Holah*
117 **Detecting allergens in food** *Edited by S. Koppelman and S. Hefle*
118 **Improving the fat content of foods** *Edited by C. Williams and J. Buttriss*
119 **Improving traceability in food processing and distribution** *Edited by I. Smith and A. Furness*
120 **Flavour in food** *Edited by A. Voilley and P. Etievant*
121 **The Chorleywood bread process** *S. P. Cauvain and L. S. Young*
122 **Food spoilage microorganisms** *Edited by C. de W. Blackburn*
123 **Emerging foodborne pathogens** *Edited by Y. Motarjemi and M. Adams*
124 **Benders' dictionary of nutrition and food technology Eighth edition** *D. A. Bender*
125 **Optimising sweet taste in foods** *Edited by W. J. Spillane*
126 **Brewing: new technologies** *Edited by C. Bamforth*
127 **Handbook of herbs and spices Volume 3** *Edited by K. V. Peter*
128 **Lawrie's meat science Seventh edition** *R. A. Lawrie in collaboration with D. A. Ledward*
129 **Modifying lipids for use in food** *Edited by F. Gunstone*
130 **Meat products handbook: practical science and technology** *G. Feiner*
131 **Food consumption and disease risk: consumer-pathogen interactions** *Edited by M. Potter*
132 **Acrylamide and other hazardous compounds in heat-treated foods** *Edited by K. Skog and J. Alexander*
133 **Managing allergens in food** *Edited by C. Mills, H. Wichers and K. Hoffman-Sommergruber*
134 **Microbiological analysis of red meat, poultry and eggs** *Edited by G. Mead*
135 **Maximising the value of marine by-products** *Edited by F. Shahidi*
136 **Chemical migration and food contact materials** *Edited by K. Barnes, R. Sinclair and D. Watson*
137 **Understanding consumers of food products** *Edited by L. Frewer and H. van Trijp*
138 **Reducing salt in foods: practical strategies** *Edited by D. Kilcast and F. Angus*

139 **Modelling microorganisms in food** *Edited by S. Brul, S. Van Gerwen and M. Zwietering*
140 **Tamime and Robinson's Yoghurt: science and technology Third edition** *A. Y. Tamime and R. K. Robinson*
141 **Handbook of waste management and co-product recovery in food processing Volume 1** *Edited by K. W. Waldron*
142 **Improving the flavour of cheese** *Edited by B. Weimer*
143 **Novel food ingredients for weight control** *Edited by C. J. K. Henry*
144 **Consumer-led food product development** *Edited by H. MacFie*
145 **Functional dairy products Volume 2** *Edited by M. Saarela*
146 **Modifying flavour in food** *Edited by A. J. Taylor and J. Hort*
147 **Cheese problems solved** *Edited by P. L. H. McSweeney*
148 **Handbook of organic food safety and quality** *Edited by J. Cooper, C. Leifert and U. Niggli*
149 **Understanding and controlling the microstructure of complex foods** *Edited by D. J. McClements*
150 **Novel enzyme technology for food applications** *Edited by R. Rastall*
151 **Food preservation by pulsed electric fields: from research to application** *Edited by H. L. M. Lelieveld and S. W. H. de Haan*
152 **Technology of functional cereal products** *Edited by B. R. Hamaker*
153 **Case studies in food product development** *Edited by M. Earle and R. Earle*
154 **Delivery and controlled release of bioactives in foods and nutraceuticals** *Edited by N. Garti*
155 **Fruit and vegetable flavour: recent advances and future prospects** *Edited by B. Brückner and S. G. Wyllie*
156 **Food fortification and supplementation: technological, safety and regulatory aspects** *Edited by P. Berry Ottaway*
157 **Improving the health-promoting properties of fruit and vegetable products** *Edited by F. A. Tomás-Barberán and M. I. Gil*
158 **Improving seafood products for the consumer** *Edited by T. Børresen*
159 **In-pack processed foods: improving quality** *Edited by P. Richardson*
160 **Handbook of water and energy management in food processing** *Edited by J. Klemeš, R.. Smith and J.-K. Kim*
161 **Environmentally compatible food packaging** *Edited by E. Chiellini*
162 **Improving farmed fish quality and safety** *Edited by Ø. Lie*
163 **Carbohydrate-active enzymes** *Edited by K.-H. Park*
164 **Chilled foods: a comprehensive guide Third edition** *Edited by M. Brown*
165 **Food for the ageing population** *Edited by M. M. Raats, C. P. G. M. de Groot and W. A Van Staveren*
166 **Improving the sensory and nutritional quality of fresh meat** *Edited by J. P. Kerry and D. A. Ledward*
167 **Shellfish safety and quality** *Edited by S. E. Shumway and G. E. Rodrick*
168 **Functional and speciality beverage technology** *Edited by P. Paquin*
169 **Functional foods: principles and technology** *M. Guo*
170 **Endocrine-disrupting chemicals in food** *Edited by I. Shaw*
171 **Meals in science and practice: interdisciplinary research and business applications** *Edited by H. L. Meiselman*
172 **Food constituents and oral health: current status and future prospects** *Edited by M. Wilson*
173 **Handbook of hydrocolloids Second edition** *Edited by G. O. Phillips and P. A. Williams*
174 **Food processing technology: principles and practice Third edition** *P. J. Fellows*

175 **Science and technology of enrobed and filled chocolate, confectionery and bakery products** *Edited by G. Talbot*
176 **Foodborne pathogens: hazards, risk analysis and control Second edition** *Edited by C. de W. Blackburn and P. J. McClure*
177 **Designing functional foods: measuring and controlling food structure breakdown and absorption** *Edited by D. J. McClements and E. A. Decker*
178 **New technologies in aquaculture: improving production efficiency, quality and environmental management** *Edited by G. Burnell and G. Allan*
179 **More baking problems solved** *S. P. Cauvain and L. S. Young*
180 **Soft drink and fruit juice problems solved** *P. Ashurst and R. Hargitt*
181 **Biofilms in the food and beverage industries** *Edited by P. M. Fratamico, B. A. Annous and N. W. Gunther*
182 **Dairy-derived ingredients: food and neutraceutical uses** *Edited by M. Corredig*
183 **Handbook of waste management and co-product recovery in food processing Volume 2** *Edited by K. W. Waldron*
184 **Innovations in food labelling** *Edited by J. Albert*
185 **Delivering performance in food supply chains** *Edited by C. Mena and G. Stevens*
186 **Chemical deterioration and physical instability of food and beverages** *Edited by L. H. Skibsted, J. Risbo and M. L. Andersen*
187 **Managing wine quality Volume 1: viticulture and wine quality** *Edited by A. G. Reynolds*
188 **Improving the safety and quality of milk Volume 1: milk production and processing** *Edited by M. Griffiths*
189 **Improving the safety and quality of milk Volume 2: improving quality in milk products** *Edited by M. Griffiths*
190 **Cereal grains: assessing and managing quality** *Edited by C. Wrigley and I. Batey*
191 **Sensory analysis for food and beverage quality control: a practical guide** *Edited by D. Kilcast*
192 **Managing wine quality Volume 2: oenology and wine quality** *Edited by A. G. Reynolds*
193 **Winemaking problems solved** *Edited by C. E. Butzke*
194 **Environmental assessment and management in the food industry** *Edited by U. Sonesson, J. Berlin and F. Ziegler*
195 **Consumer-driven innovation in food and personal care products** *Edited by S. R. Jaeger and H. MacFie*
196 **Tracing pathogens in the food chain** *Edited by S. Brul, P.M. Fratamico and T.A. McMeekin*
197 **Case studies in novel food processing technologies: innovations in processing, packaging, and predictive modelling** *Edited by C. J. Doona, K. Kustin and F. E. Feeherry*
198 **Freeze-drying of pharmaceutical and food products** *T.-C. Hua, B.-L. Liu and H. Zhang*
199 **Oxidation in foods and beverages and antioxidant applications Volume 1: understanding mechanisms of oxidation and antioxidant activity** *Edited by E. A. Decker, R. J. Elias and D. J. McClements*
200 **Oxidation in foods and beverages and antioxidant applications Volume 2: management in different industry sectors** *Edited by E. A. Decker, R. J. Elias and D. J. McClements*
201 **Protective cultures, antimicrobial metabolites and bacteriophages for food and beverage biopreservation** *Edited by C. Lacroix*
202 **Separation, extraction and concentration processes in the food, beverage and nutraceutical industries** *Edited by S. S. H. Rizvi*

203 **Determining mycotoxins and mycotoxigenic fungi in food and feed** *Edited by S. De Saeger*
204 **Developing children's food products** *Edited by D. Kilcast and F. Angus*
205 **Functional foods: concept to product Second edition** *Edited by M. Saarela*
206 **Postharvest biology and technology of tropical and subtropical fruits Volume 1: fundamental issues** *Edited by E. M. Yahia*
207 **Postharvest biology and technology of tropical and subtropical fruits Volume 2: açai to citrus** *Edited by E. M. Yahia*
208 **Postharvest biology and technology of tropical and subtropical fruits Volume 3: cocona to mango** *Edited by E. M. Yahia*
209 **Postharvest biology and technology of tropical and subtropical fruits Volume 4: mangosteen to white sapote** *Edited by E. M. Yahia*
210 **Food and beverage stability and shelf life** *Edited by D. Kilcast and P. Subramaniam*
211 **Processed meats: improving safety, nutrition and quality** *Edited by J. P. Kerry and J. F. Kerry*
212 **Food chain integrity: a holistic approach to food traceability, safety, quality and authenticity** *Edited by J. Hoorfar, K. Jordan, F. Butler and R. Prugger*
213 **Improving the safety and quality of eggs and egg products Volume 1** *Edited by Y. Nys, M. Bain and F. Van Immerseel*
214 **Improving the safety and quality of eggs and egg products Volume 2** *Edited by F. Van Immerseel, Y. Nys and M. Bain*
215 **Animal feed contamination: effects on livestock and food safety** *Edited by J. Fink-Gremmels*
216 **Hygienic design of food factories** *Edited by J. Holah and H. L. M. Lelieveld*
217 **Manley's technology of biscuits, crackers and cookies Fourth edition** *Edited by D. Manley*
218 **Nanotechnology in the food, beverage and nutraceutical industries** *Edited by Q. Huang*
219 **Rice quality: a guide to rice properties and analysis** *K. R. Bhattacharya*
220 **Advances in meat, poultry and seafood packaging** *Edited by J. P. Kerry*
221 **Reducing saturated fats in foods** *Edited by G. Talbot*
222 **Handbook of food proteins** *Edited by G. O. Phillips and P. A. Williams*
223 **Lifetime nutritional influences on cognition, behaviour and psychiatric illness** *Edited by D. Benton*
224 **Food machinery for the production of cereal foods, snack foods and confectionery** *L.-M. Cheng*
225 **Alcoholic beverages: sensory evaluation and consumer research** *Edited by J. Piggott*
226 **Extrusion problems solved: food, pet food and feed** *M. N. Riaz and G. J. Rokey*
227 **Handbook of herbs and spices Second edition Volume 1** *Edited by K. V. Peter*
228 **Handbook of herbs and spices Second edition Volume 2** *Edited by K. V. Peter*
229 **Breadmaking: improving quality Second edition** *Edited by S. P. Cauvain*
230 **Emerging food packaging technologies: principles and practice** *Edited by K. L. Yam and D. S. Lee*
231 **Infectious disease in aquaculture: prevention and control** *Edited by B. Austin*
232 **Diet, immunity and inflammation** *Edited by P. C. Calder and P. Yaqoob*
233 **Natural food additives, ingredients and flavourings** *Edited by D. Baines and R. Seal*
234 **Microbial decontamination in the food industry: novel methods and applications** *Edited by A. Demirci and M.O. Ngadi*
235 **Chemical contaminants and residues in foods** *Edited by D. Schrenk*
236 **Robotics and automation in the food industry: current and future technologies** *Edited by D. G. Caldwell*

237 **Fibre-rich and wholegrain foods: improving quality** *Edited by J. A. Delcour and K. Poutanen*
238 **Computer vision technology in the food and beverage industries** *Edited by D.-W. Sun*
239 **Encapsulation technologies and delivery systems for food ingredients and nutraceuticals** *Edited by N. Garti and D. J. McClements*
240 **Case studies in food safety and authenticity** *Edited by J. Hoorfar*
241 **Heat treatment for insect control: developments and applications** *D. Hammond*
242 **Advances in aquaculture hatchery technology** *Edited by G. Allan and G. Burnell*
243 **Open innovation in the food and beverage industry** *Edited by M. Garcia Martinez*
244 **Trends in packaging of food, beverages and other fast-moving consumer goods (FMCG)** *Edited by N. Farmer*
245 **New analytical approaches for verifying the origin of food** *Edited by P. Brereton*
246 **Microbial production of food ingredients, enzymes and nutraceuticals** *Edited by B. McNeil, D. Archer, I. Giavasis and L. Harvey*
247 **Persistent organic pollutants and toxic metals in foods** *Edited by M. Rose and A. Fernandes*
248 **Cereal grains for the food and beverage industries** *E. Arendt and E. Zannini*
249 **Viruses in food and water: risks, surveillance and control** *Edited by N. Cook*
250 **Improving the safety and quality of nuts** *Edited by L. J. Harris*
251 **Metabolomics in food and nutrition** *Edited by B. Weimer and C. Slupsky*
252 **Food enrichment with omega-3 fatty acids** *Edited by C. Jacobsen, N. Skall Nielsen, A. Frisenfeldt Horn and A.-D. Moltke Sørensen*
253 **Instrumental assessment of food sensory quality: a practical guide** *Edited by D. Kilcast*
254 **Food microstructures: microscopy, measurement and modelling** *Edited by V. J. Morris and K. Groves*
255 **Handbook of food powders: processes and properties** *Edited by B. R. Bhandari, N. Bansal, M. Zhang and P. Schuck*
256 **Functional ingredients from algae for foods and nutraceuticals** *Edited by H. Domínguez*
257 **Satiation, satiety and the control of food intake: theory and practice** *Edited by J. E. Blundell and F. Bellisle*
258 **Hygiene in food processing: principles and practice Second edition** *Edited by H. L. M. Lelieveld, J. Holah and D. Napper*

Foreword by J. Hyman

My working definition of innovation is: new ways of adding value to *all* of our business activities. Needless to say there are a lot of alternative and equally good definitions. In my definition *new* can mean: *new* to me, *new* to the industry, *new* to the world or even *new* to the universe. And the newness can be small and incremental or large and radical.

Although my definition includes *all* of our business processes, my own work has focused on new product development (NPD), so it is this experience I will share with you here.

The term 'open innovation' was coined by Henry Chesbrough at the turn of this century. Chesbrough, based in California, exhorted US companies to look a bit further afield than their own workforces when seeking expertise. What he described was already common practice in many countries, but because the USA had a heritage of vertical integration where car companies would extract their own iron ore, and grow their own rubber trees, Chesbrough had to prove that outsourcing could be the way forward if the collaboration was managed carefully and the parties trusted each other. His idea was not rocket science, but it was very well expressed.

Since then we have become a 'webbed-up-world' where experience and know-how is a click away and intellectual property (IP) is safely traded on corporate versions of ebay.

In order to understand and explain my thoughts I have grouped what I know into four fluid headings. Many of the items could show up under more than one of these titles:

1. corporate innovation;
2. NPD process innovation;
3. channel innovation; and
4. consumer innovation.

Corporate innovation

In today's business space, where everything can be subcontracted, the one thing that must be done by the top team is the development of the corporate strategy, which is done by blending global and local trends and fusing them, artfully, with the company's past reputation and future intentions. Even here, agencies and consultants assist with providing information and guidance. The top team also provides the leadership, which entails setting and maintaining standards. The company's marketing strategy develops from the corporate strategy and provides the framework for the NPD strategy.

However, let me put a cautionary word in here. Because the NPD strategy should be looking three to five years ahead, it must be compatible with the corporate and marketing plans, but it cannot be dependent on an annual planning process. Hardwiring the NPD plan to the annual marketing plan is the big mistake a lot of well-informed companies continue to make and it leads to a sudden need for fast-track NPD, rather than a slower-track, more considered, more inventive, more capital hungry, more protectable and more profitable NPD. Most importantly for this book, the top team sets the outsourcing policy which determines the amount of open innovation a company can and will do.

The current business fashion is for mergers and acquisitions, and this means there are fewer companies managing larger numbers of products and brands. These larger companies eventually trim their product tails and focus on their power brands. Their smaller brands are then sold to specialist companies which can revive them by adding back the resource and focus they need.

NPD process innovation

The past twenty years has seen companies establishing their own processes for innovation. On closer examination these 'processes' are project templates with some formalised reviews along the way which means that good projects are flagged through whereas bad projects are killed quickly.

Here is a typical process, which is more-or-less linear although there is sometimes rework to do when things go wrong:

Strategy development → product and brand strategy → consumer insight finding → idea generation → concept development → prototyping → scale up → launch → review → continuous improvement

For a company to maintain competitive advantage the challenge is to make each phase of the project better, quicker and cheaper, and open innovation, has a lot to offer here. However, it is actually humans who run the processes, and therefore it is people who determine the success or failure of projects. There are four essential human elements to be remembered, not just for NPD but for a company's approach to innovation generally.

The first element is to include everyone. I believe it is essential to avoid having innovation specialists and champions. It is better to have a whole company regard itself as innovative and, thankfully, it is easy to teach the basics.

The second element is to include everything. This means the company becomes used to improving the way it does everything from source to destination. The Japanese call this 'Kaizen' meaning continuous improvement in all we do. The company culture becomes accustomed to supplier innovation, transport innovation, accounting innovation and selling innovation. This is why it is pointless giving companies 'innovation' awards based only on their visible NPD. For all we know the underlying processes might be archaic.

The third element is to have an easy method for innovation. The best I've come across is Alex Osborn's creative problem solving (CPS) process, originally published in 1953, but which has stood the test of time. (*Applied imagination: principles and procedures of creative problem solving*, 1979). This process works on any innovation challenge and consists of six segments:

- objective finding
- fact finding
- challenge finding
- idea finding
- solution finding
- acceptance finding.

There is a vast amount of literature on the CPS process and the fundamental methodology is in the public domain. Using CPS transforms an organisation's ability to innovate.

The fourth element needed if our organisations are to survive and flourish is a climate for innovation. This is not a soft measure, it is a hard metric and, in 1971, Swedish Professor Goran Ekvall identified a number of dimensions which he has correlated with financial success (*Creativity at the place of work: a study of suggestors and suggestion systems in the Swedish mechanical industry*, 1971). Roughly translated Ekvall's dimensions are:

- trust and openness
- challenge and involvement
- freedom
- idea time
- playfulness and humour
- low levels of conflict
- idea support
- healthy levels of debate
- risk-taking.

Finally, how do we know whether we are doing the right amount of NPD? My rule-of-thumb, based on my work with The Deming Institute is that a mature company should have 10% of its turnover coming from products that are less than 24 months old. Any lower than this and the company might be stagnating; any more than this and the core might be decaying too quickly.

Channel innovation

Most of my work has been with the UK grocery trade. My universe has been fast-moving-consumer-goods (FMCG) sold through a handful of major companies with massive outlets and huge concentrations of power. However, every trend has its counter-trend and nowadays large out-of-town hypermarkets are not getting the planning permissions they used to because local authorities are favouring the high streets again. This is causing big retailers to create high-street-friendly formats and collaborate more with local suppliers. At the same time on-line grocery shopping is growing and buying from the web will have an impact on product development in ways we cannot yet imagine.

Therefore, we need to consider our options without going through a shopping list. Here are six key questions which will help us consider the subject:

- What ... are we supplying? This covers our product list and how we should be de-listing the slow-sellers as we add new items.
- Where ... are our customers going to want it. This covers in-home, in-office, in-car and in-restaurants.
- When ... are our customers going to eat or drink it? In a 24/7 society, the idea of day-parts makes planning easier. Here are some obvious segments to consider: morning, mid-morning, lunch, afternoon, evening meal, pre-bed time. These segments vary between mid-week and weekend.
- How ... are we going to manufacture and transport it and how are the end-users going to eat it? Formats include: fresh, ambient, chilled, frozen.
- Why ... should our consumers want it? This is probably the most important question. And even more importantly why should they want our product over that of our competitors?
- Who ... are our target shoppers, cooks and eaters? In an increasingly individualised yet globalised world, these questions are becoming harder to answer.

Consumer innovation

To paraphrase fifty years' of marketing wisdom, the job of business is to get and keep customers and the job of marketing and NPD is to understand consumers' present and future needs. In our information-drenched world, consumers are busily developing their own strategies to outfox suppliers. Users have become savvy to the extent that they can easily be engaged all along the NPD process to help us develop the right propositions. However, this is also bad news, as consumers are now very sceptical of what we produce!

Today's consumer research cannot rely on old-style, overt questioning. Instead, we have to use covert behavioural observation. Although behavioural economics has become an essential marketing skill. Watching people in their own homes through ethnographic filming and self-reporting (using smartphones) is becoming a given. An observation here is that marketing is the one department where outsourcing has always been the norm. Marketers were the first to recognise that they needed specialist non-staffers to provide research analysis and creative thinking. As a result, marketers are good at outsourcing and are therefore a good example for studying how to do open innovation.

Now that we've finished the preliminaries let's delve into the book. It is an excellent read and full of fascinating case-studies on how and when to consider open innovation, how to get it right and how to avoid getting it wrong. There are also some very robust observations leading to some strong innovation insights. Also because of its academic approach there is a rich list of citations and sources for further reading. For a marketer like me, the book is a lesson in precision.

Interestingly the book itself is an excellent example of open innovation. The publisher commissioned the editor to commission the authors, and in many cases no one has actually met face-to-face. In my case, I downloaded the chapters, read them and wrote this foreword; I will upload my work to a shared web box and the editor will review and revise if necessary.

Thus, with goodwill and collaboration you now have a book on open innovation in the food and drink industry. I hope you enjoy it as much as I have.

Jeffrey Hyman
Chairman
The Food & Drink Innovation Network|www.fdin.org.uk
Sharing innovation best practice in the UK food & drink industry

Foreword by W. H. Noordman and E. M. Meijer

Open innovation should be an integral part of any innovative company's innovation strategy and its way of working in innovation. Open innovation delivers significant benefits. Opportunities and challenges differ between companies of different size, industry and strategic orientation.

Changing nature of innovation

Changes in the business environment in the food and beverage industry impact the innovation needs of companies and how companies organize innovation. The world is becoming more and more connected and competition is fierce; consumer trends travel quickly and product lifecycles shorten. Moreover, research and technology has a global footprint and access to leading scientists and their capabilities becomes more and more competitive. In addition, legislation is becoming increasingly strict, limiting the room for innovation in the food industry.

In Europe's mature market, which faces a severe economic downturn, the innovation demands on companies change quickly. Furthermore, the food and beverage industry faces real challenges in the coming decades relating to sustainability, feeding the increasing world's population, addressing the needs of an aging population, and scarcity of resources. Food and beverage companies cannot address these challenges on their own using the competences on which they relied in the past. External drivers are, therefore, required for breakthrough innovations or for entering adjacent market segments (Bröring, Chapter 3).

And then there are the sheer numbers: only a fraction of the world's technological potential works for a given company and many employees do not stay with the same company for their whole career. Therefore, in this globally connected world, knowledge cannot be managed internally. This underscores the need for opening up the innovation process and for collaboration. Excelling in collaboration can foster innovation and become a differentiating factor; it is up to individual companies to determine how to set up innovative systems and make them fit for their company's purpose.

Open innovation in the food and beverage industry

The food and beverage industry in general is seen as a low-tech industry with a dominance of incremental rather than radical innovation and a relatively low innovation rate. However, open innovation is widely adopted. In maturity, food and beverage companies might be lagging slightly behind several other industry sectors such as electronics and automotive. However, the challenges faced by food and beverage companies with respect to open innovation are the same challenges as in other industries.

Changing nature in open innovation

Open innovation is not new on the agenda. A clear trend can be observed from closed innovation in the 'golden age of corporate R&D labs' (the post-war period up to the 80s) to increasingly 'open' in later part of the 20th century. In that period, many companies, following anticipated benefits and logical promises, rushed into open innovation without clearly defining their priorities. This has led to considerable disappointment. Failure rates of open-innovation alliances are known to be high, maybe up to a staggering 40 to 70%. Reasons include lack of strategic alignment, insufficient communication, cultural mismatch, high expectations that cannot be met, an inability to generate results, or even rivalry. To counter these factors calls for a thorough identification of business priorities (Garcia Martinez, Chapter 8) and, subsequently, for careful partner selection and partnership management. Therefore, the nature of open innovation currently is changing towards more intense and longer term partnerships, embedded in networks and ecosystems with a substantial degree of cohesion with respect to strategic drivers.

Building intense partnerships requires significant investments, monetary as well as in (management) time, looking further than the success of the (first) joint development project. It is extremely important to be clear on the partnering objective: is it to generate knowledge, to absorb competences, or to co-innovate? Although strategic and operational fit is essential, it is as important to focus on the softer topics such as how partners work together, and how to stimulate collaborative behavior. Because companies try to build intense partnerships, the open-innovation market becomes more exclusive and, hence, more competitive.

By building partnerships, companies become part of an innovation network. An important type of formal innovation network is the public private partnership (PPP), such as the Dutch Technological Top Institute (TTI) for Food and Nutrition. Such PPPs allow companies to build discriminating power for the future, and they perform the function of training potential future staff. Governments, companies and knowledge institutes (together called 'the golden triangle') join forces to invest in strategic research that enables exploitation of identified economic and social growth opportunities.

Industry and governments alike recognize that players have to be brought together to foster innovation. Food Valley (Chapter 10) and Flemish organizations like Flanders FOOD (Chapter 11) are examples of effective organizations that serve this purpose. Although virtual ecosystems were popular a decade ago, exemplified by consortia formed for example by European Framework programs, we currently see a trend towards colocation in so-called physical ecosystems. Bringing parties geographically close together maximizes informal interaction and boosts creativity across institutional borders. Industry is consolidating its own R&D centers (Unilever, Danone, FrieslandCampina), opening up their main sites for other companies to create open-innovation campuses (e.g. the Philips high-tech campus in Eindhoven or other examples mentioned in Chapter 13) or locate their R&D site in a relevant ecosystem. For example, FrieslandCampina is building its new innovation center on the campus of Wageningen University and Research Center, in the heart of the 'ecosystem' Food Valley.

Differences in open innovation

Companies of different size, industry and strategic orientation have different open-innovation needs and face different opportunities and challenges. Most companies, including Kraft, General Mills and Unilever (chapters 13 and 14) see the prime importance of customer–supplier partnerships to fulfil innovation demands and have ample experience in managing this type of partnership. Many companies feel the need to involve consumers more intimately, particularly earlier in their innovations, and they establish a co-creation partnership with them (Kemp, Chapter 7, Garcia Martinez, Chapter 8). Food companies are experimenting how best to do this.

However, there are large differences in the different steps in the innovation process. At one end of the spectrum, in the implementation phase, there is direct valorization by consortia, which often consist primarily of small to medium-sized enterprises (SMEs). SMEs are mainly active in the valorization of knowledge. They may require help in network formation with (international) partners, technology transfer, and finding access to knowledge and subsidies (Omta and Fortuin, Chapter 10; Kuhne, Chapter 11). Governments that aim to stimulate economic development need to pay attention to these needs. In The Netherlands, SMEs make up 80% of the AgroFood industry, and, to a large extent, determine the innovation power of this sector as a whole. Large companies are also active in valorization, but, in addition, they have a need to develop leading knowledge in emerging science areas to pave the way for subsequent innovations. For this reason, large companies join forces in the area of strategic research, the upstream part of the innovation process. The challenge here, from the perspective of a company, is the clear formulation of a demand and the absorption of generated knowledge. This requires a relatively deep internal competence base and, therefore, this type of open innovation is almost exclusively the domain of large companies.

Another main distinction is seen in the strategic orientation of a business: is it playing offense or defense? Even amongst business groups within one company these differences are apparent, indicating different priorities with respect to the intensity and breadth of open-innovation partnerships.

Implementation of open innovation

Companies need to find their way on how, in practice, to implement open innovation within their organization. First, it is to necessary define what is fit for purpose in the context of the company in question and, then, to set up a tailored open-innovation strategy or framework. It is all about determining the priorities. The challenge, thereafter, is to implement the innovation. Change needs to happen on various levels: organization, processes and tools, and, most importantly, mindset.

On the organizational level, it is clear that open innovation should not be located within the domain of an R&D department in isolation. Open innovation should be embraced by the company, and, therefore, top management support is required. The involvement of all disciplines involved in innovation is instrumental. Only firms where R&D and commercial disciplines are fully aligned will be successful in setting up partnerships with customers, suppliers or competitors. Many companies choose to assign a central open-innovation function or team. However, the role of such a team varies widely, as can be seen from the diversity in ways Mars (Chapter 17), Kraft (Chapter 13), General Mills (Chapter 13), and Heinz (Chapter 20) set up such a team. Some companies assign this team with specific strategic tasks such as strategic management or co-ordination of the network, others with specific operational tasks such as executing or outsourcing scouting activities for the business groups or handling incoming ideas (e.g., Heinz, Chapter 20). Others make this function responsible for both strategic and operational topics, or give this team a facilitating role, including gathering and sharing best practices and training.

Organizations can profit from adopting specific open-innovation tools and processes, not only to make open innovation more efficient, but mainly to communicate, adopt and improve the best practices, as also indicated by Garcia Martinez in Chapter 17. For instance, the 'want, find, get, manage' model has been adopted by many companies, including Mars (Chapter 17), Unilever and FrieslandCampina, as a framework to internally as well as externally align the steps to take. Using such a model, companies can develop specific tools that help decision making, for instance, decisions on 'make, buy or ally' or on partner selection. Companies may want to use processes or tools to manage their business and technology needs and share them with partners or eventually with the whole world, or structure their scouting process. Successful organizations develop their ability to attract best ideas from the network and to ensure proper follow up. For this reason, companies may set up internet portals to attract ideas from outside, or use third parties' help. Internal training programs also help to create a common language and awareness. No matter how sophisticated the processes and tools are, we believe that open-innovation needs to be seen in the perspective of the total innovation capabilities of the company. In other words, open innovation will not fly if the internal innovation process does not function properly.

A change in mindset probably is the most difficult part in successfully implementing open innovation; from 'not invented here' to 'proudly found elsewhere'; from being closed to controlled sharing of innovation strategies and roadmaps and knowledge with partners; from individual internal R&D groups working in isolation to internally connecting to make the broad network available and share best practices. With respect to mind set, the big corporations might face a bigger challenge than smaller companies (SMEs) owing to the fact that they have been used to find solutions internally because their internal capabilities are huge.

This book gives an excellent overview of the state of the art regarding open innovation in the food and beverage industry and is a great source of guidance for those involved in open-innovation dynamics.

W. H. Noordman, E. M. Meijer
FrieslandCampina, Wageningen, The Netherlands

Part I

The changing nature of innovation in the food and drink industry

1

Trends in the acquisition of external knowledge for innovation in the food industry

M. Acosta, D. Coronado and E. Ferrándiz, University of Cadiz, Spain

DOI: 10.1533/9780857097248.1.3

Abstract: Within the framework of 'open innovation', a number of key issues related to the acquisition of external knowledge in food technology are addressed. The methodology relies on patents as a measure of the generation of new knowledge, and patent citations as a proxy for external knowledge acquisition towards innovation in food technology. The main results reveal different patterns of knowledge acquisition depending on the sector, and strong differences across countries. In addition, in relative terms, our data showed few differences in knowledge acquisition patterns between large firms (large in terms of their capacity to develop many patented technologies) and smaller firms.

Key words: food technology, food patents, food firms, non-patent and patent citations.

1.1 Introduction

Scientific and technological knowledge acquisition from such organizations as other firms, universities and research institutes is a mechanism that may help firms to create, develop or improve their own innovations. This process of incorporating new knowledge from other institutions can be framed as the open-innovation paradigm. According to the open-innovation model, firms incorporate external as well as internal ideas, and internal and external paths to market, as they look to advance their technology (Chesbrough, 2003, p. 24). The original concept of open innovation coined by Chesbrough has been refined by the same author to denote the use of purposive inflows and outflows of knowledge, to accelerate internal innovation and expand the markets for external use of innovation, respectively (Chesbrough, 2006).

The open-innovation model assumes that competitive advantage through the development of new technologies or improvement of current ones can stem from the discoveries or ideas of others. This means that companies need not and should not rely exclusively on their own R&D (Chesbrough, 2006). Since Chesbrough's seminal work, a considerable number of studies have analysed the open-innovation process at various levels, including at firm, industry and region levels (see van de Vrande *et al.*, 2009 for a review), and new trends and directions have been identified (see, for example, Gassmann *et al.*, 2010). Most analyses of the topic agree that the paradigm of open innovation is of particular interest to food sectors because of their many chain and network ties, but this potential has not been fully utilized (Saguy, 2011). Furthermore, there is scant evidence of the use of this potent tool in this sector (Fortuin and Omta, 2009).

A defining characteristic of the open-innovation model is the 'revealing' of inventions, findings, discoveries and knowledge (von Hippel and von Krogh, 2003, von Hippel and von Krogh, 2006, Gassmann *et al.*, 2010). 'Revealing' occurs when the available useful knowledge is transferred to others through compensation (e.g., licensing) or through a 'free revealing' process of spillovers for which there is no compensation to the innovator (e.g., published patents or papers, or informal contacts with researchers). In this chapter, we analyse the particular dimension of the open-innovation process by which firms incorporate new knowledge developed by others without paying any compensation. The focus is the acquisition of freely available knowledge by food firms to create or develop their own patented technologies. To explore this aspect of the open-innovation process in food technology, we address the following four key questions:

1. What are the sources of knowledge for patented food technology?
2. What kind of scientific and technological knowledge is the most widely incorporated in food technology?
3. Are there differences across countries or firms?
4. Has there been a growing trend towards open innovation in patented food technology?

The chapter is organized as follows. Section 1.2 reviews the relevant literature on the importance of the acquisition of knowledge in food sectors. Section 1.3 describes the methodology to measure open innovation. Section 1.4 explains the data and sources. Section 1.5 provides answers to the main research questions. Section 1.6 summarizes the main results and notes some policy implications.

1.2 Reasons for open innovation in the food industry

A comprehensive analysis of the entire open-innovation process in food sectors is beyond the scope of this chapter, but it is worth noting that as a general framework, the open-innovation paradigm provides an interesting theoretical umbrella. In the following paragraphs, we summarize several reasons why food firms rely on external scientific and technological knowledge to support their own innovation.

First, firms in the food industry have several well-known characteristics that distinguish them from firms in other industries:

1. They operate in a mature and relatively low-technology sector (Christensen *et al.*, 1996, Grunert *et al.*, 1997).
2. Very often, food innovations are initially developed in smaller companies that may lack the know-how needed to commercialize them (Fryer and Versteeg, 2008).
3. Patented food technology is highly concentrated in a few firms, most of which are multinational companies (Alfranca *et al.*, 2004).
4. Innovation in food sectors is usually incremental rather than radical, mostly comprising improvements or variations of existing products (Galizzi and Venturini, 1996, Grunert *et al.*, 1997).
5. Therefore, the food industry brings to the marketplace the benefits of research conducted further upstream (Garcia Martinez and Briz, 2000, Traill and Meulenberg, 2002).

Second, as a result of the characteristics summarized above, the food industry is thought to draw not only on R&D but also on interactions with other actors, during which process the technological and scientific knowledge produced outside the firm plays an important role. The New Food Economy (Kinsey, 2001, Weaver, 2008) partially explains this process. In the New Food Economy, innovation is generated in networks in which value is created through productive working relationships or collaboration (Avermaete and Viaene, 2002). Several studies have pointed out that enterprises in food sectors are highly dependent on external sources of information for innovation and hence must open their innovation process to their network (Avermaete *et al.*, 2004, Sarkar and Costa, 2008). External sources of knowledge include business relationships, access to research providers and collaboration infrastructure (Avermaete and Viaene, 2002, Bröring, 2008). This enhances the capacity of firms to capture knowledge from their environment and incorporate new knowledge, which could explain the number of innovations in food sectors. The rare empirical research on this topic in food sectors confirms that firms in this industry benefit more from spillovers from outside the industry than from those within it (see Ramani *et al.*, 2008).

Third, previous empirical research has shown that the knowledge useful for food firms originates in multiple scientific and technological sectors, such as pharmaceuticals, chemicals and agriculture (Galizzi and Venturini, 1996, Wilkinson, 1998, Connor, 1988), machinery, tools and electrical product sectors (Johnson and Evenson, 1999) and emergent scientific fields, such as nanotechnology (Sanguansri and Augustin, 2006, Kalpana Sastry *et al.*, 2009) or biotechnology (Levidow and Bijman, 2002, Carew, 2005).

Despite the importance of previous scientific and technological knowledge for innovation in food firms, there is still scant empirical evidence on the science–technology links in food sectors. Alfranca *et al.* (2004) analysed knowledge flows within multinational food companies, concluding that food and chemical capabilities within this industry are significantly related. The paper by Acosta and Coronado (2003) examined the science–technology flows and revealed multiple relationships between industrial sectors, including between the food industry and scientific fields. This previous research has shed light on some important issues of innovation and knowledge diffusion in food sectors, but to our knowledge, only Acosta *et al.* (2011) have used patent citations (PCs) to analyse the use of scientific knowledge by food firms.

Finally, the progressive pathway to new models of innovation in which the relationships between various institutions (such as firms, universities, laboratories and governments) are emphasized is strongly related to the open-innovation process in food firms. The science–technology relationships in food sectors are strongly linked to debates about the efficient organization of innovation processes. Interest in the knowledge base of innovations in food sectors relies on the expansion of the systemic approach to improve the organization of the production and exchange of scientific and technological knowledge. The assumptions necessary to understand the systemic approach are derived from modern theories of innovation based on evolutionary propositions, as originally described by Nelson and Winter (1982). According to this perspective, knowledge as a resource and interactive learning are both regarded as fundamental aspects of the process (Lundvall and Johnson, 1994). New organizational forms have appeared throughout the institutional sphere, research centres, industry and government, which demonstrate the importance of knowledge and the flow of learning in processes of economic growth and social transformation. These forms include national, regional or sectoral systems of innovation (Lundvall, 1992, Nelson, 1993, Malerba, 2002). In this regard, a number of papers have used concepts from the systems-of-innovation framework to shed light on agri-food research and innovation processes (Arias-Aranda and Romerosa-Martínez, 2010, Klerkx and Leeuwis, 2008, Sumberg, 2005). In line with the systemic approach, these studies highlighted the importance of relationships among the components of the system (such as governments, firms, universities and research centres, or intermediaries) to enhance food innovation in both developing and developed countries. However, the implementation and organization of the system are not easy tasks, and it is not always possible to extrapolate experiences to yield more generalized insights because of such factors as differences in culture, specialisation patterns or knowledge bases.

1.3 Measuring open innovation in the food industry

Dahlander and Gann (2010) developed an analytical framework by structuring the process of open innovation in two dimensions: inbound/outbound (Chesbrough, 2006, Gassmann and Enkel, 2004) and pecuniary/non-pecuniary. Inbound open innovation is an outside–in process and involves opening the innovation process to knowledge exploration. External knowledge exploration refers to the acquisition of knowledge from external sources. By contrast, outbound open innovation is an inside–out process and includes opening the innovation process to knowledge exploitation (Lichtenthaler, 2011). Dahlander and Gann's framework is a good theoretical starting point for analysing the process of acquisition of external knowledge in food firms using the information included in patents. In particular, the methodology we follow to analyse the scientific and technological base of innovations in food sectors relies on the analysis of PCs and non-patent citations (NPCs) in patent documents; the latter can include such sources as scientific publications, conference proceedings, books or technical manuals.

Many studies have shown that the analysis of patents is a valid and objective method for analysing the processes of innovation and technology transfer. However, patents have several drawbacks as a technological indicator. For example, firms differ in their propensity to patent their technologies (revealed as differences in the number of patents per unit of R&D expenditure, or in raw numbers of patent applications). Not all inventions are patentable, and some specific types of technology are not necessarily patented (see Griliches, 1990 for a review). To this general argument, we must add the particular characteristics of the sector under analysis. As is well known, the food industry is not especially prone to patenting, and most patented technology is developed by large firms (Alfranca *et al.*, 2002, 2004). Despite these limitations, patent data provide a useful indicator of firms' new technological developments, and the technological and scientific information included in patents reveals how food firms incorporate this internal and external knowledge to facilitate innovation.

Following the same assumption as in other recent studies, we use citations in patent documents as a proxy variable to represent a particular unit or item of scientific/technological knowledge that is required or considered useful for the development of patented technology (McMillan *et al.*, 2000, Meyer, 2000, Meyer, 2002, Tijssen, 2002, Verbeek *et al.*, 2002).

Patented technology generally incorporates previous public and private knowledge. Consequently, in patent documents, as in scientific articles, it is usual to provide references or citations to describe the antecedents to, or 'state of the art' before, the invention. This state of the art includes other patents on which the invention has built, but also bibliographical references to scientific literature and technical publications. These citations are indicators of the potential contribution of the published knowledge (articles or patents) to the patented inventions. However, the essential question to justify the procedure followed in this study is whether the citations in patent documents capture the incorporation of scientific and technological knowledge by food firms. On the one hand, PCs have been widely used as indicators, for example to track technological relationships among inventions (Lee and Su, 2011) and knowledge flows among firms (Jaffe *et al.*, 1993). On the other hand, scientific citations in patent documents, or non-patent citations (NPCs), are more likely to be carefully selected, because of the patent examiners' control and for legal reasons, than are references in scientific papers (Collins and Wyatt, 1988, Verbeek *et al.*, 2002). Although there are other motives affecting the presence or frequency of NPCs in patents that are not related to interactions with science, such as the different citation practices of national patent offices, the aggregate studies seem to confirm that these citations measure the intensity of the science supporting the innovations patented. For a review of the NPC procedure, see Acosta and Coronado, 2003.

On the basis of the above arguments, we conclude that citations in patent documents are useful but partial indicators for analysing how food firms capture external knowledge. Scientific citations or PCs only reflect the part of codified knowledge that is utilized as a source of ideas, analytical methods and data, but this methodology does not consider other interesting sources.

1.4 Sources and types of data

The empirical data used in this study comprise a set of 19 511 European patent applications to the European Patent Office (EPO) and are included in the EPO Worldwide Patent Statistical Database (PATSTAT, September 2009).[1] We have used patent applications to the EPO to measure innovation output in food sectors for two main reasons. First, European patents are expected to be more technologically significant than are those registered with national patent offices (Del Barrio-Castro and García-Quevedo, 2005) because they involve greater costs. Additionally, although applications for European patents can also be filed with other international and national offices, we limit our sample to European patents filed with the EPO. Our motivation is to circumvent the well-known differences in citation practices across patent offices (Bacchiocchi and Montobbio, 2010) in order to facilitate international comparisons.

The procedure we applied to trace the acquisition of knowledge supporting patented food technologies followed the steps below.

First, we obtained patent applications to the EPO related to food sectors with a priority date in the period 1998 to 2006. In this step, we retrieved those patents with International Patent Classification (IPC) codes included in either the agriculture, food chemistry or agricultural and food processing, machinery and apparatus sectors, following the technology classification of ISI-OST-INPI (Hinze *et al.*, 1997). The appendix includes all the technology in these sectors and their IPC equivalences. Then, we linked these patents to information on applicants and inventors and limited the sample to those patents for which at least one of the applicants or inventors was a firm. For this purpose, we used the keywords suggested in van Looy *et al.* (2006). It is worth mentioning that the OECD's Patent Statistics Manual recommends using the 'patent family' (the set of patents filed in several countries that are related to each other by one or several common priority filings) instead of individual patents for this type of analysis. However, this could lead to misleading conclusions because older patent families are likely to have more applications and related documents, and thus more citations; more applications imply more citations because sometimes the review of the state of the art is repeated in each patent of each family. To avoid this bias, we have considered only the first European patent of the family. Furthermore, we controlled for self-citations.

[1]For a review of the NPC procedure, see Acosta and Coronado (2003) and Acosta *et al.* (2005).

Table 1.1 Number of first European patent applications in agri-food patent families by year (1998–2006)

	1998	1999	2000	2001	2002	2003	2004	2005	2006	Total (98–06)
Sector 14	1013	1055	1085	1075	1175	1127	1183	1225	1055	9993
Sector 25	855	1009	1057	1028	977	959	914	956	906	8661
Sector 14–25 jointly	97	129	114	121	106	72	85	72	61	857

Source: Own elaboration based on Patstat data.

Second, we linked these patents to their scientific and technological backward citations, and identified whether these citations were added by the applicant or the examiner, or during other phases of the application procedure (e.g., during the opposition phase). For the purpose of this work, we have focused on the information provided only for the applicants. The patents cited by the applicant were also classified into technological fields. For that purpose, we matched these cited patents to their IPC codes in PATSTAT. For those patents whose IPC codes were not available in PATSTAT, we retrieved the patent details from Espacenet, which is a free online patent database offered by the EPO. Then, we grouped these IPC codes into six broader technological fields according to the technology classification of the ISI-OST-INPI, updated in February 2005 (see Schmoch, 2008 for details).

Finally, we assigned each patent to its corresponding company/companies in order to obtain patent statistics at the firm and country levels. Owing to the lack of normalization of company names and codes in PATSTAT, we harmonized the names of companies by hand. As a result, we reduced the number of firms in our sample from the 15 595 initially obtained to 6597; that is, 57.7% of firms were incorrectly identified as unique in PATSTAT. The firms in our sample are related to 69 countries.

According to the above criteria, our sample contains 19 511 European patents: 9993 from sector 14, 8661 from sector 25, and 857 from both sectors 14 and 25 (Table 1.1). Figures are steady from year to year, but with a small decrease in 2006 because of the new incorporation of data in each patent office and in the PATSTAT (Table 1.1).

1.5 Results of the open-innovation study

The sources and types of knowledge incorporated in food technology are now examined and differences between countries and firms are discussed.

1.5.1 What are the sources of knowledge for patented food technology?

In food patents, both PCs and NPCs show the extent of the acquisition of scientific and technological knowledge through a process of spillover*s* for which there is no

Table 1.2 Origin of citations in agri-food patents by sectors (1998–2006)

	Patents (P)	Patent citations (PC)		Non-patent citations (NCP)		Total citations	Citations for each 100 patents	
		No.	%	No.	%		PC/100P	NPC/100P
Sector 14	9993	866	45.15	1052	54.85	1918	9	11
Sector 25	8661	3360	90.54	351	9.46	3711	39	4
Sector 14–25	857	65	89.04	8	10.96	73	8	1
Total	19 511	4291	75.25	1411	24.75	5702	22	7

Source: Own elaboration based on Patstat data.

compensation. The following paragraphs present the main figures concerning the external knowledge background underpinning the development of patented inventions; the use of the inventor's own knowledge was excluded from the analysis (self-citations were removed from the sample).

According to Table 1.2, food patents include 5702 citations made by the applicants, 75% of which are PCs, whereas the remainder are scientific citations. On average, patents in food sectors include 22 PCs and seven scientific citations for every 100 patents. The distinctions between sectors show the differences in the type of knowledge incorporated in each type of technology; for example, patents in sector 14 include nine PCs and 11 scientific citations for every 100 patents, whereas those in sector 25 include 39 PCs and four scientific citations for every 100 patents.

The approximate ratio of external technological knowledge to scientific knowledge is three (i.e., three PCs for each NPC). This suggests that the main source of external knowledge acquisition in food sectors is other technologies rather than scientific knowledge. However, considering sectors 14 and 25 separately yields a different picture. In sector 14, this ratio is practically one to one (one PC per scientific citation), whereas in sector 25 the ratio is 10:1 (10 PCs for each NPC). This difference in knowledge acquisition between sectors 14 and 25 (14–25) is not surprising: sector 14 includes many technologies related to the chemical components in food processing, for which advances published in scientific journals may be the main source of knowledge. By contrast, sector 25 mainly includes machinery and engineering, for which the main supporting source of external knowledge is frequently previous inventions of a similar nature.

1.5.2 What kind of technological knowledge is the most widely incorporated in food technology?

To analyse the technological sources of external knowledge, we have classified the PCs into seven broad technological fields (Table 1.3); we are considering only PCs, not NPCs, as a proxy for the acquisition of technological knowledge. The main source of technological knowledge for food inventions is the mechanical engineering and machinery sector; the second most significant source is process

Table 1.3 Applicant citations by sector and cited technological fields (1998–2006)

	Sector 14	Sector 25	Sector 14–25
I. Electrical engineering	13	109	2
II. Instruments	188	255	3
III. Chemistry and pharmaceuticals	396	110	5
IV. Process engineering and special equipment	603	512	44
V. Mechanical engineering and machinery	113	2816	34
VI. Consumption	35	210	7
Other	3	13	0
Unknown cited patent IPC	20	83	5
Total	1371	4108	100

Source: Own elaboration based on Patstat data.

engineering and special equipment; the third, chemistry and pharmaceuticals. Although almost all of the technological knowledge sectors are used in only one of the three food sectors (14, 25 and 14–25), technologies from process engineering and special equipment are a source for all three.

Table 1.3 shows that in sector 14, 44% of PCs stem from technologies in the field of process engineering and special equipment; 30%, from chemistry and pharmaceuticals; and 14%, from instruments. In sector 25, 69% of PCs are from mechanical engineering and machinery; and 12%, from process engineering and special equipment. Sectors 14–25 include citations of process engineering and special equipment (44%) and mechanical engineering and machinery (33%).

1.5.3 Are there differences across countries?

To find out if there were differences across countries, we grouped the patents in our sample by country according to the location of the innovative firm and the sector to which the patent is linked (Table 1.4). From these data, we can observe the following. First, there is a high spatial concentration of European food patents in a few countries: note that three countries account for more than 50% of European patents in sectors 14, 25 and 14–25. In addition, it must be highlighted that the USA leads the production of European patents in food sectors, with a share of patents in each of the three sectors ranging from 24 to 27%. Similarly, Germany is among the top three innovators in food sectors, with a share of patents ranging from 11 to 22% in sectors 14, 25 and 14–25. Second, there are strong disparities in the share of PCs and NPCs across countries and sectors. In sector 14, the share of PCs across the top five countries (i.e., those with the largest number of patents) ranges from 11% for Switzerland to 21% for the USA. Meanwhile, the share of NPCs ranges from 3% for Switzerland to 32% for the USA. In sector 25, the share of PCs across the top five countries varies across a wider range, from 5% for the Netherlands to 32% for Germany. However, the differences in the share of NPCs are smaller than in sector 14, ranging from 8% for the USA to 20% for Germany and the Netherlands. Finally, focusing on patents in sectors 14 and 25 combined, the share of PCs across the top five countries ranges from 3% for Switzerland to 35% for the USA; the

Table 1.4 Patents and applicant citations by country. Sectors 14, 25 and 14–25 (1998–2006)

Sector/country	No. of patents	Share of patents	Patent citations	NPL citations
Sector 14				
USA (US)	2702	23.98	256	374
Japan (JP)	1736	15.40	44	52
Germany (DE)	1233	10.94	216	274
The Netherlands (NL)	1167	10.35	151	205
Switzerland (CH)	873	7.75	102	36
United Kingdom (GB)	706	6.26	26	42
France (FR)	426	3.78	39	22
Denmark (DK)	410	3.64	4	
Italy (IT)	404	3.58	16	15
Belgium (BE)	192	1.70	39	101
Others	1421	12.61	63	49
Total	11 270	100	956	1170
Sector 25				
USA (US)	2218	24.18	816	32
Germany (DE)	1986	21.65	1129	74
Japan (JP)	865	9.43	203	43
The Netherlands (NL)	570	6.21	283	74
Italy (IT)	532	5.80	159	57
United Kingdom (GB)	451	4.92	44	11
France (FR)	435	4.74	159	36
Switzerland (CH)	393	4.28	133	7
Sweden (SE)	297	3.24	29	2
Belgium (BE)	243	2.65	215	6
Others	1184	12.91	337	33
Total	9174	100	3507	375
Sector 14 and 25 jointly				
USA (US)	272	27.76	25	2
Germany (DE)	118	12.04	19	1
Switzerland (CH)	108	11.02	2	0
The Netherlands (NL)	97	9.90	12	1
Japan (JP)	90	9.18	3	0
Italy (IT)	63	6.43	5	1
United Kingdom (GB)	61	6.22	2	0
France (FR)	35	3.57	1	2
Belgium (BE)	18	1.84	0	1
Australia (AU)	14	1.43	0	0
Others	104	10.61	3	0
Total	980	100	72	8

Source: Own elaboration based on Patstat data.

share of NPCs ranges from 13% for Germany and the Netherlands to 25% for the USA.

Overall, these results confirm that there are strong differences across countries in terms of patenting activity and the propensity to incorporate previous scientific and technological knowledge. However, these results may be strongly related to the location of firms with a large capacity to innovate. In the next section, we disaggregate our data at firm level.

1.5.4 Are differences across firms?

Table 1.5 presents the distribution of firms, patents and citations by firms. The sample is sufficiently large to allow us to draw conclusions about firms' propensity to patent in each sector, and about patterns of firms' knowledge acquisition.

First, from 7287 firms in our sample, 44.7% are in sector 14, which accounts for 52.6% of all patents; 47.6% are in sector 25, with 42.8% of patents; and 7.7% are in sectors 14–25, with 4.6% of patents. Note that the count of firms in sector 14, 25 and 14–25 is slightly higher than the number of firms in our sample that was previously given (6597). This can be explained because a firm that has patents in more than one agri-food sector would be double counted in this table. Dividing the number of patents in each sector by the number of firms, we obtain the finding that on average, sector 14 accounts for 3.5 patents per firm; sector 25, for 2.7 patents per firm; and sectors 14–25, for 1.7 patents per firm. These figures suggest differences in the innovative capacity of firms depending on the sectoral focus of their innovations: firms in sector 14 are more prone to patent new technologies than are firms in sectors 25 and 14–25.

Second, firms in sector 14 accumulated 21.1% of PCs and 75.3% of scientific citations, and firms in sector 25 accumulated 77.3% of PCs and 24.1% of scientific citations. These results are slightly different from those in Table 1.2 because the unit of analysis is now the firm, and in some cases, we account for two or more firms per patent. The figures corroborate the view that different types of knowledge acquisition occur in different sectors: firms in sector 14 produce technologies utilizing a strong external science base, whereas firms in sector 25 produce technologies with strong support from external technological knowledge.

Third, there is a strong concentration of patents in few firms: 10 firms, all multinational, account for approximately 20% of all patents in each sector. However, the number of firms in each sector is not the same: in sectors 14 and 25, the 10 firms with the highest patent activity account for 0.3% of all firms in the sector, whereas this figure is 2% in sectors 14–25, indicating a higher concentration of patent activity in sectors 14 and 25.

Finally, to ascertain whether firm size affects the incorporation of external knowledge, the last two columns in Table 1.5 present the number of citations divided by the number of patents. Note that on average, large firms in sectors 14 and 14–25, with the highest number of patents, do not use scientific and technological external knowledge more frequently than do other firms (there are two exceptions: Nestec SA and Kraft Foods Holdings Inc.). However, the acquisition of external technological knowledge by large firms in sector 25 duplicates the use of technological knowledge by smaller firms. Overall, this suggests that there is no size effect in the incorporation of scientific knowledge. Size is relevant only for the acquisition of technological knowledge in sector 25.

Table 1.5 Top patenting companies and citations by sector (1998–2006)

	Country	No. patents (P)	Share of patents (%)	Accumulated share of patents (%)	No. patent citations (PC)	No. scientific citations (NPC)	PC/P	NPC/P
Sector 14								
Unilever NL[2]	NL	388	3.44	3.44	22	42	0.057	0.108
Unilever GB[2]	GB	376	3.34	6.78	22	42	0.059	0.112
Nestle S.A.	CH	300	2.66	9.44	9	2	0.030	0.007
Nestec S.A.[2]	CH	222	1.97	11.41	75	23	0.338	0.104
Dsm Ip Assets B.V	NL	158	1.4	12.81	5	3	0.032	0.019
Ajinomoto Co. Inc.	JP	151	1.34	14.15	2	3	0.013	0.020
Kraft Foods Holdings Inc.	US	143	1.27	15.42	27	39	0.189	0.273
Basf Plant Science Gmbh	DE	128	1.14	16.56	2	1	0.016	0.008
Novozymes A/S	DK	122	1.08	17.64	0	0	0.000	0.000
Healthpro Brands Inc.	US	112	0.99	18.63	2	0	0.018	0.000
Top Ten		2100	18.63	18.63	166	155	0.079	0.074
Others		9170	81.37	81.37	786	1015	0.086	0.111
Total (number of firms 3256)		11 270	100	100	956	1170	0.085	0.104
Sector 25								
Deere & Company	US	589	6.42	6.42	540	22	0.917	0.037
Claas Saulgau GMBH	DE	285	3.11	9.53	213	2	0.747	0.007
Amazonen-Werke H. Dreyer Gmbh & Co. Kg	DE	173	1.89	11.42	50	14	0.289	0.081
Lely Enterprises Ag	CH	167	1.82	13.24	84	1	0.503	0.006
Delaval Holding Ab	SE	166	1.81	15.05	18		0.108	0.000
Cnh Belgium N.V.	BE	147	1.6	16.65	171	3	1.163	0.020
Kuhn S.A.	FR	122	1.33	17.98	44	17	0.361	0.139
Shimano Inc	JP	113	1.23	19.21	37	4	0.327	0.035
Maschinenfabrik Bernard Krone Gmbh & Co. Kg	DE	90	0.98	20.19	30	0	0.333	0.000
Kalle Nalo Gmbh & Co. Kg	DE	69	0.75	20.94	33	0	0.478	0.000

Top Ten		1921	20.94	20.94	1220	63	0.635	0.033
Others		7253	79.06	79.06	2287	312	0.315	0.043
Total (number of firms 3468)		9174	100	100	3507	375	0.382	0.041
Sector 14 and 25								
Nestle S.A.	CH	43	4.39	4.39	1	0	0.023	0.000
Mars Inc.	US	23	2.35	6.74	0	0	0.000	0.000
Unilever GB[2]	GB	23	2.35	9.09	1	0	0.043	0.000
Unilever NL[2]	NL	21	2.14	11.23	1	0	0.048	0.000
Frito-Lay Inc.	US	15	1.53	12.76	0	0	0.000	0.000
Nestec S.A.[2]	CH	15	1.53	14.29	0	0	0.000	0.000
Bähler AG	CH	14	1.43	15.72	0	0	0.000	0.000
Kraft Foods Holdings Inc.	US	11	1.12	16.84	5	0	0.455	0.000
Friesland Brands B.V.	NL	10	1.02	17.86	3	1	0.300	0.100
Firmenich S.A.	CH	9	0.92	18.78	0	0	0.000	0.000
Top Ten		184	18.78	18.78	11	1	0.060	0.005
Others		796	81.22	81.22	61	7	0.077	0.009
Total (number of firms 563)		980	100	100	72	8	0.073	0.008

Source: Own elaboration based on Patstat data.
[1]We use the term 'patent' referring to Inpadoc patent families.
[2]Nestec is the central advanced research facility of Nestle S.A. Unilever is a British–Dutch multinational.

1.5.5 Has there been a growing trend towards open innovation in patented food technology?

In this section, we examine the trend in the evolution of firms' acquisition of external knowledge. This analysis has two main limitations. First, the available time series is relatively short. Second, the data in our sample only cover the period before 2006; therefore, they do not include the past five years. We have data from 1998 to 2008, but very few patents are recorded in our database for 2007 and 2008 because of the time lag in the updating of the PATSTAT database. Despite these limitations, we can draw some conclusions from our data. Figures 1.3–1.5 present the average number of citations (PCs and NPCs) per patent for each food sector from 1998 to 2006.

In absolute terms, the amount of technological knowledge (PCs) incorporated in food patents shows an upwards trend in sector 14 and a stable pattern in sectors 25 and 14–25 (Fig. 1.1). Furthermore, there seems to be a wider adoption of technological knowledge in the last three years of the study period. If we compare Fig. 1.1 and 1.2, food patents make less frequent use of scientific knowledge than of technological knowledge; more specifically, the only sector in which the use of scientific knowledge has significant weight is sector 14. Note that the erratic evolution of NPCs in the first two years of the sample potentially reflects the influence of a single patent in 1998 that included an unusually large number of NPCs. Therefore, excluding the year 1998, the most frequent incorporation of scientific knowledge occurs in the last three years of the study period.

However, when focusing on the average knowledge incorporated into each food patent, a different pattern emerges. The evolution of PCs and that of scientific citations are very similar in sector 14 (Fig. 1.3). In sectors 25 and 14–25 (Fig. 1.4 and 1.5), we see more frequent incorporation of technological rather than scientific knowledge, but cannot draw a clear conclusion about whether this is more pronounced in recent years.

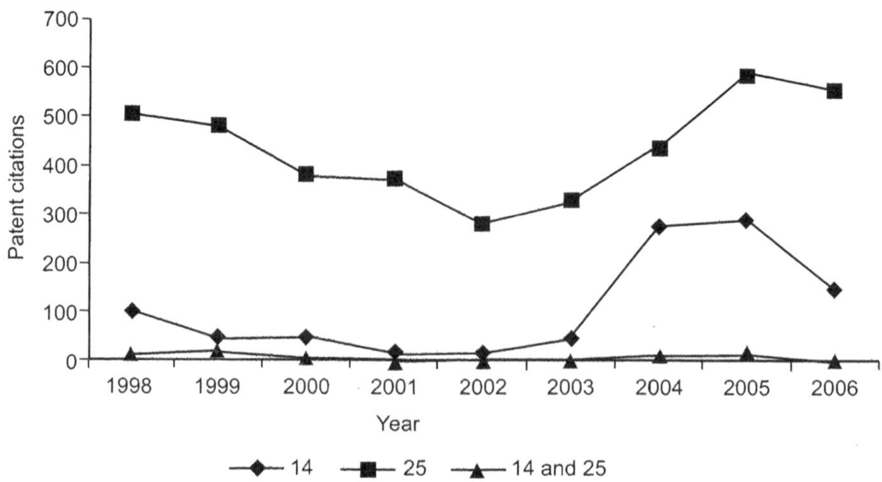

Fig. 1.1 External knowledge incorporation in agri-food sectors. *Source:* own elaboration based on Patstat data.

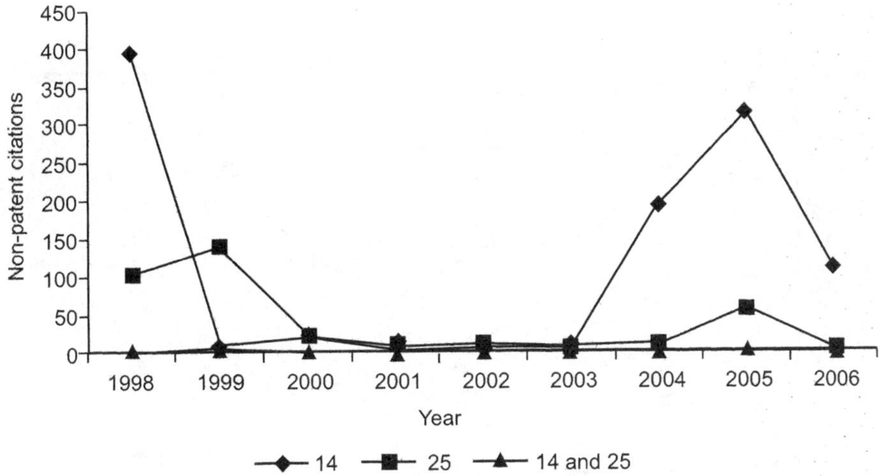

Fig. 1.2 External knowledge incorporation in agri-food sectors. *Source:* own elaboration based on Patstat data.

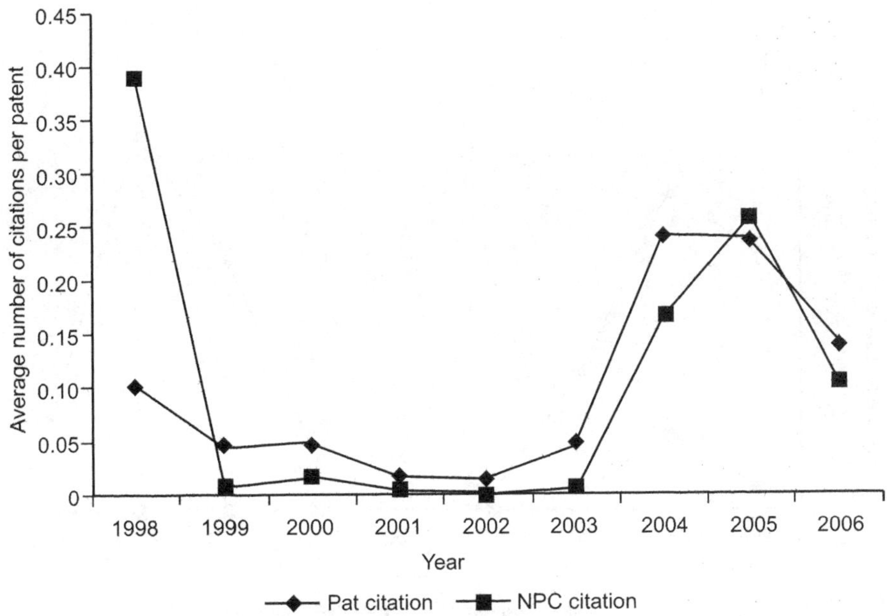

Fig. 1.3 External knowledge incorporation in sector 14. *Source:* own elaboration based on Patstat data.

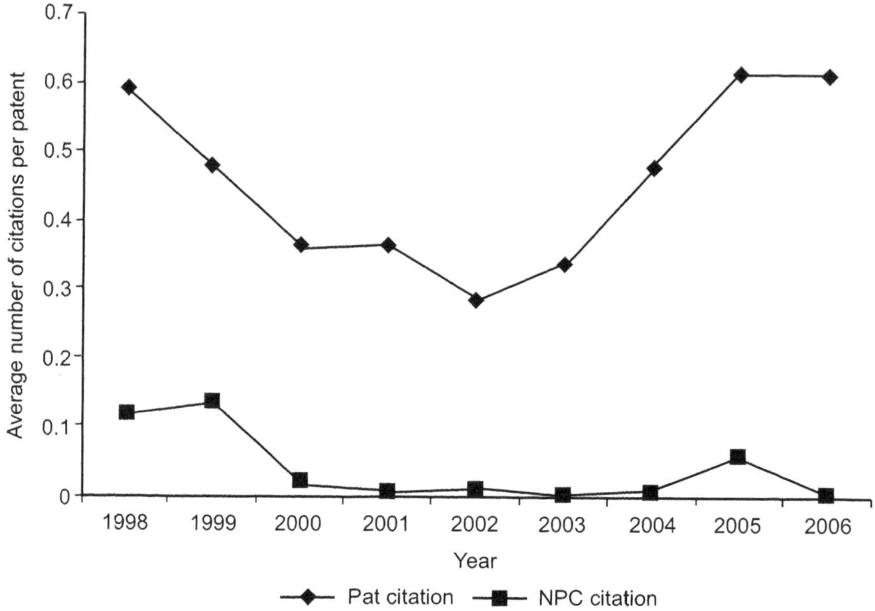

Fig. 1.4 External knowledge incorporation in sector 25. *Source:* own elaboration based on Patstat data.

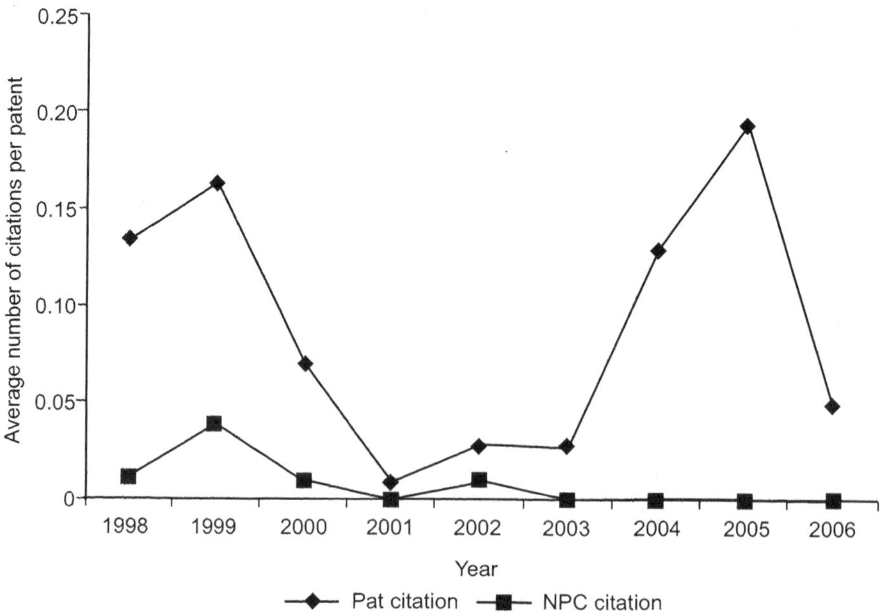

Fig. 1.5 External knowledge incorporation in sectors 14–25. *Source:* own elaboration based on Patstat data.

1.6 Conclusions

According to the open-innovation model, firms seek to advance in their development and improvement of technologies by acquiring knowledge that may originate in the discoveries or ideas of others. This chapter has explored this topic using quantitative indicators: *patents* as a proxy for firm technologies, and citations of external sources in patents as a proxy for knowledge acquisition. In the following paragraphs, we highlight the main findings:

1.6.1 Sources of external knowledge

Food firms incorporate external scientific and technological knowledge to support their own inventions. Sector 14 is the most prone to incorporate scientific knowledge rather than technological knowledge. By contrast, sector 25 supports technological advances essentially with technological rather than scientific knowledge. Inventions in sectors 14–25 incorporate both scientific and technological knowledge, but such knowledge incorporation is less frequent overall than in sector 14 or sector 25.

1.6.2 Types of technological knowledge

Sector 14 takes the majority of its external knowledge base from technologies related to chemistry, pharmaceuticals and process engineering and special equipment, whereas the technological bases in sector 25 are linked to mechanical engineering and machinery. Sectors 14–25 capture external knowledge from both process engineering and special equipment and mechanical engineering and machinery, although the number of citations is not as significant as in other food sectors.

1.6.3 Distribution across countries

There is a strong national concentration in the production of food technology: the USA, Japan and Germany account for more than 50% of all European patents. The USA leads in the use of external scientific and technological knowledge. The data also reveal clear differences across countries in the acquisition of knowledge; for example, patent applications from USA firms incorporate scientific and technological knowledge more frequently than do those from other countries in sector 14, whereas in sector 25, Germany leads the field.

1.6.5 Differences across firms

In absolute terms, a small group of multinational firms generates most food technology and acquires most external knowledge in each sector. In relative terms (dividing the number of citations by the number of patent applications for each firm), there is no size effect on the incorporation of scientific knowledge, except for firms in sector 25. This fact could be explained by the greater requirements, dimension and technological capacity of large firms to produce and commercialize patents. In other words, size is important in sector 25 because firms require a greater capacity to incorporate complex external technological knowledge to support their technological developments.

These results are subject to several limitations in the data set and in our reliance on patents and the citations therein to capture innovation and the acquisition of knowledge, respectively. We have considered here only a partial perspective, and clearly more research is needed to identify the sources of external knowledge and the ways in which food firms incorporate it to improve their technological capacities. Nevertheless, to encourage better exploitation of external scientific and technological knowledge in food technology, a few policy implications can be put forward.

Given that external knowledge acquisition in food technology has its origin mainly in the technological/scientific knowledge developed by other firms, government incentives should promote the incorporation of technologists or scientists depending on the characteristics of the technology and the sector. For example, firms in sector 14 should receive support for the incorporation of both scientists and technologists. By contrast, firms in sector 25 have shown an increase in the use of technological knowledge; therefore, they would gain more advantage by incorporating technologists rather than scientists.

Lastly, to make such incentives more efficient, there is no need to differentiate between firms according to their past experience in developing useful technology or their size. The figures presented above show a concentration of patents in a few multinational firms; however, with the exception of the more intensive use of technological knowledge by a group of 10 large firms in sector 25, we found practically no difference in the use of external knowledge in sector 14 or in sectors 14–25 between those multinational firms and other, smaller firms.

1.7 Acknowledgements

The authors are grateful for the financial assistance provided by Junta de Andalucía, Consejería de Innovación, Ciencia y Empresa (P06-SEJ-02087 and P08-SEJ-03981).

1.8 References

ACOSTA, M. AND CORONADO, D. (2003). Science–technology flows in Spanish regions: An analysis of scientific citations in patents. *Research Policy*, **32**, 1783–1803.

ACOSTA, M., CORONADO, D. AND TORIBIO, M. R. (2011). The use of scientific knowledge by Spanish agrifood firms. *Food Policy*, **36**, 507–516.

ALFRANCA, O., RAMA, R. AND VON TUNZELMANN, N. (2002). A patent analysis of global food and beverage firms: The persistence of innovation. *Agribusiness*, **18**, 349–368.

ALFRANCA, O., RAMA, R. AND VON TUNZELMANN, N. (2004). Combining different brands of in-house knowledge: Technological capabilities in food, biotechnology, chemicals and drugs in agri-food multinationals. *Science and Public Policy*, **31**, 227–244.

ARIAS-ARANDA, D. AND ROMEROSA-MARTÍNEZ, M. M. (2010). Innovation in the functional foods industry in a peripheral region of the European Union: Andalusia (Spain). *Food Policy*, **35**, 240–246.

AVERMAETE, T. AND VIAENE, J. (2002). On innovation and meeting regulation: the case of the Belgian food industry. DRUID Summer Conference on Industrial Dynamics of the New and Old Economy: Who is Embracing Whom?, 6–8 June 2002 Copenhagen.

AVERMAETE, T., VIAENE, J., MORGAN, E. J., PITTS, E., CRAWFORD, N. AND MAHON, D. (2004). Determinants of product and process innovation in small food manufacturing firms. *Trends in Food Science and Technology*, **15**, 474–483.

BACCHIOCCHI, E. AND MONTOBBIO, F. (2010). International knowledge diffusion and home-bias effect: do USPTO and EPO patent citations tell the same story? *Scandinavian Journal of Economics*, **112**, 441–470.

BRÖRING, S. (2008). How systemic innovations require alterations along the entire supply chain: The case of animal-derived functional foods. *Journal of Chain and Network Science*, **8**, 107–119.

CAREW, R. (2005). Science policy and agricultural biotechnology in Canada. *Review of Agricultural Economics*, **27**, 300–316.

CHESBROUGH, H. (2003). *Open innovation: the new imperative for creating and profiting from technology*, Boston, MA., Harvard Business School Press.

CHESBROUGH, H. W. (2006). Open innovation: a new paradigm for understanding industrial innovation. *In:* Chesbrough, H. W., Vanhaverbeke, W. and West, J. (eds.) *Open innovation: researching a new paradigm.* Oxford: Oxford University Press.

CHRISTENSEN, J. L., RAMA, R. AND TUNZELMANN, N. V. (1996). Study on innovation in the European food products and beverages industry. *EIMS/SPRINT.* Brussels: The European Commission.

COLLINS, P. AND WYATT, S. (1988). Citations in patents to the basic research literature. *Research Policy* **17**, 65–77.

CONNOR, J. M. (1988). *Food processing: an industrial powerhouse in transition*, Lexington, MA, Lexington Books.

DAHLANDER, L. AND GANN, D. M. (2010). How open is innovation? *Research Policy*, **39**, 699–709.

DEL BARRIO-CASTRO, T. AND GARCIA-QUEVEDO, J. A. (2005). Effects of university research on the geography of innovation. *Regional Studies*, **39**, 1217–1229.

FORTUIN, F. T. J. M. AND OMTA, S. W. F. (2009). Innovation drivers and barriers in food processing. *British Food Journal*, **111**, 839–851.

FRYER, P. J. AND VERSTEEG, C. (2008). Processing technology innovation in the food industry. *Innovation: Management, Policy and Practice*, **10**, 74–90.

GALIZZI, G. AND VENTURINI, L. (1996). Product innovation in the food industry: Nature, characteristics and determinants. *In:* Galizzi, G. and Venturini, L. (eds.) *Economics of Innovation: The Case of Food Industry.* Heidelberg: Physica-Verlag.

GARCIA MARTINEZ, M. AND BRIZ, J. (2000). Innovation in the Spanish food and drink industry. *International Food and Agribusiness Management Review*, **3**, 155–176.

GASSMANN, O. AND ENKEL, E. (2004). Towards a theory of open innovation: three core process archetypes. R&D Management Conference, July 6–9, 2004 Lisbon, Portugal.

GASSMANN, O., ENKEL, E. AND CHESBROUGH, H. (2010). The future of open innovation. *R&D Management*, **40**, 213–221.

GRILICHES, Z. (1990). Patent statistics as economic indicators: A survey. *Journal of Economic Literature*, **28**, 1661–1707.

GRUNERT, K., HARMSEN, H., MEULENBERG, M., KUIPER, E., OTTOWITZ, T., DECLERCK, F., TRAILL, B. AND GÖRANSSON, G. (1997). A framework for analysing innovation in the food sector. *Product and Process Innovation in the Food Industry*, 1–37.

HINZE, S., REISS, T. AND SCHMOCH, U. (1997). Statistical analysis on the distance between fields of technology. *Report for European Commission. TSER Project.* Brussels.

JAFFE, A., TRAJTENBERG, M. AND HENDERSON, R. (1993). Geographic localization of knowledge spillovers as evidenced by patent citations. *Quarterly Journal of Economics*, **79**, 577–598.

JOHNSON, D. K. N. AND EVENSON, R. E. (1999). R&D spillovers to agriculture: Measurement and application. *Contemporary Economic Policy*, **17**, 432–436.

KALPANA SASTRY, R., RASHMI, H. B., RAO, N. H. AND ILYAS, S. M. (2009). Integrating nanotechnology into agri-food systems research in India: a conceptual framework. *Technological Forecasting and Social Change*, **77**, 639–648.

KINSEY, J. D. (2001). The new food economy: Consumers, farms, pharms, and science. *American Journal of Agricultural Economics*, **83**, 1113–1130.

KLERKX, L. AND LEEUWIS, C. (2008). Matching demand and supply in the agricultural knowledge infrastructure: Experiences with innovation intermediaries. *Food Policy*, **33**, 260–276.

LEE, P.-C. AND SU, H.-N. (2011). Quantitative mapping of scientific research. The case of electrical conducting polymer nanocomposite. *Technological Forecasting and Social Change*, **78**, 132–151.

LEVIDOW, L. AND BIJMAN, J. (2002). Farm inputs under pressure from the European food industry. *Food Policy*, **27**, 31–45.

LICHTENTHALER, U. (2011). Open innovation: past research, current debates, and future directions. *Academy of Management Perspectives*, **25**(1), 75–93.

LUNDVALL, B. A. (1992). *National systems of innovation: towards a theory of innovation and interactive learning*, London, Frances Pinter.

LUNDVALL, B. A. AND JOHNSON, B. (1994). The learning economy. *Journal of Industrial Studies*, **1**, 23–42.

MALERBA, F. (2002). Sectoral systems of innovation and production. *Research Policy*, **31**, 247–264.

MCMILLAN, G. S., NARIN, F. AND DEEDS, D. L. (2000). An analysis of the critical role of public science in innovation: the case of biotechnology. *Research Policy*, **29**, 1–8.

MEYER, M. (2000). Does science push technology? Patents citing scientific literature. *Research Policy*, **29**, 409–434.

MEYER, M. (2002). Tracing knowledge flows in innovation systems. *Scientometrics*, **54**, 193–212.

NELSON, R. R. (1993). *National Systems of Innovation*, Oxford, Oxford University Press.

NELSON, R. R. AND WINTER, S. G. (1982). *An evolutionary theory of economic change*, Cambridge, Massachusetts, Harvard University Press.

RAMANI, S. V., EL-AROUI, M. A. AND CARRÈRE, M. (2008). On estimating a knowledge production function at the firm and sector level using patent statistics. *Research Policy*, **37**, 1568–1578.

SAGUY, I. S. (2011). Paradigm shifts in academia and the food industry required to meet innovation challenges. *Trends in Food Science and Technology*, **22**, 467–475.

SANGUANSRI, P. AND AUGUSTIN, M. A. (2006). Nanoscale materials development – a food industry perspective. *Trends in Food Science and Technology*, **17**, 547–556.

SARKAR, S. AND COSTA, A. I. A. (2008). Dynamics of open innovation in the food industry. *Trends in Food Science and Technology*, **19**, 574–580.

SCHMOCH, U. (2008). Concept of a technology classification for country comparisons. *Final Report to the World Intellectual Property Organisation (WIPO).*

SUMBERG, J. (2005). Systems of innovation theory and the changing architecture of agricultural research in Africa. *Food Policy*, **30**, 21–41.

TIJSSEN, R. J. W. (2002). Science dependence of technologies: evidence from inventions and their inventors. *Research Policy*, **31**, 509–526.

TRAILL, W. B. AND MEULENBERG, M. (2002). Innovation in the food industry. *Agribusiness*, **18**, 1–21.

VAN DE VRANDE, V., DE JONG, J. P. J., VANHAVERBEKE, W. AND DE ROCHEMONT, M. (2009). Open innovation in SMEs: Trends, motives and management challenges. *Technovation*, **29**, 423–437.

VAN LOOY, B., DU PLESSIS, M. AND MAGERMAN, T. (2006). Data production methods for harmonized patent statistics: patentee sector allocation. *FETEW MSI Research Report 0606*. Leuven: K.U. Leuven.

VERBEEK, A., DEBACKERE, K., LUWEL, M., ANDRES, P., ZIMMERMANN, E. AND DELEUS, F. (2002). Linking science to technology: using bibliographic reference in patents to build linkage schemes. *Scientometrics*, **54**, 339–420.

VON HIPPEL, E. AND VON KROGH, G. (2003). Open source software development and the private-collective innovation model: issues for organization science. *Organization Science*, **14**, 208–223.

VON HIPPEL, E. AND VON KROGH, G. (2006). Free revealing and the private-collective model for innovation incentives. *R&D Management*, **36**, 295–306.

WEAVER, R. D. (2008). Collaborative pull innovation: Origins and adoption in the new economy. *Agribusiness*, **24**, 388–402.

WILKINSON, J. (1998). The R&D priorities of leading food firms and long-term innovation in the agrofood system. *International Journal of Technology Management*, **16**, 711–720.

1.9 Appendix: concordance between agri-food technological sectors and International Patent Classification (IPC)

1.9.1 Technological sector 14. Agriculture, food chemistry

IPC	Description
A01H	New plants or processes for obtaining them
A21D	Treatment, e.g. preservation, of flour or dough for baking, e.g. by addition of materials; baking; bakery products; preservation thereof
A23B	Preserving, e.g. by canning, meat, fish, eggs, fruit, vegetables, edible seeds; chemical ripening of fruit or vegetables; the preserved, ripened, or canned products.
A23C	Milk, milk products; milk substitutes; manufacturing; pasteurizing; sterilizing and preserving
A23D	Butter substitutes; edible oils and fats
A23F	Coffee; tea; their substitutes; manufacture, preparation, or infusion thereof
A23G	Cocoa; chocolate; confectionery; ice-cream
A23J	Protein compositions for foodstuffs
A23K	Feeding-stuffs specially adapted for animals; methods specially adapted for production thereof
A23L	Foods, foodstuffs, or non-alcoholic beverages; their preparation or treatment, e.g. cooking, modification of nutritive qualities, physical treatment
C12C	Brewing of beer
C12F	Recovery of by-products of fermented solutions; denaturing of or denatured, alcohol
C12G	Wine; other alcoholic beverages; preparation thereof
C12H	Pasteurisation, sterilisation, preservation, purification, clarification, ageing of alcoholic beverages or removal of alcohol therefrom.
C12J	Vinegar; its preparation
C13D	Production and purification of sugar juices
C13F	Preparation and processing of raw sugar, sugar, and syrup

C13J Extraction of sugar from molasses
C13K Glucose; invert sugar; lactose; maltose; other sugars

1.9.2 Technological sector 25. Agriculture and food processing, machinery and apparatus

IPC **Description**
A01B Soil working in agriculture or forestry; parts, details, or accessories of agricultural machines or implements, in general
A01C Planting; sowing; fertilizing
A01D Harvesting; mowing
A01F Processing of harvested produce; hay and straw presses; devices for storing agricultural or horticultural produce
A01G Culture of vegetables, flowers, fruit, vines, and hops; forestry; watering
A01J Manufacture of dairy products (devices, apparatus)
A01K Animal husbandry; care of birds, fishes, insects; fishing; rearing or breeding animals, not otherwise provided for (housing animals, devices, etc.)
A01L Shoeing of animals
A01M Catching and trapping of animals; apparatus for the destruction of noxious animals or noxious plants
A21B Bakers' ovens; machines or equipment for baking
A21C Machines or equipment for making or processing dough; handling baked articles made from dough
A22 Butchering; meat treatment; processing poultry or fish
A23N Machines or apparatus for treating harvested fruit, vegetables in bulk, not provided for elsewhere; apparatus for preparing animal feeding-stuffs
A23P Foods or foodstuffs; their treatment, not covered by other classes
B02B Preparing grain for milling; refining granular fruit to commercial products by working the surface
C12L Pitching and depitching machines; brewing devices; cellar tools
C13C Cutting mills; shredding knives; pulp presses
C13G Evaporation apparatus; boiling pans
C13H Cutting machines for sugar; combined cutting, sorting and packing machines for sugar

2

The tension between traditional innovation strategies and openness: Lindt's controlled open innovation approach

V. Lazzarotti and R. Manzini, Cattaneo University – LIUC, Italy

DOI: 10.1533/9780857097248.1.25

Abstract: In this chapter, an exploration is presented of the tension generated in companies following a very traditional approach to managing R&D and innovation (mostly based on internal resources, competences and activities) even when globalization and the rapid pace of innovation driven by market forces, leads them to involve external actors in their R&D processes. The problem this tension generates, namely balancing the need for openness with the strategic relevance of internal R&D activities and core capabilities, which the company wishes to remain internal and private, is one faced by leading chocolate manufacturer, Lindt. Evidence from this case study illustrates how even 'closed' companies can adopt a 'controlled open approach' to innovation so as to exploit its potential benefits without radically modifying their business model.

Key words: closed innovation, open innovation, chocolate industry, trade secrets.

2.1 Introduction

In this chapter we analyse the experience of a well-known company, Lindt & Sprüngli Maitre Chocolatier. This company is in a phase of transition; moving towards adopting 'open' practices in its innovation process, whilst at the same time maintaining some behaviours which are evidently 'closed'. The case is interesting because it allows us to grasp how opposite forces can be at work in stimulating or suppressing openness. This chapter reviews innovation and technology management studies on open innovation (OI) and contextual factors, details the methodology and context, and describes Lindt's experience, discussing this in the context of previous studies. Finally, the chapter draws conclusions and suggests further research along a similar model on companies in other areas of industry.

2.2 Literature review

Since Chesbrough published his book in 2003, the concept of OI has received a considerable amount of attention from practitioners and researchers. Today it is largely accepted that an artificial dichotomy between closed and open approaches does not represent how things really are (Dahlander and Gann, 2010). Thus, exploring the different degrees and types of openness, which may be seen to lie along a continuum, provides a richer and more interesting way forward (Chesbrough, 2003). In particular, this approach allows a deeper and more realistic investigation into company behavior, including the nature and context of innovation sources (Dahlander and Gann, 2010; Gassmann, 2006). An increasing number of studies adopt this approach, for example Laursen and Salter (2004) with their concept of breadth (i.e. number of sources used for OI) and depth (i.e. intensity of collaboration with each source) of external sources, pave the way for more recent reinterpretations by Keupp and Gassmann (2009). Lazzarotti and Manzini (2009) attempt to develop a concept of 'openness degree' in a preliminary work by integrating two variables: 1) the number and type of external partners (i.e. partner variety); and 2) the number and type of phases in the innovation process opened to external contributions, in and/or out (i.e. innovation phase variety) so as to collaborate and, hopefully, co-create through the innovation funnel.

Adopting this notion, previous research in innovation and technology management literature has also attempted to study relationships among different degrees of OI, and the various contextual factors involved. This research is driven by the notion that such factors influence the companies' choices in terms of how open their innovation process should be. Contextual factors can be both external variables representing the environment (including: industry's R&D intensity; technological turbulence; prevalent type of innovation – breakthrough or radical versus incremental; and globalization trends) and internal, representing the specific situation of a firm, though evidently also matching the larger technological and competitive context (including: internal R&D intensity; innovation strategy; and appropriability strategy). The underlying idea is that firms must adapt to their contextual environment (Lawrence and Lorsch, 1967). Firms make the most suitable choices in terms of degree of openness in the innovation process based on context.

It is possible to identify two types of factors, i.e. external and internal or firm-specific ones.

2.2.1 External factors

In terms of external factors, the state of technology in an industry, was studied in terms of intensity, turbulence and convergence. Technological intensity (Fortuin and Omta, 2008) seems to force organizations toward co-operation in that they are not able to cope with technology using only their own strengths: companies in high-tech industries (e.g., semiconductors) show a higher propensity to co-operate, using external sources to support product development, in an environment characterized by rapid technological change. At the same time, costs, time and risks of R&D should be reduced (Calantone and Stanko, 2007). Above all, examples can be found in the biotech and IT industries.

Technological turbulence appears to exert a similar impact (Ozman, 2008) because it positively impacts on acquiring technology. Rapidly changing technology conditions will increase the motivation of firms to be involved in a large number of interactions in order to explore new knowledge residing outside the firm, and to be informed about new developments. Radical innovations using new technologies, and creating new markets, are also more likely.

Lastly, technological convergence, although reshaping the competitive arena and altering the borders between industries, presses firms to collaborate: high levels of interdisciplinary research mean that a single company is not able to innovate succesfully using its own capability alone (Fortuin and Omta, 2008).

Another external contextual factor is the move towards globalization (Chesbrough, 2003; Fortuin and Omta, 2008). Globalization appears to foster OI because it is characterized by: higher mobility of capital, labor, and knowledge; lower logistics costs; more efficient ICT; and increased market homogeneity across different countries. This lowers entry barriers for new international competitors, providing a competitive advantage for those companies that innovate more quickly and are able to better adapt.

2.2.2 Firm-specific factors

Studies of internal factors are based on the view that degree of OI is mainly determined by an individual company's strategic choice, rather than by industry characteristics, even though general consistency between a company's behavior and its technological and competitive environment is a reasonable assumption. For example, Lichtenthaler (2008) identifies groups of firms that pursue similar degrees of openness, and characterizes them in terms of internal contextual variables representative of the innovation strategy, such as R&D intensity and emphasis on radical rather than incremental innovation. The higher the R&D intensity (expenditure on R&D as a share of sales), the greater the acquisition of technology. This finding supports the contention that firms pursue external technology acquisition as a complement to internal R&D rather than a substitute. The degree of openness also appears to rise with an emphasis on radical innovation, both in the degree of external technology commercialization, and external acquisition. This is necessary to facilitate acceptance on the market, as well as the creation of a standard, and because firms which emphasize radical innovation are not able to develop all salient knowledge internally, but must also rely on complementary external sources.

The appropriability factor is important for the single firm and manifests in a specific intellectual property (IP) strategy. Given a certain appropriability regime in the industry, a single company can play the IP game to facilitate (or not) its OI strategy (Alexy *et al.*, 2009). It is clear that where secrecy is excessive or even where the use of patents are inadequate (a company would not collaborate with another without at least one patent in place: 'no patent, no talk'), OI is seriously curtailed. However, we should also recognize that it is not always a simple matter to find the right balance in using IP tools to protect proprietary knowledge and technology, and thus limit spill-over risks. Some types of technologies, know-how and innovations, are not effectively patentable. For example: mature technologies; tacit know-how; process; and incremental innovations (Bogers, 2011; Cohen *et al.*, 2002; de Faria and Sofka, 2010; Hertzfeld *et al.*, 2006; Kingston, 2001). In such cases secrecy, and thus a lower degree of openness, may be necessary.

In summary, this review details factors that may influence the degree of openness in innovation and highlights the opposing forces often involved. The following case study provides a relevant example of how these forces can work in the food and drink industry. There is certainly a tendency towards openness (i.e. technological and structural changes), but some factors orient the industry towards more traditionally closed approaches to innovation. Indeed, evidence from the food and drink industry finds a low degree of R&D intensity; non-technology based innovations, mainly of an incremental type; critical success factors based on quality, reputation, and tradition; and appropriability strategies primarily based on trade secrets.

2.3 Research method for Lindt case study

The research method adopted here is the single case study. Despite the widely acknowledged limitations of this approach, especially in terms of reliability and validity (Ginsberg and Abrahamson, 1991; Yin, 2002), the case study method is able to capture the full complexity of a phenomenon, including its 'softer' aspects. Given that the aim of the study is to investigate OI practices in depth, this advantage was critical in selecting the research approach.

The company Lindt & Sprüngli Maitre Chocolatier was chosen as the subject of the study. As a matter of fact, in an era characterized by OI, the company is immediately distinguished by its essentially closed innovation approach. As the unit of analysis we chose the Italian branch of the company, located at Induno Olona. The main reason for this was twofold. Firstly, as part of a very cohesive group, the branch may represent the group's collective philosophy. Secondly, the Italian branch plays a leadership role in the group's wider R&D activities, so its involvement with innovation processes is evident.

Information was collected through direct interviews with Lindt's Italian R&D management. Internal documents were also consulted. The applied research methodology had two main limitations. Firstly, because the empirical base was mainly built from direct personal interviews with the company's top managers, the results are susceptible to bias arising from distorted and subjective interpretations and rationalizations. Secondly, as in most single case studies, the empirical research does not permit any systematic generalization. However, our aim was not to generalize from a single case study, but rather to offer a detailed description of the phenomenon in order firstly to put reported theories to the test (Eisenhardt, 1989), and secondly to offer new insight towards future investigations aimed at generalizing results.

2.4 Open and closed innovation at Lindt

Lindt & Sprüngli is recognized as a leader in the market for premium quality chocolate, offering a large selection of products in more than 100 countries around the world. The company has six production sites in Europe, two in the USA and distribution and sales companies on four continents. The beginnings of Lindt & Sprüngli are in 1845, when father and son, in partnership, manufactured solid chocolate in their small confectionery firm Sprüngli & Son for the first time. Over the past 165 years of Lindt & Sprüngli's life, the company has become famed as one of the most innovative and creative companies making high quality chocolate, transforming itself into a multinational group through progressive integration of licensees and strategic acquisitions (such as the Italian Caffarel and the Austrian Hofbauer). Lindt & Sprüngli has been listed on the Swiss stock exchange (SIX) since 1986. The group has production companies in Switzerland, Germany, France, Italy, the USA and Austria, distribution and sales companies in England, Hong Kong, Spain, Poland, Canada, Australia, Sweden, Mexico and the Czech Republic, as well as sales offices in Dubai and Ireland. In addition, Lindt & Sprüngli sells its products through a vast network of local independent trade partners.

The two Italian subsidiary companies, Lindt & Sprüngli SpA and Caffarel SpA, reported consolidated sales for 2010 of EUR 205.1 million (previous year adjusted: EUR 214.5 million). Lindt proved to be the most dynamic brand in the Italian chocolate market, increasing its market share and further enhancing its leadership in this premium segment. The Lindor chocolate balls took center stage, benefiting from a high cost marketing plan.

2.4.1 The innovation process at Lindt

The innovation process at Lindt consists of four elements:

1. the recipe, which is probably the most difficult innovation, and is usually incremental in its development;
2. the product concept, which includes expectations as to product use, the specific target it addresses, and the atmosphere it should create;
3. product packaging; and
4. the machines used in production, since new products may require modifications to manufacturing techniques.

Marketing	R&D	Operation
• Concept definition • Packaging definition	• Concept definition • Product recipe realization • Selecting suppliers	• Packaging realization • Design

Fig. 2.1 Organizational units that deal with innovation at Lindt and their competences.

At Lindt, the process of innovation involves several different units (Fig. 2.1) and is composed of the following phases:

- Concept generation: the marketing section proposes a new or updated product concept, following suggestions from the analysis of consumer needs;
- Concept development: the creation of the first prototypes in the R&D laboratories;
- Project elaboration: elaborating a first attempt at the project, in which relationships with marketing are of key importance;
- Development loop: production of the first samples by R&D laboratories, including implementing changes and modifications, aimed at understanding the effectiveness of the concept. The output of this phase is a sample of three recipes;
- Sensory tests are conducted on the prototypes from the selection of recipes by the sample. In this phase, a selection can be based on a panel of consumers, selected to match the target market.
- Packaging is developed in parallel with the sensory test by the operation departments.
- Production tests are conducted through a combination of technology and R&D, aimed at understanding whether the new product can be realized with existing equipment, or whether new tools are required.

In parallel, marketing analyzes the desired image the product should project. An executive committee, composed of the heads from each function, follows each step through monthly meetings where the latest news is presented and discussed.

The innovation funnel has an average two year duration. The percentage of ideas that pass the concept phase is equal to 80%; 80% of the concepts come to the marketing stage, and then 60% of these may go on to the next step. In one year, 600 product variants are developed, corresponding to 30 ideas, and 7/8 launches.

2.4.2 Open innovation at Lindt

The Lindt approach to OI is very cautious. The idea is 'to open the process when it is not possible to go on by ourselves', and is motivated by the fact that Lindt considers its experience and competences to be excellent and unique in the world of chocolate and, as a consequence, that technological collaborations are not necessary to create successful new products. Its 'maitre chocolatiers', working in eight production plants dispersed globally, are the repository of a specific and distinctive know-how, accumulated over 150 years of experience, and not imitable by competitors. However, Lindt does in some cases need to interact with other actors in order to develop innovative products. It usually chooses a 'controlled open approach'. These collaborations are explored in more detail in the following subsections.

New concept generation

New concept generation is kept entirely internal, but follows the interactions of the marketing unit with customers, (i.e. with focus groups), fairs, and relationships with vendors. Suppliers of raw materials work with Lindt on the basis of precise technical specifications (e.g. the size of hazelnuts).

Development of new recipes

Lindt remains as closed as possible because recipes are protected only by means of trade secrets (Lindt does not use any intellectual property tool to protect its recipes) and hence containing know-how is critical. Even within the company, very few people know the recipes exactly. The need to interact with other companies emerges from a necessity. In order to innovate recipes, Lindt needs the appropriate ingredients. The relationship is particularly collaborative with companies working on aroma. Lindt has identified a list of excellent suppliers for each type of aroma, and with these Lindt works in iterative experimentation until the appropriate mix is achieved to match with the product concept. A non-disclosure agreement is signed with these partners to protect all information exchanged during collaborations.

Experimentation

Once new concepts and the related new recipes are identified, these must be tested. The first rough prototypes made with readily available materials are tested. This experimentation phase (known as sensorial testing) is conducted in collaboration with a panel of selected expert consumers, chosen to represent the target market, in order to evaluate how product fits to concept. Experts have a strong and long term relation with Lindt, and they are selected from among frequent consumers of Lindt chocolate.

Packaging

It is vital for Lindt that packaging is able to preserve the product; allow easy and safe transport; and effectively communicate the product concept. As a consequence, several graphic design agencies collaborate with Lindt on this phase, as well as consumer focus groups.

Manufacturing

New products sometimes require modifications in the technology and machinery used in production. In these cases, a few suppliers of machinery are involved to evaluate the feasibility of proposed modifications, and to implement these.

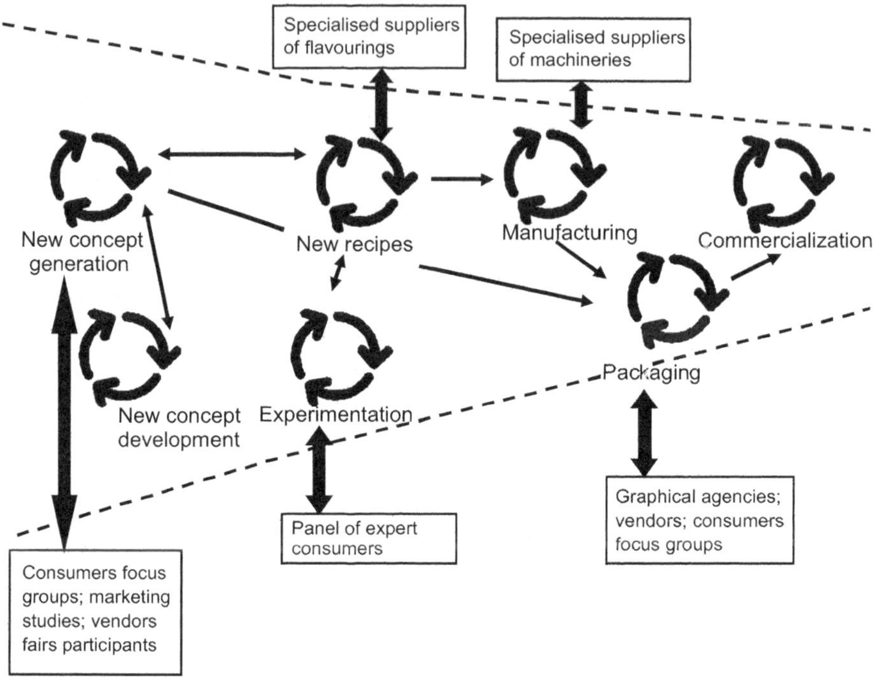

Fig. 2.2 Lindt's open innovation funnel.

The 'controlled approach' to OI adopted at Lindt can be visualized as a funnel as in Fig. 2.2.

2.5 Lindt's open-innovation approach in practice: the innovation project Noccior

The product which is now identified as Noccior is the result of a complex innovation process. The first version of the product was born 50 years ago, and had no specific name: it was simply an easter chocolate, with a rock effect deriving from chopped hazelnuts on the surface, and realized with the use of Rocher technology. In the sixties it was produced by hand, and only sold through the pastry channel: this channel accounts for 50% of total sales of the easter range and 15% of sales in this product (we will call it 'Before-Noccior').

In the nineties, less effort was put into Before-Noccior: the strategy of this period being focused on the product known as Gold Bunny. Before-Noccior represented 14% of total sales for the company in 2008. During this period, Perugina and Ferrero, among the most relevant competitors, especially in Italy, sought to increase their market share by focusing on aesthetics over quality. At the same time, Lindt realized that Before-Noccior was too expensive. Hence, an interest began in renewing Before-Noccior. Given the relevance of Before-

Noccior to the Lindt portfolio, the innovation process was initiated, involving R&D, marketing and operation units. The aim was to renew all salient features of Before-Noccior.

The first step was taken in communication, aimed at reinforcing the message that this was a high quality product. The quality of the product was tested and compared with competitors using objective parameters such as roundness, percentage of nuts, and taste. In executing this comparison, research and development was crucial, providing the sales force and customers with the parameters by which the quality of Before-Noccior could be measured. A tasting kit was prepared by R&D and marketing, and given to a panel of customers, with whom Lindt created an intense relationship, allowing the company to verify and, it may be said, to 'certify' the superiority of the product to that of the competition, along every parameter.

Hence, the innovation of Before-Noccior did not concern the recipe, which was already developed and did not require significant changes. Instead, innovation focused on several areas: the name; packaging aesthetic and functional design; product characteristics; and production process.

The process started by deriving a name able to evoke the new product concept. To this aim, Lindt collaborated with a company specialized in naming. After five months work, the name derived was in fact the result of an idea generated internally: Noccior, given by the combination of 'nocciole' (hazelnuts) with 'oro' (gold). Another external company checked the name with a sample of customers, who judged the name easy, explicit, and direct.

The next step was to redesign the packaging. From 1999 to 2008 many types of packaging had been used. Before-Noccior used to have functional, closed packaging; it was felt that a better presentation was needed to renew the product. The packaging review resulted from collaborations with actors outside the company, particularly with packaging suppliers and a marketing agency based in Milan and Naples, the latter city being key for sales. For the choice of colors and materials: the papermaking research and particularly the graphic design, were aimed at representing a premium product whilst conveying festivity. Developing the new packaging was costly, impacting the total packaging cost. Given that quality and taste were already highly developed, product innovation focused on size. Reduced size was essential in order to put a lower priced and hence more affordable product on the market.

The reduced size gave customers a product with a 30% reduction both in weight and cost and, later, with the addition of small chocolates made with the Noccior recipe. This extreme downsizing allowed Lindt to realize a line extension of the brand, consisting in a 'big' product for families, dedicated to times of festivity such as Easter, but also an every-day product aimed at the same loyal customers.

Innovating size resulted in several problems in the production process, especially in weight testing and the production line. The new sizes meant the creation of a dedicated facility. There were thus two phases resulting from the innovative design of Noccior. Firstly the creation of the plant. Secondly, the implementation of new packaging.

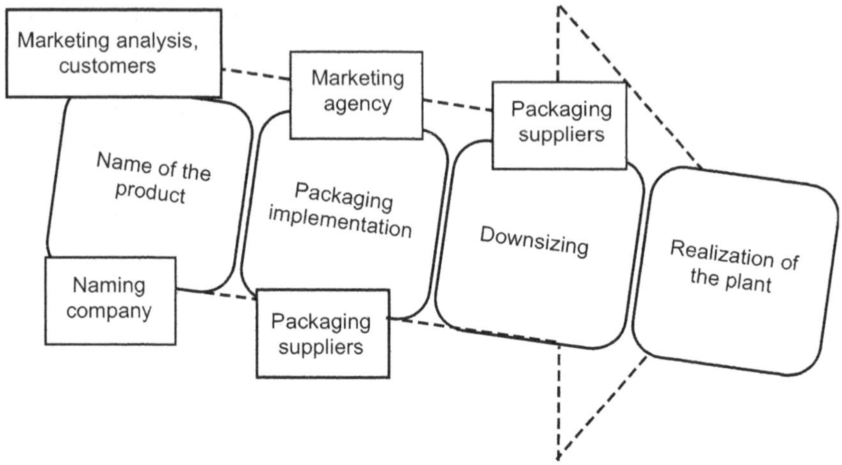

Fig. 2.3 Phases of the Noccior project and the involved partners.

Until the mid-nineties the production process of Before-Noccior was done completely by hand (from the realization of the basic product, the chocolate coating, and each stage), but the volumes required by the newly innovated product were clearly incompatible with this type of production. Mechanization was required. The problem was therefore the creation of a plant to manufacture this complex product. Lindt, in order to avoid problems of spill over, did not consider collaboration with external partners. The risk of losing control over critical information concerning the recipe and the treatments, protected only by means of trade secrets, was too great. Particularly because there was a very limited number of companies manufacturing these types of plant worldwide, and all Lindt's competitors were in close contact with them. Therefore this part of the innovation funnel was entirely closed. The engineering unit designed the machines. The new machinery coming about as a reimagining of a snack production machine.

Implementing the new packaging posed two critical difficulties: the first was the inclusion of the little eggs in the package, being fragile and empty inside; and the second concerned the use of a material suitable for the irregular shape, typical of the rocher mode of production. Again a partnership was necessary, in this case with the packaging supplier; the innovation protected only by trade secret.

The open-innovation process adopted for the Noccior is shown in Fig. 2.3.

2.6 Results of controlled open innovation in the Lindt case

The case of Lindt and of Noccior provides an interesting and paradigmatic example of the tensions that can arise within companies that traditionally adopt a closed innovation model, and are now forced by competition and technological change to move towards a more open approach. The influence of external context factors, (among those listed in our literature review), is not pushing Lindt towards

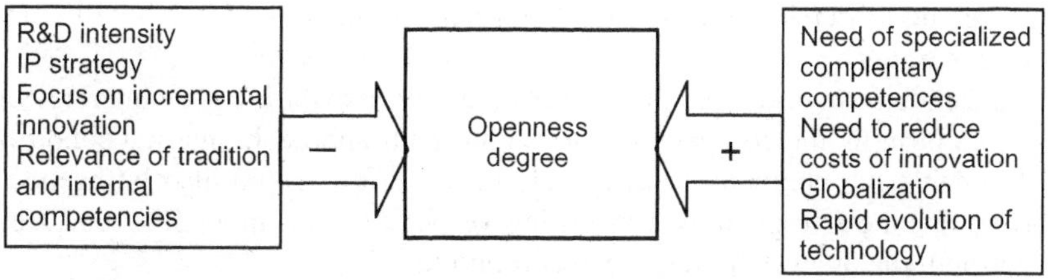

Fig. 2.4 Opposing forces in Lindt drive the choice of the openness degree.

completely OI. In fact, the level of R&D intensity is not very high (for example with respect to other industries, such as pharmaceuticals or electronics); the appropriability regime does not facilitate collaborations (because IP protection is mainly based upon trade secrets); and the type of innovation is far more incremental than radical. Tradition and proprietary competences are central to the company's strategy (Fortuin and Omta, 2009; Garcia Martinez and Briz, 2000; Hardy, 2009; Lagnevik, 2003; Sarkar and Costa., 2008). At the same time, several increasingly relevant forces stimulate the Lindt innovation process to open up: the need to access specialized competences (such as those for packaging or identifying a new aroma); the need to reduce the costs of innovation; the globalization of markets; and the rapid evolution of technology, especially IT, which can significantly improve the efficiency and effectiveness of the innovation process (Bellairs, 2010; Bérézai, 2010; Garcia Martinez, 2011; Garcia Martinez and Briz, 2000; Goers, 2009; Thomas, 2011; Traitler *et al.*, 2011).

Figure 2.4 shows the opposing forces determining the OI approach chosen by Lindt.

The case of Noccior represents an example in which all these forces were active at the same time. As a consequence, a closed and open process of innovation was observed, in which some phases were open to external contributions, whereas others remained completely closed. Furthermore, even the most open phases were characterized by a highly 'controlled' approach. Given the trends in opening and closing influences, we can expect that Lindt will adopt an increasingly open approach in future. The question is whether Lindt will be able to capture this opportunity (or adapt to this need). To answer this question, a number of internal factors must be considered, including the organizational and cultural context, particularly absorptive capacity and human resource management (Thuc Anh *et al.*, 2006; Jiménez-Barrionuevo *et al.*, 2011; Volberda *et al.*, 2010). Studies have shown that a high degree of internal competence, which in turn determines a high level of absorptive capacity, can ease the way for the opening up of innovation processes. Without any doubt Lindt has great internal competences in chocolate, both from a technological and a marketing perspective. Lindt emphasizes team work and relationships between and among functional units (e.g., purchasing with R&D and marketing, operations with R&D), a fact recognized in the literature as a factor enabling OI. Human resource management style can also be relevant. Lindt always encourages, recognizes and rewards individual innovation, personal initiative and leadership, throughout the organization, with a large investment in lifelong learning.

In the end, it can be argued that if context evolves towards a growing relevance of factors pushing the company towards a more OI approach, internal conditions would enable the company to adapt. However, effort would also be needed to improve openness, especially concerning IP policy, with more frequent use of protection tools other than trade secrets required.

2.7 Conclusions

The case presented in this chapter may be considered paradigmatic: We can frequently recognize opposing forces in companies that have traditionally adopted a closed approach to innovation. In cases where the ultimate source of competitive advantage lies in tradition, excellent tacit know-how, and trade secrets, the shift towards OI is risk-laden and complex. The 'controlled' approach to OI adopted by Lindt can thus be seen as a solution suitable for application in similar cases. Such an approach is characterized by two fundamental decisions:

- identifying the phases of the innovation process that can, in fact, be open: being open does not necessarily mean opening the whole innovation funnel, but it is possible to identify those phases for which the risks of openness are limited and, at the same time, the benefits potentially achievable are high with technological collaboration;
- identifying the partners who are to be involved in the innovation process, limiting openness to only those actors with whom a long term and trusting relationship can be built and maintained.

However, even a controlled approach to OI should be conceived dynamically: the choice of which phases to open, and of the partners in collaboration should be continuously monitored, and in some cases revised, in order to adapt to the evolving context, both internal and external.

2.8 References

ALEXY O, CRISCUOLO P AND SALTER A (2009), 'Does IP strategy have to cripple open innovation?', *MIT Sloan Management Review*, **51**(1), 71–77.

BELLAIRS J (2010), 'Open innovation gaining momentum in the food industry', *Cereal Foods World*, **55**(1), 4–6.

BÉRÉZAI P (2010), 'Future R&D strategies in food & drinks: evolution from orthodox approaches to open innovation models', March, *Business Insights*, 74–82.

BOGERS M (2011), 'The open innovation paradox: knowledge sharing and protection in R&D collaborations', *European Journal of Innovation Management*, **14**(1), 93–117.

CALANTONE RJ AND STANKO MA (2007), 'Drivers of outsourced innovation: an exploratory study', *Journal of Product Innovation Management*, **24**, 230–241.

CHESBROUGH H (2003), *Open innovation: the new imperative for creating and profiting from technology*, Harvard Business School Press, Boston.

COHEN WM, GOTO A, NAGATA A, NELSON RR AND WALSH JP (2002), 'R&D spillovers, patents and the incentives to innovate in Japan and the United States', *Research Policy*, **31**, 1349–1367.

DAHLANDER L AND GANN D (2010), 'How open is innovation', *Research Policy*, **39**(6), 699–709.

DE FARIA P AND SOFKA W (2010), 'Knowledge protection strategies of multinational firms: a cross-country comparison', *Research Policy*, **39**, 956–968.

EISENHARDT KM (1989), 'Building theories from case study research', *The Academy of Management Review*, **14**(4), 532–550.

FORTUIN FTJM AND OMTA SWF (2008), 'The Dark Side of Open Innovation. A Survey of Failed Inter-company Cooperation', *Proceedings of the 8th international conference on management in Agrifood chains and networks*, p. 10.

FORTUIN FTJM AND OMTA SWF (2009), 'Innovation drivers and barriers in food processing', *British Food Journal*, **111**(8), 839–851.

GARCIA MARTINEZ M AND BRIZ J (2000), 'Innovation activities in the Spanish food and drink industry', *International Food and Agribusiness Management Review*, **3**, 155–176.

GARCIA MARTINEZ M (2011), 'Future R&D strategies in food & drinks: evolution from orthodox approaches to open innovation', IFAMA, Annual World Symposium. 20–21 June, Frankfurt, Germany.

GASSMANN O (2006), 'Opening up the innovation process: towards an agenda', *R&D Management*, Special Issue Open Innovation, **36**, 223–228.

GINSBERG A AND ABRAHAMSON E (1991), 'Champion of change and strategic shift: the role of internal and external change advocates', *Journal of Management Studies*, **28**(2), 173–190.

GOERS SK (2009), 'Open innovation: driving growth by applying principles of open innovation', Kraft Foods, open access at http://openinnovation.berkeley.edu/speaker_series/Goers_Berkeley_02_23_09.pdf.

HARDY N (2009), *Future innovations in food and drinks to 2015. NPD, trend convergence and emerging business opportunities*, Business Insights, http://store.business-insights.com.

HERTZFELD HR, LINK A AND VONORTAS NS (2006), 'Intellectual property protection mechanisms in research partnerships', *Research Policy*, **35**, 825–838.

JIMÉNEZ-BARRIONUEVO, MM, GARCÍA-MORALES, VJ AND MOLINA LM (2011), 'Validation of an instrument to measure absorptive capacity' *Technovation*, **31**, 190–202.

KEUPP MM AND GASSMANN O (2009), 'Determinants and archetype users of open innovation', *R&D Management*, **39**(4), 331–341.

KINGSTON W (2001), 'Innovation needs patent reform', *Research Policy*, **30**, 403–423.

LAGNEVIK M (2003), *The dynamics of innovation clusters: a study of the food industry*. Northampton, MA: Edward Elgar Pub.

LAURSEN K AND SALTER A (2004), 'Searching high and low: what type of firms use universities as a source of innovation?', *Research Policy*, **33**(8), 1201–1215.

LAZZAROTTI V AND MANZINI R (2009), 'Different modes of open innovation: a theoretical framework and an empirical study', *Journal of Innovation Management*, **13**(4), 1–22.

LAWRENCE PR AND LORSCH JW (1967), *Organization and environment*, Cambridge, MA, Harvard University Press.

LICHTENTHALER U (2008), 'Open innovation in practice: an analysis of strategic approaches to technology transactions', *IEEE Transactions on Engineering Management*, **55**(1), 148–157.

OZMAN M (2008), 'The two faces of open innovation: network externalities and learning', *Document de Travail*, 24.

SARKAR S AND COSTA A (2008), 'Dynamics of open innovation in the food industry', *Trends in Food Science & Technology*, **19**(11), 574–580.

THOMAS K (2011), *Hot trends in food and drinks innovation. Six key strategies for driving innovations and marketing in the next decade*, Business Insights.

THUC ANH PT, BAUGHN C, MINH HANG NT AND NEUPERT KE (2006), 'Knowledge acquisition from foreign parents in international joint ventures: an empirical study in Vietnam', *International Business Review*, **15**(5), 463–487.
TRAITLER H, WATZKE HJ AND SAGUY IS (2011), 'Reinventing R&D in an open innovation ecosystem', *Journal of Food Science*, **76**(2), 62–68.
VOLBERDA HW, FOSS NJ AND LYLES MA (2010), 'Perspective – absorbing the concept of absorptive capacity: how to realize its potential in the organization field', *Organization Science*, **21**(4), 931–951.
YIN R (2002), *Case study research*, Sage Publications.

3

The role of open innovation in the industry convergence between foods and pharmaceuticals

S. Bröring, University of Applied Sciences Osnabrück, Germany and Wageningen University and Research Centre, The Netherlands

DOI: 10.1533/9780857097248.1.39

Abstract: The extent to which the food industry shows tendencies to converge with the pharmaceutical industry are explored in the context of the implications for innovation management and the role of open innovation in particular. Convergence is identified as the blurring of boundaries between industries by converging value propositions, technologies and markets. The effects of increasing health care costs as well as a greater consumer interest in preventing diseases by making distinctive food choices are shown to be central reasons for an ongoing process of industry convergence at the intersection of the pharmaceutical and the food industry.

Key words: industry convergence, open innovation, technology management, functional foods, food supplements, nutraceuticals.

3.1 Introduction

Convergence is a phenomenon to be observed in many industries such as telecommunications, computing and consumer electronics (Katz, 1996; Duysters and Haagedoorn, 1997; Prahalad, 1998; Bröring, 2005). In general, it can be defined as the blurring of boundaries between different previously distinct industries (Choi and Valikangas, 2001). Likewise the Organisation for Economic Co-operation and Development (OECD) defines convergence as follows 'the blurring of technical and regulatory boundaries between sectors of the economy' (OECD, 1992, p. 13).

Convergence leads to emerging interindustry segments. These offer a plethora of opportunities for new fields of business, but firms have to employ knowledge and technologies not within their traditional expertise (Bierly and Chakrabarti, 1999). Thus, firms require many critical but unfamiliar knowledge areas for engaging in innovation (Bröring, 2010b). This leads to the question of how companies of different industries as well as different resource backgrounds and competence profiles can engage in the emerging interindustry segments and to what extent do different companies open up their innovation process.

The food industry is no stranger to this phenomenon as it seems to show tendencies to partly converge with the pharmaceutical industry leading to the emerging sector of nutraceuticals and functional foods. Accordingly, this chapter addresses two research questions:

1. To what extent can we observe tendencies of industry convergence between foods and drugs?
2. What is the role of open innovation in the convergence between food and drugs?

Thus, by analysing different cases of convergence, the chapter seeks to contribute to management research by linking the relatively unexplored context of industry convergence in the food industry to open innovation.

The remainder of this chapter is organised as follows. Section 3.2 looks at the growing number of studies of industry convergence to give an overview on drivers and patterns of this phenomenon. Subsequently, in Section 3.3, the challenges of industry convergence for innovation management are described in order to derive the need for open innovation. The specific case of industry convergence between foods and pharmaceuticals is analysed in Section 3.4 answering the first research question, to what extent convergence can be witnessed between the two sectors. In line with research question two, Section 3.5 explores the role of open innovation in the converging segment of nutraceuticals and functional foods in particular. On the basis of this discussion, conclusions and implications for theory and practice are highlighted in Section 3.6. To provide the reader with some thoughts for the future, this chapter closes with an outlook on potential future developments of the interindustry segment of 'nutraceuticals' given in Section 3.7.

3.2 A brief literature review on industry convergence

Aspects of industry convergence including patterns of convergence and technology and market drivers for convergence are now explored.

3.2.1 Different patterns of convergence

Industry convergence itself, is an evolving field of research in strategic management (Kim *et al.*, 2006; Pennings and Puranam, 2001). When investigated, the focus has predominantly been on computing, communication and consumer electronics (Dong *et al.*, 2004; Duysters and Hagedoorn, 1997; Greenstein and

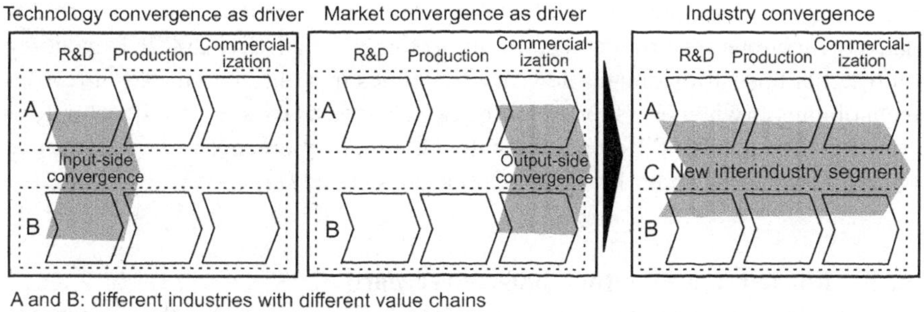

Fig. 3.1 Input-side and output-side convergence resulting in industry convergence. *Source:* Bröring, 2010a.

Khanna, 1997; Katz, 1996; Lee *et al.*, 2005; Rockenhäuser, 1999; Stieglitz, 2004; Wirtz, 2001). These existing studies on convergence commonly show that, in general, there are different levels and sources of convergence, occurring in almost any combination. Convergence can take place at the product–market level, at the firm level or at the industry level, and it is the latter that is the most profound and that, thus, is the subject of this chapter. Moreover, convergence can be induced by interindustry networks of new technology platforms, which are invisible from a consumer's view, or convergence of customer demands, regulation and industry standards. In most cases market or technological drivers of industry convergence present an interwoven development, but they can as well occur separately (Gambardella and Torrisi, 1998; Stieglitz, 2004). This means that forces driving technological convergence do not always initially lead to a convergence in end markets. In general, there exist various types of industry convergence which differ with regard to their consequences to the involved firms. This contribution follows Malhorta and Gupta (2001) in distinguishing industry convergence into input-side and output-side convergence as detailed in Fig. 3.1.

3.2.2 Technology-driven input-side convergence

Before any trends of convergence emerge, patterns of technological developments as well as technological competencies vary significantly between industries (Cantwell and Paniccia, 1998). This situation may change if new technology areas arise which are applied across industries (Bierly and Chakrabarti, 1999). In this context input-side refers to the converging trends of technologies and technology-platforms (Malhorta and Gupta, 2001; Pennings and Puranam, 2001). For example, the convergence of telecommunication, computing and consumer electronics has been induced by the emergence of the digital era. The technological advancements that have led to the possibilities of digital transmission did not distinguish between bits reflecting videos, voices or texts. The same phenomenon can be seen in the revolution of biotechnology having the DNA as a common code, which likewise resembles a common technology platform applied to different markets. Hence, technology platforms grow together and fuse (Kodama, 1992). This process occurs because technological platforms have an integrative character, linking formerly discrete functions and product features into a single system (Lei, 2000). Depending on whether this fusion creates new functionalities in addition to the two previous technologies or whether it substitutes the two previously existing ones, technology convergence can be differentiated into being complementary or substitutive (Greenstein and Khanna, 1997; Lei, 2000; Stieglitz, 2004).

3.2.3 Market-driven output-side convergence

Output-side convergence, on the other hand, is triggered by the converging demand structures of different industries, and it occurs when customers treat products of different industries in the same way, i.e. products that originally are not in competition with each other start to become substitutes (Malhorta and Gupta, 2001; Pennings and Puranam, 2001). An example for such a market-driven convergence which was initially triggered by technological convergence can be seen in personal computers and televisions. Both products are increasingly used as substitutes for one another. Laptop computers with a DVD player, for instance, combine the classical functions of a data processing computer with consumer electronics and entertainment. They are increasingly popular for watching DVDs or even TV programmes – product functions traditionally not served by the computer industry but by the consumer electronics sector (Rockenhäuser, 1999). Therefore, market convergence can be seen as a result of the consumer trend towards convenience and one-stop shopping (Graack, 1996), where consumers try to satisfy different needs in one transaction (Pennings and Puranam, 2001). Moreover, it has to be noted that, in most cases, technology convergence is the ultimate trigger for industry convergence, whereas market convergence is reinforcing it but not driving convergence in the first place. However, this work follows Pennings and Puranam (2001) in viewing market-related convergence resulting from a 'growing similarity of needs' as a separate force driving the process of convergence.

3.2.4 Regulation as a driver for industry convergence

Furthermore, a third dimension of convergence can be seen in the convergence of regulations, which can be the result of industry convergence, or can trigger it. In the telecommunication sector regulation has blocked convergence initially. Only with deregulation and privatisation of this sector has convergence been possible (Lei, 2000). In other cases, because technological developments are rapid, regulation lags behind. This is the case in modern biotechnology, where product technologies were developed years before any legal classification emerged (Castle *et al.*, 2006). Because there are no industry standards regulation only evolves over time. On the one hand, this can produce a broader set of options, but it may also lead to confusion and a decisive decrease in planning reliability. Thus innovations placed on the market when there is a lack of regulation may, later on have to be adapted in order to comply with emerging legislation (Bröring, 2005).

Table 3.1 Industry convergence as a special situation for innovation

Convergence of technologies	Convergence of markets	Regulation and standards
• Application of new technologies across industry boundaries • Fusion of existing technologies owned in different industries to form a common one	• Demand structures converge • Substitute products arise from another industry	• Missing industry standards • Regulation for the new 'converged' sector is only about to emerge
New areas of technological knowledge become relevant for innovation	New areas of market knowledge become relevant for innovation	Legal uncertainty in defining the options for innovation

Source: Bröring, 2010a.

There are, therefore, many dynamic characteristics of industry convergence that influence innovation and these are classified as given in Table 3.1.

3.2.5 Complementary vs. substitutive forms of industry convergence

The occurrence of both technology-driven input-side convergence and market-driven output-side convergence leads to the development of a new interindustry segment. Owing to convergence of the input- and the output-side, a new value chain emerges (Dong *et al.*, 2004). An important question in this context is the degree to which a new industry segment leads to substitution of the old segments. According to Greenstein and Khanna (1997), convergence can either lead to a total phasing out of the two formerly separate industries (1+1=1) or it can also trigger the emergence of a new industry segment, which complements the two formerly existing ones (1+1=3). As shown in Fig. 3.1, the question is whether the newly emerged value chain C replaces A and B or is complementary. One explanation for the different outcome of convergence being either substitutive or complementary can be seen in the respective combination of technology and market convergence: only if both technology *and* market convergence can be characterised as substitutive, industry convergence as such is substitutive (Bröring, 2010a). This means that two value chains are merging, leading to a total phase out of the two hitherto separate ones. Another explanation might be seen in the distance of the two merging markets, as market convergence is a precondition for the convergence of two industry sectors (Curran *et al.*, 2010). However, current research in the field of convergence suggests that in most cases there are rather a number of many small sub-segments than one large merged industry (Stieglitz, 2004). Furthermore, substitutive convergence may change the importance of single industries; value creation may be shifted to a sub-segment, which only gradually substitutes the *ex ante* segments (Lind, 2005).

3.3 Convergence-related challenges and the role of open innovation

At this point one may ask: what does industry convergence have to do with open innovation? To answer the question, this section first of all explores the challenges for innovation management in a context of convergence. These challenges stress the need for a more open approach to innovation. Following this rationale, the second part of this section examines the open-innovation approach as a means to cope with the challenges resulting from industry convergence.

3.3.1 Challenges for innovation in industry convergence

One can distinguish various challenges arising from competency gaps, and over uncertainty to problems resulting from missing standards and lack of absorptive capacity. These challenges are now examined.

Missing competences and path-dependency in industry convergence

The necessary areas of knowledge and accompanying competences are traditionally owned by different industries. That these essential knowledge and competence sources are industry specific (Fai and von Tunzelmann, 2001) can be explained by the fact that they are cumulative and path-dependent. This is because competences only develop over time following specific development paths (Diericks and Cool, 1989; Greener, 2002; Leonard-Barton, 1992; Teece *et al.*, 1997). Hence, it seems that no player, notwithstanding from which industry affiliation he originates, possesses all the competences needed for innovating successfully in times of convergence.

The degree to which competence gaps arise is contingent upon the pattern of convergence. With respect to innovation management the differentiation into substitutive and complementary convergence seems necessary in order to specify how pressing trends of convergence are. For total substitution of previous sectors, innovation seems to be imperative to keep up with trends of convergence, whereas, for complementary convergence, firms may seize the opportunities of industry convergence or just focus on the existing 'historical' industry sector not requiring any adaptation.

Uncertainty due to missing standards and lack of absorptive capacity

As already discussed, new areas of knowledge may become relevant for engaging in innovation. Concerning regulation and industry standards, Day and Schoemaker (2000) state that there are different rules in emerging industries, because neither industry standards, nor regulation have yet been developed. Following the classification of Sanchez (1996), this contribution understands interindustry segments arising from convergence as evolving markets. Evolving markets can be characterised with a higher degree of uncertainty. Uncertainty arises for two reasons. First, the evolving market itself is not characterised by clear preferences or defined technological designs. Second, firms involved in this market may struggle because they lack the required competences and are sometimes also insecure about which competences to develop, because the specific competences and resources cannot be identified with higher precision. Firms seem to be insecure about their position in an emerging value chain.

One reason for the inability to identify opportunities resulting from industry convergence can be seen in limited absorptive capacity. Cohen and Levinthal (1990) claim that for the most external knowledge areas to be used effectively, an organisation must already possess a considerable level of competence in the area so that it is able to recognise the value of new, external information, assimilate it, and apply it to commercial ends (Cohen and Levinthal, 1990). They refer to this notion as absorptive capacity which only evolves over time and is path-dependent. The reasoning for path-dependency originates from the influence of past activities on future ones. Cognitive processes are cumulative and idiosyncratic and, therefore, accumulated experiences determine the capability of a firm to absorb the external knowledge needed for idea generation. For example, firms that build a distinct technological competence are more likely to build on existing resources (e.g., Dosi, 1982; Diericks and Cool, 1989; Helfat, 1994).

Different 'industry recipes' and the distance between converging sectors
Successful innovation requires organisations to combine many critical knowledge areas (Bierly and Chakrabarti, 1999). The theoretical grounding of this contribution is the resource-based view (RBV) of the firm as it understands competences and resources as influences for firm differences and behaviour (Penrose, 1959; Wernerfeld, 1984; Barney, 1991). It explains that there are industry-specific competences (Fai and von Tunzelmann, 2001) and industry specific 'recipes' for innovation (Spender, 1989). Therefore, another important characteristic in analysing industry convergence and its impact for innovation and technology management is the dissimilarity of the previously separate industry sectors. This can be seen in the extent to which the two converging sectors have been alike in terms of their competence bases before convergence. In this connection Kim *et al.* (2006) refer to conventional convergence phenomena and heterogeneous interindustry convergence. The latter indicates a greater distance between the two converging sectors as a result of different qualifications of the two sectors. Being theoretically anchored in the RBV, in this chapter we examine heterogeneity in terms of differences in resources and competences of the two converging industries. The greater the difference between competences and experiences, and, therefore, the more the industry-specific development path differs, the more profound are the problems firms have in realising their opportunities in an emerging interindustry segment (Bröring *et al.*, 2006) owing to lack of absorptive capacity as described in the second challenge above. In addition, the greater the distance, the more difficult interindustry collaboration can be, owing to different resource profiles, but, most importantly, also because of different approaches (industry recipes) of how to conduct R&D (Bröring, 2005). Firms seem to need excellent combinative capabilities for linking internal and external R&D (Kogut and Zander, 1992) to overcome these challenges.

3.3.2 Open innovation as an approach to cope with industry convergence related challenges

For industry convergence, which is based on both convergence of technologies as well as convergence of markets, firms face knowledge, capability and resource gaps. Chesbrough (2003) has coined the term 'open innovation' to indicate the interactive and open nature of innovation and technology management which allows for an active exchange of knowledge between the innovating company and its environment. This means that different steps from idea generation to product launch are not necessarily performed by the company itself but also through specific partnerships. The opposite of this open approach can be seen in a closed innovation design where a company carries out every step from idea generation to development and commercialisation internally, using its own resources and competences (Chesbrough, 2003). Consequently, open innovation can be defined as 'the use of purposive inflows and outflows of knowledge to accelerate internal innovation, and expand markets for external use of innovation…' (Chesbrough, 2006, p.1). This definition indicates two flows of open innovation: inbound (inflows) and outbound (outflows) open innovation (see Chesbrough and Crowther, 2006). To further characterise these different flow patterns Gassmann and Enkel (2006) distinguish three core processes of open innovation:

1. the outside-in process to internalise external knowledge;
2. the inside-out process to externalise (commercialise) internal knowledge via licensing etc.; and
3. the coupled-process involving co-creation and an exchange between two different parties.

At first sight, all three processes seem relevant for innovation in the context of convergence. However, the outside-in process, in particular, seems relevant to this contribution. This rationale follows the overall intention to understand how companies can use open innovation in times of industry convergence to close competence gaps, because firms involved in convergence grapple with new bodies of technological knowledge not typically found in their own industry. Therefore, external technology sourcing to close competence gaps in technology development is crucial for convergence in new product development (NPD). In this context many researchers have focused on R&D networks to develop new platform technologies such as biotechnology (see e.g., Chiesa and Toletti, 2004), licensing (for example, see Atuahene-Gima and Patterson, 1993; Tidd and Trewhalla, 1997; Arora *et al.*, 2002), joint R&D agreements (for example, see Arora and Gambardella, 1990; Ritter and Gemünden, 2003; Steensma and Corley, 2000), joint ventures (for example, see Schildt *et al.*, 2005; Tidd *et al.*, 2005), or acquisitions (for example, see Arora and Gambardella, 1990; Hagedoorn and Duysters, 2002; Pisano, 1991), stressing their respective (dis)advantages in NPD.

It is beyond the scope of this contribution to further explore the respective advantages and disadvantages of different forms of collaborative arrangements. However, as reports suggest, the chosen form of collaboration in NPD depends on the extent of innovativeness of a certain innovation (Bröring *et al.*, 2006). One can assume that the more existing competences are challenged by a NPD project, the higher is the need for complementary knowledge sourced-in from outside the firm (see Powell *et al.*, 1998; Bröring *et al.*, 2006; Rothaermel and Deeds, 2006).

Furthermore, the more important it is to the firm to establish the appropriate alternative partnership with the appropriate partner at the appropriate time of the NPD process. For instance, access to market information can be a relevant criterion for setting up a strategic alliance in NPD in emerging value chains. Thus, interorganisational projects targeting resource complementarities become key to organisational design (Hagedoorn, 1993).

3.4 Evidence for industry convergence between foods and pharmaceuticals

Having elaborated on the theoretical basis of industry convergence and open innovation this section now tries to answer the first research question, to determine to what extent industry convergence can be witnessed in the nutraceuticals and functional foods (NFF) sector.

3.4.1 Trends of convergence between nutrition and pharmaceuticals to form 'nutraceuticals'

Health and well-being is one of the most influential food trends opening up a well-spring of innovation opportunities for the food, specialty chemicals and pharmaceutical industries. As a result since the beginning of the 1990s the new interindustry segment of nutraceuticals and functional foods has developed at the borderline of foods and pharmaceutical products (Bröring, 2005). As depicted in Fig. 3.2 one can detect multiple trends of convergence and a gradual overlap of, in particular, the food and pharmaceutical industries leading to a new interindustry segment at the border between foods and drugs.

That this new interindustry segment is no longer 'just' an academic playing field to analyse competitive behaviour, and innovation strategies of functional foods has become even more evident in the industry itself. Nestlé's recent announcements in September 2010 on the creation of 'Nestlé Health Science S.A.' and the 'Nestlé Institute of Health Sciences' to confidently '... pioneer a new industry between food and pharma...' show that these opportunities will trigger more R&D and innovations in the future. According to Robert Redgwell, Senior Research Scientist at Nestlé Research Centre, the company strives to become a leader in the emerging field of personalised nutrition (Redgwell, 2011).

Looking at the value chain of the emerging NFF sector, one can observe both: input-side convergence and output-side convergence as depicted in Fig. 3.3. Moreover, the term 'nutraceutical' illustrates the coalescence of the nutrition and the pharmaceutical industries. It integrates different technologies and consumer trends but does not lead to a phasing out of the two formerly distinct industries. Therefore, the NFF sector presents a case of complementary industry convergence driven by complementary technological and market convergence as depicted in Fig. 3.3.

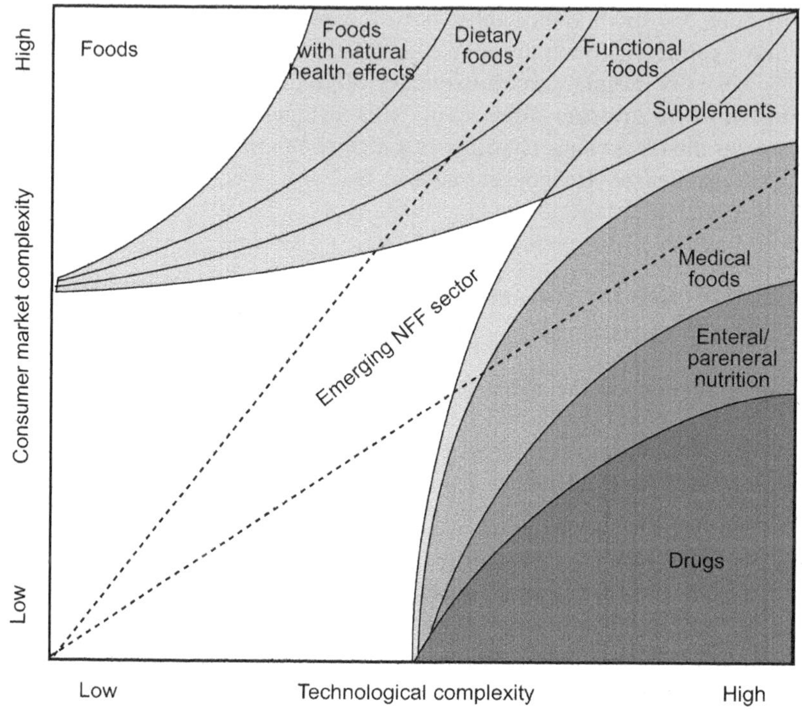

Fig. 3.2 The emerging interindustry segment of NFF between foods and drugs. *Source:* Bröring, 2005.

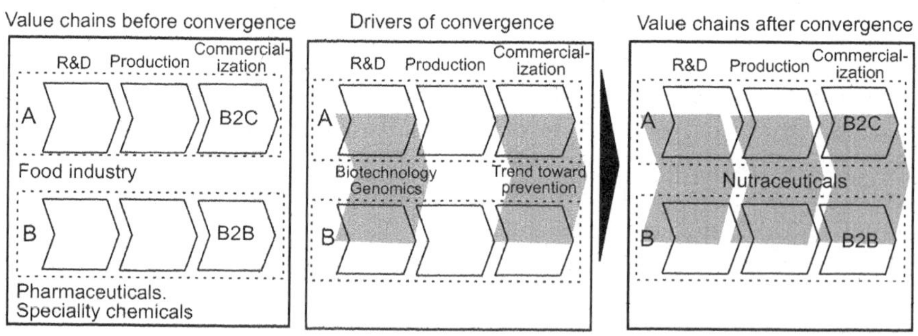

Fig. 3.3 Convergence of food and pharmaceuticals. *Source:* Bröring and Cloutier, 2008.

3.4.2 Technology-driven input-side convergence in the neutraceuticals and functional foods (NFF) sector

New technologies and their rapid diffusion across industry boundaries are the main drivers for industry convergence leading to interindustry segments. This increase in overlap of market and technological structures is triggered by advances made in the broad field of biotechnology and, most notably, genomics (Bröring, 2005;

Bröring and Cloutier, 2008; Siedlock *et al.*, 2010). Thus, technology-driven input-side convergence can be seen as a driver for convergence between foods and drugs, as both sectors increasingly draw on similar technological platforms. For instance, the progress of biotechnology plays a special role for the development of bioactive food or pharmaceutical ingredients. Furthermore 'nutrigenomics' as the connection of nutritional sciences and genomics represents a new area with huge potential for innovation. Scientific knowledge from the fields of genomics, crop science and genetic engineering as well as modern medicine is combined with the goal to develop individualised nutritional products for medical prevention and treatment (Zhang *et al.*, 2010). Thus, 'the gap between food and pharma is narrowing' as Georgiou *et al.* (2011, p. 2) put it. In the *Journal of Pharmacology*, different areas of the interaction between pharmacology and nutritional science are examined by using a joint approach as for instance in the fight against obesity and its metabolic complications. Dietary measures have been shown to be equally as effective as drug treatment in preventing, stopping or reverting the development of type 2 diabetes. Hence, there is an increasing realisation that the continuum between health and disease exists without clear boundaries and requires technological competences from nutritional as well as pharmacological sciences.

3.4.3 Market-driven output-side convergence in the NFF sector

Market drivers such as an increasing consumer interest in preventing diseases with distinctive food choices along with societal triggers as the increasing health care costs have led to the emergence of a new interindustry segment between foods and drugs (Bröring, 2010a). Convergence can, therefore, also be induced by the fusion of demand structures and the combination of previously distinct product features into one hybrid product (Pennings and Puranam, 2001; Sääksjärvi, 2004). Hence, this form of complementary technological integration is accompanied by the convergence of product functions (e.g., nutritive and preventive functions). As illustrated in Fig. 3.3, the driver for market-related output-side convergence can be

Table 3.2 Market trends driving the convergence between foods and drugs

Driving trends	Implications
Aging population	Increased life expectancy, interest in anti-aging products
Rising health care costs	Economic burden, limited reimbursement, increasing self-medication
Awareness of diet/health relationship	Increased demand for healthier foods
Proactive behaviour	Increased demand for products which will prevent disease
Willingness to pay for health benefit	Increased interest of food companies to increase margins in functional foods in times where retail has induced severe price wars

Source: Bröring, 2005.

Table 3.3 Product functions of nutraceuticals and functional foods versus drugs

Functions of nutraceuticals & functional foods: prevention	Functions of drugs: treatment and prevention
Future benefit	Immediate effect
Broad-brush approach	Targeted population
Safe	Benefit > risk

Source: Clydesdale (1998).

seen in a growing trend towards health and well-being, which provides constantly growing sales of hybrid products such as cholesterol-lowering foods. Looking at these market trends more closely, one can observe that there are four main market/societal drivers for the convergence between food and pharma as depicted in Table 3.2.

Thus, the emerging segment of nutraceuticals and functional foods offers room for a variety of products from the food and the pharmaceutical industry (Green and van der Ouderaa, 2003). These range from the simple enrichment of foods with functional ingredients to products of personalised nutrition based on the scientific knowledge on genomics (Bröring, 2010a). Therefore, the emerging value chain resulting from market-driven output side convergence is an attractive target for players of food (e.g., Unilever, Nestlé and Danone) and pharmaceutical or chemical (e.g., Merck, Abbot Labs, DSM and BASF) industry origin (Bröring, 2005). However, one common distinction of the two product areas is their intended product function.

According to Clydesdale (1998), products of the food-like NFF sector are consumed for preventive uses. This implies that they deliver only a future benefit as detailed in Table 3.3. Drugs, on the other hand, deliver an immediate effect. However, this distinction of food and drugs by product function does not always hold: for instance phytosterol-enriched foods bring an immediate benefit of reducing the blood-cholesterol level (Chadwick *et al.*, 2003). On the other hand the pharmaceutical industry is also aiming at the segment of prevention. Looking at the technological field of genomics, both foods and drugs may be able to deliver benefits to a targeted segment of the population. Hence, product functions are blurring, which is even more reinforced by the consumer trend of seeking to combine health benefits in food products. Thus, market convergence next to technological convergence seems to be a second driver for industry convergence.

3.4.4 Regulation as a driver for the convergence between foods and pharmaceuticals

Next to the convergence of technology platforms and consumer needs, regulation presents a third driver for the ongoing process of industry convergence between foods and drugs. Legal definitions and regulations are important to innovating firms, because they determine what is possible. Hence, a detailed *ex ante* analysis of the relevant regulatory environment is a crucial prerequisite for planning the

NPD process and estimating the costs for tests regarding product safety and approval. In established markets these external conditions may be defined and well known. In the NFF sector, the external conditions are still evolving, thus causing additional costs (Wijnands *et al.*, 2007). However, legal definitions for what is a functional food or a nutraceutical are not available in most countries, let alone their harmonisation on the EU or on a supranational level. At the same time, a thorough understanding of the evolving regulatory environment is ranked as very important by manufacturers (Hilliam, 1998). However, regulation as for instance the legislation regarding health claims on food products (see EC 1924/2006) is evolving only very slowly. The regulation on nutrition and health claims made on foods, adopted by the council and parliament in 2006, lays down harmonised rules across the EU for the use of nutrition claims such as 'low fat', 'high fibre' or health claims such as 'reducing blood chloresterol'. However, it exemplifies how convergence increasingly manifests itself on the regulatory level. Why is that? There are increasingly similar regulatory requirements for foods and drugs not only in the EU but also in North America (Herath *et al.*, 2006). This is the case for the substantiation of health benefits as required by the European Food Safety Authority (EFSA) as part of the assessment of health claims (Flynn, 2010). Functional foods and supplements are confronted with clinical trials having similar high ('gold') standards (randomised, controlled, double-blind) as required in the design of phase II or III clinical studies involved with the development of new pharmaceuticals. This imposes heavy challenges in terms of clinical research expertise and investment for food companies striving to enter into the NFF sector.

To conclude and, thus, answer research question one: it is possible to observe trends of industry convergence between foods and drugs. This convergence can not only be observed in the rise of so-called hybrid products such as the cholesterol-lowering margarine but also increasingly manifests itself in the application of common technology platforms on the one hand and common regulatory approaches on the other hand. However, one has to note, food and pharmaceutical industries 'just' show a complementary case of convergence building a new interindustry segment. That means there is clearly not going to be a total phasing out process as in other industries, where hybrid products increasingly substitute the old classical ones. In other words, companies can seize the opportunities from convergence, but they certainly do not have to because the old former classical industries will remain.

3.5 Open innovation in order to cope with convergence in the neutraceuticals and functional foods (NFF) sector

As discussed in Section 3.3, industry convergence brings three main challenges for innovation:

1. missing competences,
2. uncertainty owing to missing industry standards, and

3. different approaches owing to different 'industry recipes' for innovation. In the following subsection, the first challenge 'missing competences' is the focus as it is closely linked to the need for open innovation.

3.5.1 Missing competences of players in the NFF sector

Products of the nutraceutical sector like cholesterol-lowering food products require technological competences from the pharmaceutical and chemical industries as well as strong consumer market competences including consumer insights and preferably strong assets such as well-known brands. It seems that both sectors with their specific competence profiles are complimentary to each other (compare Table 3.4). Speciality chemicals companies are strong in technology development and food companies are consumer-oriented and already employ marketing competences (Bröring, 2005; Rothaermel and Deeds, 2006).

Table 3.4 Competence differences of actors in the NFF Sector

	Industry background	
	Food/nutrition	Pharma/speciality chemicals
Technological competences		
R&D intensity	• Low (< 1% of sales)	• High (>14% pharma/ >4% chemicals)
R&D cycles	• Short (1–3 years)	• Long (10–20 years in pharma)
IP strategies	• Trademarks, consumer brands	• Patents (preferably on the product/compound)
New technologies	• Traditionally external sourcing, no own development skills	• Foster internal development of technologies
Market competences		
Market insights	• Consumer market (B2C)	• Industrial clients (B2B)
Access to distribution channels	• Mass-market distribution via retail chains	• B2B wholesalers, pharmacies, nutritional speciality stores
Communication	• Oriented to end-consumer	• Science, industrial partner–oriented
Regulatory competences		
Regulatory approval processes	• Little or no skills	• Own regulatory approval business functions
Familiarity with safety/efficacy testing	• No experience in clinical product evaluation	• Long-term experiences with clinical trials • Via Contract Research Organizations (CROs) or own facilities
Monitoring of the emerging regulation	• Limited access to changes in regulation via industry associations	• To some degree active in developing standards (biomarkers) relevant for health claims

Source: Bröring (2005).

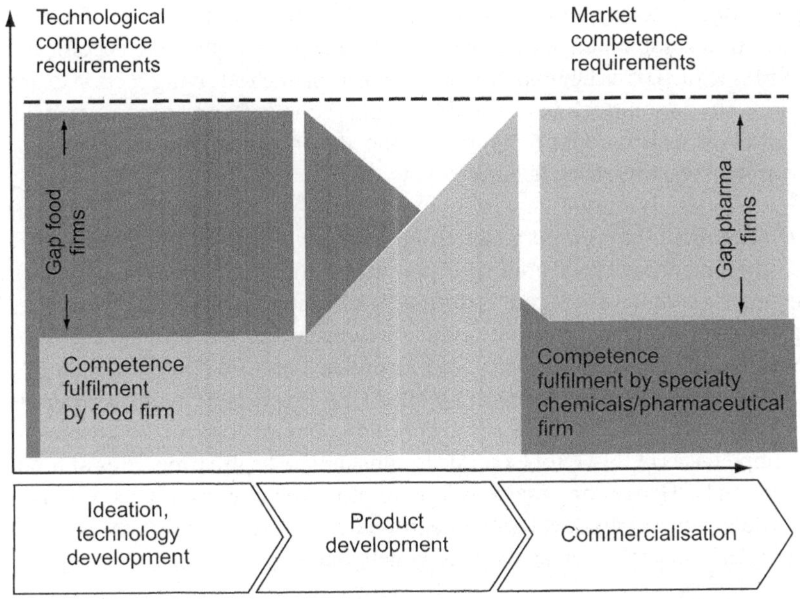

Fig. 3.4 Competence gaps arising in NPD in the converging NFF sector. *Source:* Bröring and Cloutier, 2008.

However, competence gaps are obvious: one can observe food companies producing functional food products without having undergone any clinical trials to substantiate their claimed health benefits. At the same time, there are pharmaceutical companies trying to apply their technological knowledge to the emerging sector but failing owing to limited adaptation of the product to the consumer market. Hence, there exist different forms of competence gaps in the NFF sector. Before analysing how open innovation can help to close competence gaps, these gaps need to be described in detail.

The schematic overview given in Fig. 3.4 resembles the situation of a technology-intense consumer product requiring both strong technological and strong consumer market competences. Following the preceding rationale of different competence profiles for the two different industries, the situation depicted in Fig. 3.4 creates technological gaps (e.g., clinical trials) for food companies and business-to-consumer (B2C)-market-related competence gaps for pharmaceutical/specialty chemical companies.

3.5.2 Open innovation as a means to close competence gaps

Firms who seek to enter the interindustry segment of NFF realise the need for partners and therefore often engage in R&D collaborations of different scope and intention. The question now is: what stage of the NPD process requires what type of open-innovation processes? In terms of the three processes of open innovation as described by Gassmann and Enkel (2006), it can be proposed that the outside–

in processes to source-in missing competences, but also the coupled process to benefit from complementary competences seem appropriate in the specific case of the NFF sector. To further explore this proposition, the following section examines previous studies on (technology and/or market) competence sourcing (outside–in processes) as well as coupling processes and also includes 'minicases' to illustrate these processes on a more detailed level.

R&D networks for developing new technology platforms for the NFF sector

R&D competences of food companies as such are limited, because only a few food companies maintain a separate or formal R&D department. In the evolving NFF sector several R&D networks have emerged during the last decade. For instance, in North America, the Institute of Nutraceuticals and Functional Foods (INAF) located at the Université Laval in Quebec City presents a classical example for pooling resources and competences. This network enables food companies to use the competences at INAF for example to substantiate health properties of foods via clinical trials. Hence, one major role is to close technology-related competence gaps. New technology development activities, however, in most cases, require a larger R&D network such as the NuGO network.

Minicase nutrigenomics organisation (NuGO)

The NuGO network can be regarded as a R&D network focusing on radical nutrigenomic-based innovation, which requires a close interdisciplinary collaboration. NuGO evolved from a European-funded Network of Excellence, the full title of which was 'The European Nutrigenomics Organisation: linking genomics, nutrition and health research'. Today, NuGo is an association of Universities and Research Institutes focusing on jointly developing the research area of nutrigenomics and, thereby, advancing this new technology platform.

This case illustrates the interdisciplinary character of new technology development in R&D networks. In a coupled open-innovation process, different partners share their competences reciprocally to further develop an emerging technology platform. Even if these developments are taking place very much at the public research level, they increasingly integrate company R&D. Thus, these networks are about to build the basis for another subsegment between food and pharmaceutical industries: personalised nutrition.

Acquisitions in the NFF sector as a means for competence sourcing

In line with the different options of how to close competence gaps, acquisitions clearly present an important vehicle. According to Frost & Sullivan (2011) 119 mergers and acquisitions (M&A) transactions have been closed in the global food ingredients (antioxidants, functional bio-ingredients, etc.) segment between 2007 and 2010. This study has also analysed the purpose of acquisitions showing that some acquisitions where striving for forward integration (3.3%) whereas most of the acquisitions intended to gain synergies (40%) and strengthen the product/technology portfolio (32%). For instance, DSM acquired Martek to benefit from Martek's strong technological and market position for omega-3 fatty acids

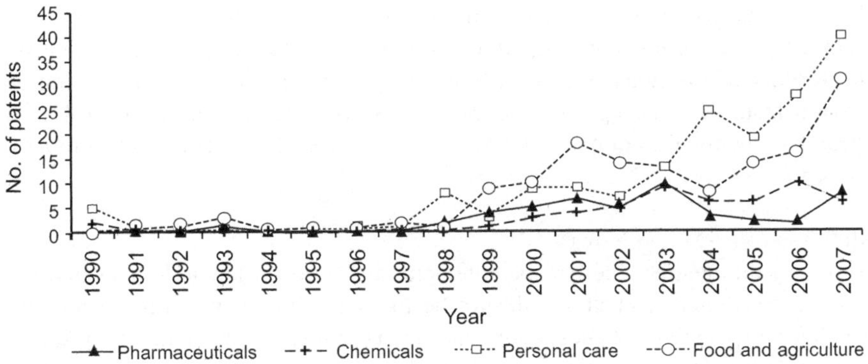

Fig. 3.5 Number of patents according to the industrial background of the filing companies. *Source:* Bornkessel *et al.*, 2011.

ingredients (Frost & Sullivan, 2011). However, more relevant cross-industry acquisitions further strengthening the convergence argument have been performed by Nestlé.

Minicase Nestlé Health Sciences SA

Nestlé being traditionally rooted in the food industry with the new establishment of the Nestlé Health Science SA in September 2010 shows increasing 'cross-industry' M&A activities. Coming from the food industry Nestlé seems to reshape its competence base by acquiring different start-ups which belong to the pharmaceutical industry, for instance CM&D Pharma, to foster its medical nutrition developments.

This case indicates that Nestlé seems to engage in technology sourcing to further develop its medical nutrition business. In doing so, it seems that the food company is building new competences for the emerging NFF sector. In this regard, acquisitions can be seen as an inbound/outside-in process of open innovation to source new competences.

R&D collaborations in the NFF sector

However, most competence gaps are not closed by acquisitions or cross-ownerships, but by different forms of interindustry R&D collaborations. This holds particularly true for SMEs representing the majority of all food companies (Capitanio *et al.*, 2009). In general, competence gaps in the NFF sector lead to numerous collaborations in the NFF sector. To give an example, a recent patent analysis in the field of probiotics scrutinising the development of new probiotics ingredients shows an increasing number of collaborations as depicted in Fig. 3.5. Collaborations between Food & Agriculture and Pharmaceuticals & Chemicals have risen since the early 2000s.

Moreover, a previous study on collaborations involving a sample of 54 R&D projects in the NFF sector showed that 87% have been carried out by using either R&D and/or marketing collaborations (Bröring and Cloutier, 2008). This finding illustrates the complexity of developing functional food products requiring input from different parties, e.g., ingredient suppliers, food manufacturers, scientific institutions, and regulatory bodies. In most cases, the need to collaborate can be derived from the missing competences as has also been the case for the Dutch specialty chemicals company DSM and the Swiss Pharmaceutical company Novartis:

Minicase of DSM and Novartis

Although the health benefit of supplements is based on technological innovations, a consumer product should not be too scientific–technically positioned. The Dutch specialty chemicals company DSM realised that it had no consumer market competences during the innovation process of its sports nutrition Peptopro®, a whey protein which improves utilisation of glucose by muscle cells. Therefore, it collaborated with a consumer goods company specialised in sports nutrition, rather than commercialising the product on its own. Missing market competences can be observed in the attempt of the Swiss pharmaceutical company Novartis, which had been very actively promoting its functional food range AVIVA. The company failed and withdrew the whole range after only three years. The pharmaceutical company being very strong in the over the counter (OTC) market for pharmaceuticals and consumer health care products was missing business-to-consumer market competences (e.g. food retail experience) needed in the fast moving consumer goods (FMCG) market of foods.

The results of this minicase show that, in contrast to Novartis, DSM changed its innovation strategy towards a more open approach accommodating for the missing B2C market competences. Looking at the value chain, it becomes clear that in the case of Novartis existing core competences can become core rigidities and somewhat hamper companies to move forward in the value chain (Leonard-Barton, 1992). By contrast, DSM managed to do so by a close co-operation with a consumer goods company. In terms of the three different open innovation processes, R&D collaborations in this case present a coupled process where complementary competences (as in the Peptopro® case) are pooled.

At this point, one might ask how both specialty chemical and food industry come together, because they may not realise the value owing to lack of prior knowledge expressed by missing absorptive capacity. Here, one can observe so-called innovation intermediaries like Evaluserve, Innocentive, yet2.com or Ninesigma which act as matchmakers in helping to link parties, who alone would not identify their partners (Enkel *et al.*, 2009; Herzog and Niedergassel, 2007; Lakhani, 2008).

3.6 Conclusion

This chapter examines industry convergence between foods and pharmaceuticals and the role of open innovation in closing competence gaps. In the first challenge studied, the extent to which these two sectors converge, one can observe a case of complementary industry convergence between foods and pharmaceuticals leading to the NFF sector arising at the borderline of the two merging industries. Owing to the complimentary character of this case of industry convergence, i.e., not leading to a phasing out of the two previous sectors, the emerging NFF sector clearly has a niche market character. This means that companies may want to enter the segment, but they can also remain in their traditional industry boundaries not engaging in innovation for the NFF sector.

Moreover, for those companies deciding to enter the evolving interindustry segment, this contribution illustrates the convergence-related challenges and stresses the need for open innovation in particular. Accordingly, by studying the role of open innovation in industry convergence, it can be concluded that open innovation is very important for companies entering the new interindustry segment. This is because open innovation seems to fulfil two major tasks in industry convergence:

1. Closing competence gaps by outside-in open innovation: the outside-in process is needed to source in missing technological or market competences via acquisitions or R&D collaborations. Hence, even though there is an emerging value chain that offers vast opportunities for innovation, many companies fail to realise these owing to missing competences, as the example of Novartis has indicated. From these preliminary observations it seems that missing market competences are often underestimated.
2. Bringing complementary resources together by coupled open-innovation processes: coupled open-innovation processes are needed for technology development as the example of the NuGO R&D network to develop the nutrigenomics platform has indicated. They are also needed in interindustry collaboration between the food and pharmaceutical/specialty chemicals companies to bring together complementary competences.

To conclude, in a context of industry convergence, firms may be trapped by their past development resulting from an industry-specific set of competences and resources. As illustrated in Fig. 3.6, the further a firm leaves its traditional positioning in the value chain in convergence, the higher the extent to which it deviates from its existing market and technology-related path-dependencies. As a consequence, the company faces severe competence gaps and may fail. These gaps may be either on the technology side or on the market side. Thus, taking into account the degree to which a company deviates from its existing market or technology development path, it has to engage in open innovation by either sourcing-in missing market (see the example of DSM) or technological (see the example of Nestlé) competences via acquisitions or R&D collaborations. Moreover, new radical innovations involving the development of a new technology-platform and market require companies to collaborate in larger R&D networks combining technologies of different industries with public research as illustrated by the NuGO example (see Fig. 3.6.).

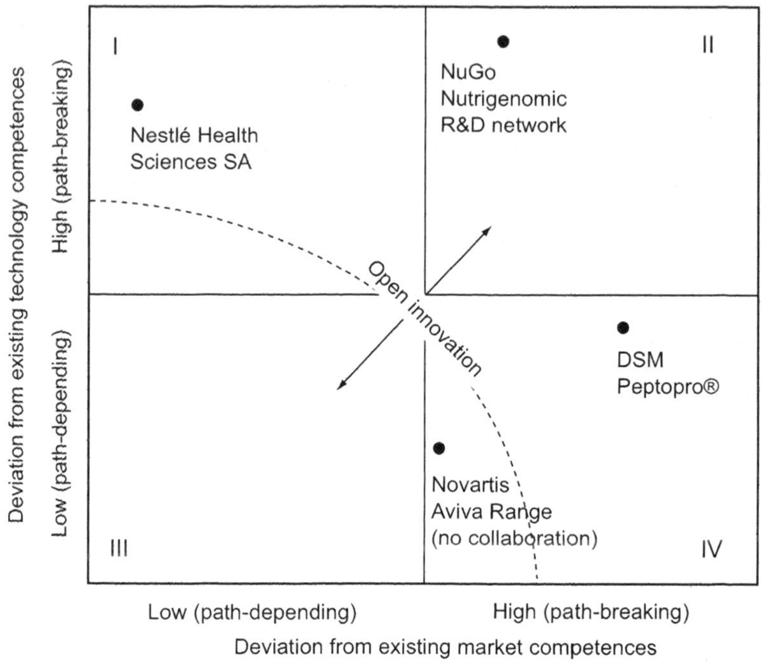

Fig. 3.6 Different types of open innovation in industry convergence. *Source:* Bröring, 2010a.

3.7 Future trends

The process of convergence, which has already led to a new interindustry segment, will not stop at the frontier of the food, pharmaceuticals and speciality chemicals industries. It will, and already does, include the AgBio industry in terms of more nutritious crops being genetically modified to alter e.g., the fatty acid composition to a more healthy profile. Beyond life sciences, other industries will become more relevant in shaping the future of functional foods, supplements and personalised nutrition (Boehlje and Bröring, 2011). For instance, competences from the IT sector will be needed to develop web-enabled and improved software tools to measure nutrient intake for personalised nutrition (see, e.g. Stumbo *et al.*, 2010). Start-up companies such as Viocare are already offering IT services around nutrition seeking to enable personalised assessment, planning, monitoring, and modification of dietary and exercise behaviours. Open innovation especially the outside-in approach of internalising external knowledge can also be witnessed here, as DSM Venturing has invested in these new technology platforms. The industry-wide application of life science (e.g., genomics) as well as IT platforms will further trigger tendencies of convergence and, thus, the evolution of inter-industry segments between the food, pharmaceutical and related industries.

3.8 References

ARORA, A. AND GAMBARDELLA, A. (1990), Complementarity and external linkages. The strategies of the large firms in biotechnology, *Journal of Industrial Economics*, **38**(4), 361–379.

ARORA, A., FOSFURI, A. AND GAMBARDELLA, A. (2002), New regulations: markets for technology in the knowledge economy, *International Social Science Journal*, **54**(1), 115–128.

ATUAHENE-GIMA, K. AND PATTERSON, P. (1993), Managerial perceptions of technology licensing as an alternative to internal R&D in new product development: an empirical investigation, *R & D Management*, **23**(4), 327–336.

BARNEY, J.B. (1991), Firm resources and sustained competitive advantage, *Journal of Management*, **17**(1), 99–120.

BIERLY, P. AND CHAKRABARTI, A.K. (1999), Managing through industry fusion, in Brockhoff, K., Chakrabarti, A.K. and Hauschildt, J. (Eds.), *The dynamics of innovation: strategic and managerial implications*, Berlin: Springer, pp. 7–30.

BOEHLJE, M. AND BRÖRING, S. (2011) The increasing multifunctionality of agricultural raw materials: three dilemmas for innovation and adoption, *International Food and Agribusiness Management Review*, **14**(2), 1–16.

BORNKESSEL, S., CURRAN, C.-S., BRÖRING, S. AND OMTA, O. (2011), Who is leading the way? Different dynamics of converging industries in nutraceuticals and functional foods, *ISPIM proceedings*, Hamburg.

BRÖRING, S. (2005), *The front end of convergence in converging industries: the case of nutraceuticals and functional foods*, Wiesbaden, Gabler-Verlag.

BRÖRING, S. (2010a) Developing innovation strategies in times of market convergence, *International Journal of Technology Management*, **49**(1/2), 272–294.

BRÖRING, S. (2010b), Innovation strategies of functional foods and supplements: challenges of the positioning between foods and drugs, *Food Science & Technology Bulletin*, **7**(8), 111–123.

BRÖRING, S. AND CLOUTIER, L.M. (2008), The role of market alliances and customer supplier interfaces in innovations in the food industry, *British Food Journal*, **110**(1), 76–97.

BRÖRING, S., CLOUTIER, L.M. AND LEKER, J. (2006), The front end of innovation in an era of industry convergence: evidence from nutraceuticals and functional foods, *R&D Management*, **36**(5), 487–498.

CAPITANIO, F., COPPOLA AND A., PASCUCCI, S (2009), Indications for drivers of innovation in the food sector, *British Food Journal*, **111**(8), 820–838.

CANTWELL, J. AND I. PANICCIA (1998), Technological change and vertical integration: analysis of international vertical integration in multinational companies, in Colombo, M. *The Changing Boundaries of the Firm: explaining evolving inter-firm relations*, New York, Routledge, pp. 158–184.

CASTLE, D., LOEPPKY, R. AND SANER, M. (2006). *Convergence in biotechnology innovation: case studies and implication for regulation*, University of Guelph, Canada.

CHADWICK, R., HENSON, S., MOSELEY, B., KOENEN, G., LIAKOPOULOS, M., MIDDEN, C., PALOU, A., RECHKEMMER, G., SCHRÖDER, D. AND VON WRIGHT, A. (2003), *Functional foods*, Heidelberg, Springer-Verlag.

CHESBROUGH, H.W. (2003), *Open innovation – the new imperative for creating and profiting from technology*, Harvard Business School Press, Boston.

CHESBROUGH, H.W. (2006), Open innovation a new paradigm for understanding industrial innovation, in Chesbrough, H.W., Vanhaverbeke W., West, J. (Eds.) *Open innovation: researching a new paradigm*, Oxford: Oxford University Press, pp. 1–12.

CHESBROUGH, H.W. AND CROWTHER, A.K. (2006), Beyond high tech: early adopters of open innovation in other industries, *R&D Management*, **36**(6), 229–236.

CHIESA, V. AND TOLETTI, G. (2004), Networks of collaborations for innovation: the case of biotechnology, *Technology Analysis & Strategic Management*, **16**, 73–96.

CHOI, D. AND VALIKANGAS, L. (2001), Patterns of strategy innovations, *European Management Journal*, **19**, 424–429.

CLYDESDALE, F.M. (1998), Science, education, and technology: new frontiers for health, *Critical Reviews in Food Science and Nutrition*, **38**(5), 397–419.

COHEN, W.M. AND LEVINTHAL, D.A. (1990) Absorptive capacity: a new perspective on learning and innovation, *Administrative Science Quarterly*, **35**(3), 128–152.

CURRAN, C.-S., BRÖRING, S. AND LEKER, J. (2010), Anticipating converging industries using publicly available data, *Technological Forecasting & Social Change*, **77**(3), 385–395.

DAY, G.S. AND SCHOEMAKER, P.J. (2000), A different game, in Day, G.S., Schoemaker, P.J. and Gunther, R.E. (Eds.) *Wharton on managing emerging technologies*, New York, Wiley, pp. 1–23.

DIERICKS, I. AND COOL, C. (1989), Asset stock accumulation and sustainability of competitive advantage, *Management Science*, **35**(12), 1504–1511.

DONG, D.-H., KIM, S., NAM, C. AND MOON, J.-S (2004), Fixed and mobile convergence and reconfiguration of telecommunications value chains, *IEEE Wireless Communications*, **11**(5), 42–47.

DOSI, G. (1982), Technological paradigms and technological trajectories: a suggested interpretation of the determinants and directions of technical change, *Research Policy*, **11**(3), 147–162.

DUYSTERS G. AND HAGEDOORN J. (1997), Technological convergence in the IT industry: the role of strategic technology alliances and technological competences, *International Journal of the Economics of Business*, **5**, 338–368.

EC (European Parliament and Council) Regulation No 1924/2006 on nutrition and health claims made on foods, in: *Official Journal of the European Union* L 404 of 30 December 2006.

ENKEL, E., GASSMANN, O. AND CHESBROUGH, H.W. (2009), Open R&D and open innovation: exploring the phenomenon, *R&D Management*, **39**(4), 311–316.

FAI F. AND VON TUNZELMANN N. (2001), Industry-specific competencies and converging technological systems: evidence from patents, *Structural Change and Economic Dynamics*, **12**(2), 141–170.

FLYNN, A. (2010), EFSA's evaluation of health claims: scientific substantiation, *EFSA technical meeting on health claims*, June 1st 2010, Parma, Italy.

FROST & SULLIVAN (2011), 2010: The year of mergers & acquisitions in the food & beverage ingredients space: econometric and market impact analyses of M&A events, Frost & Sullivan Ltd. Oxford.

GAMBARDELLA, A. AND TORRISI, S. (1998), Does technological convergence imply convergence in markets? Evidence from the electronics industry, *Research Policy*, **27**, 445–463.

GASSMANN, O. AND ENKEL, E. (2006), Open innovation: Die Öffnung des Innovationsprozesses erhöht das Innovationspotenzial, *Zeitschrift Führung und Organisation*, **3**, 132–138.

GEORGIOU, N.A., GARSSEN, J.R.G. AND WITKAMP,. R.F. (2011), Pharma-nutrition interface: the gap is narrowing, *European Journal of Pharmacology*, **651**, 1–8.

GRAACK, C. (1996), Telecom operators in the European Union: internationalization strategies and network alliances, *Telecommunication Policy*, **20**(5), 341–355.

GREEN, M. AND VAN OUDERAA, F. (2003), Nutrigenetics: where next for the food industry?, *Pharmacogenetics Journal*, **3**, 191–193.

GREENER, I. (2002), Theorizing path dependency: how does history come to matter in organsations?, *Management Decision*, **40**(6), 614–619.

GREENSTEIN, S. AND KHANNA, T. (1997), What does industry convergence mean?, in Yoffie, D. (Ed.), *Competing in the age of digital convergence*, Boston, pp. 201–226.

HAGEDOORN, J. (1993), Understanding the rationale of strategic technology partnering: interorganizational modes of cooperation and sectoral differences, *Strategic Management Journal*, **14**(5), 371–386.

HAGEDOORN, J. AND DUYSTERS, G. (2002), External sources of innovative capabilities: the preference for strategic alliances or mergers and acquisitions, *Journal of Management Studies*, **39**(2), 167–188.

HELFAT, C.E. (1994), Evolutionary trajectories in petroleum firm R&D, *Management Science*, **40**(12), 1720–1747.

HERATH, D., HENSON, S. AND CRANFIELD, J. (2006), A note on the economic rational for regulating health claims on functional foods and nutraceuticals: the case of Canada, *Health Law Review*, **15**(1), 23–32.

HERZOG, P AND NIEDERGASSEL, B. (2007) Facilitating open innovation: idea brokers in the chemical industry, *PharmaChem*, **3**, 10–12.

HILLIAM, M. (1998), The market for functional foods, *International Dairy Journal*, **8**(5–6), 349–354.

KATZ, M.L. (1996), Remarks on the economic implications of convergence, *Industrial and Corporate Change*, **5**(4), 1079–1095.

KIM, S., EUN, H.-J, KANG, Y. AND LEE, M. (2006), Developing a convergence portfolio matrix for heterogeneous inter-industry convergence, *Telecommunications Review*, **16**(1), 82–89.

KODAMA, F. (1992), Technology fusion and the new R&D, *Harvard Business Review*, **70**(4), 70–78.

KOGUT, B AND ZANDER, U. (1992) Knowledge in the firm, combinative capabilities, and the replication of technology, *Organization Science*, **3**(3), 383–397.

LAKHANI, K.R. (2008), *InnoCentive.com (A)*, Harvard Business School Case, No 608–170. Boston, Harvard Business School.

LEE, K., LIM, C. AND SONG, W. (2005), Emerging digital technology as a window of opportunity and technological leapfrogging: catching up in the digital TV by the Korean firms, *International Journal of Technology Management*, **29**(1/2), 40–63.

LEI, D.T. (2000), Industry evolution and competence development: the imperatives of technological convergence, *International Journal of Technology Management*, **19**(7/8), 699–738.

LEONARD-BARTON, D. (1992) Core capabilities and core rigidities: a paradox in managing new product development, *Strategic Management Journal*, **13**, 111–126.

LIND, J. (2005), Ubiquitous convergence: market redefinitions generated by technological change and industry life cycle, *Proceedings of the DRUID Academy Winter 2005 Conference.*

MALHORTA, A. AND GUPTA, A.K. (2001), An investigation of firms' strategic responses to industry convergence, *Academy of Management Best Paper Proceedings*, 61st annual management meeting.

OECD (1992), *Telecommunications and broadcasting: convergence or collision?*, ICCP 29, Paris, OECD.

PENNINGS J.M. AND PURANAM P. (2001), Market convergence & firm strategy: new directions for theory and research, *Proceedings ECIS Conference, The Future of Innovation Studies*, Eindhoven, Netherlands, 20–23 September 2001.

PENROSE, E. (1959), *The theory of the growth of the firm*, Oxford, Oxford University Press.

PISANO, G.P. (1991). The governance of innovation: vertical integration and collaborative arrangements in the biotechnology industry, *Research Policy*, **20**(3), 237–249.

POWELL, W.W. (1998), Learning from collaboration: Knowledge and networks in the biotechnology and pharmaceutical industries, *California Management Review*, **40**(3), 228–240.

PRAHALAD, C.K. (1998), Managing discontinuities: the emerging challenges, *Research Technology Management*, May–June, 14–22.

REDGWELL, R. (2011), *View from the top*, *World of Food Ingredients*, **4/5**, 66.

RITTER, T. AND GEMÜNDEN, H.G. (2003), Interorganizational relationships and networks: an overview, *Journal of Business Research*, **56**(9), 691–698.

ROCKENHÄUSER, J. (1999), *Digitale Konvergenz und Kompetenzmanagement*, Wiesbaden, Gabler-Verlag.

ROTHAERMEL, F.T. AND DEEDS, D.L. (2006), Alliance type, alliance experience and alliance management capability in high-technology ventures, *Journal of Business Venturing*, **21**(4), 429–460.

SÄÄKSJÄRVI, M. (2004), Consumer evaluations of hybrid innovations, *Publication of the Swedish School of Economics and Business Administration*, No. 122, Helsingfors.

SANCHEZ, R. (1996), Strategic product creation: managing new interactions of technology, markets, and organsations, *European Management Journal*, **14**(2), 121–138.

SCHILDT, H.A., MAULA, M.V.J. AND KEIL, T. (2005), Explorative and exploitative learning from external corporate ventures, *Entrepreneurship Theory and Practice*, **29**(4), 493–515.

SIEDLOK, F., SMART, P. AND GUPTA, A. (2010), Convergence and reorientation via open innovation: the emergence of nutraceuticals, *Technology Analysis & Strategic Management*, **22**(5), 571–593.

SPENDER, J.-C. (1989), *Industry recipes: an enquiry into the nature and sources of managerial judgement*, Oxford.

STEENSMA, H.K. AND CORLEY, K.G. (2000), On the performance of technology-sourcing partnerships: the interaction between partner interdependence and technology attributes, *Academy of Management Journal*, **43**(6), 1045–1067.

STIEGLITZ, N. (2004), *Strategie und Wettbewerb in konvergierenden Märkten*, Wiesbaden, Gabler-Verlag.

STUMBO, P.J., WEISS, R., NEWMANN, J.W., PENNINGTON, J.A., TUCKER, K.L., WIESENFELD, P.L., ILLNER, A.-K., KLURFELD, D.M. AND KAPUT, J. (2010), Web-enabled and improved software tools and data are needed to measure nutrient intakes and physical activity for personalized health research, *Journal of Nutrition*, **140**, 1–12.

TEECE, D.J., PISANO, G. AND SHUEN, A. (1997), Dynamic capabilities and strategic management, *Strategic Management Journal*, **18**, 509–534.

TIDD, J. AND TREWHALLA, M.J. (1997), Organizational and technological antecendents for knowledge acquisition and learning, *R&D Management*, **27**(4), 359–375.

TIDD, J., BESSANT, J. AND PAVITT, K. (2005), *Managing innovation: integrating technological, market and organizational change*. Chichester, Wiley.

WERNERFELD, B. (1984), A resource based view of the firm, *Strategic Management Journal*, **5**, 171–180.

WIJNANDS, J.H.M., VAN DER MEULEN, B. AND POPPE, K.J. (2007), *Competitiveness of the European food industry: an economic and legal assessment*, Luxembourg European Commission, DG Enterprise and Industry.

WIRTZ, B.W. (2001), Reconfiguration of value chains in converging media and communications markets, *Long Range Planning*, **34**, 489–506.

ZHANG, X., WANG, W. AND XIAO, K. (2010), Novel omics technologies in nutraceutical and functional food research, in *Genomics, proteomics, and metabolomics in nutraceuticals and functional foods* (eds D. Bagchi, F. C. Lau and M. Bagchi), Wiley-Blackwell, Oxford, UK, pp. 11–22.

4

Accelerating the innovation cycle through intermediation: the case of Kraft's melt-proof chocolate bars

R. Wielens, NineSigma, Belgium

DOI: 10.1533/9780857097248.1.63

Abstract: The successful use of open innovation by Kraft to find solutions for melt-proof chocolate bars is described. A problem broadcasting methodology enabled Kraft to reach out to over 10 000 relevant experts with this specific challenge and solutions were proposed from the open global innovation community (crowdsourcing). The key recommendations for organizations to become more successful in open innovation are summarized as clearly defined problem ownership, creating a good open-innovation environment, deciding about the span of control, creating realistic timelines and budgets, and deciding resource allocation.

Key words: crowdsourcing, expert crowds, open innovation, needs-driven innovation.

4.1 Introduction

Since the start of the company in 2000, NineSigma has run over one thousand projects in the food and drinks industry varying from melt-proof chocolate packaging to microbiological contamination detection. This chapter deals with some of the experiences, examples and results from the work done with clients in the food and drinks industry. The first part of this chapter sets the scene and shares some of the beliefs related to open innovation (OI), putting into context how OI relates to innovation in general. Next the key capabilities needed for OI are discussed and I provide a perspective on identifying latent customer needs. The third part explains the 'need driven innovation model', providing a vision on a different approach to learning about possible solutions for client needs. The case study presents the approach Kraft took when looking for a packaging solution to prevent chocolate from melting. The conclusions provide an overview of lessons learned from 12 years of OI projects and a view on the future of OI in the food and drink industry.

4.2 From research to search in company innovation

The innovation landscape and the way companies grow and innovate is changing. As shown by Henry Chesbrough, the share of R&D investments and patent ownership has shifted from big organizations to small (Chesbrough, 2011). In 1981, 70% of all R&D spend was by big corporations with over 25 000 employees and 30% by smaller companies. These days it is completely the opposite situation, 70% is done by companies with less than 25 000 employees and most remarkable, more than 20% of all R&D spent is from companies with less than 1000 employees. R&D for big corporations is slowly becoming, what I would like to call, 'Search and Development'. So what is OI? Frank Piller, director of MIT's mass customization centre and RWTH Aachen and thought leader in OI, defines it as 'The formal discipline and practice of leveraging the discoveries of non-obvious others as input for the innovation process through formal and informal relationships' (Piller and Wielens, 2011). The challenge is to develop the skills and capabilities to be able to leverage and integrate the knowledge and solutions from new and existing ecosystems that are beyond the current networks of trusted suppliers and partners. Connecting to and working with the open global innovation community requires a win–win mentality from large organizations. Winning in OI often requires a change in mentality from a big company to become the partner of choice for small organizations.

Another trend that is visible is what Lisa Gansky, author of 'The Mesh', coined the shift 'from ownership to access' (Gansky, 2010). Companies are moving away from the ownership paradigm and learning how to connect the dots, combining existing assets owned by partners. These assets might be intellectual property or production capabilities or other capabilities. Companies such as Ben Kaufmann's Quirky are challenging the big companies with agile models that tap into crowds for using an internet driven platform for ideation, concept development, product development, product design and sales and distribution. Ben Kaufmann started Quirky after learning first hand that the minimal upfront cost for a single-product company is $200 000 with a 90% chance of failure. He realized that there are thousands of people that have great product ideas every day but not the ambition, money or opportunity to start a one product company. By building a platform where he invites 'ideators' to submit concept product ideas online, he created a funnel of more than 200 concepts per week. The ideas must retail for less then $150 and should not require integrated software to keep it manageable. His community of over 100 000 members grows 20% per month and involves in addition to 'ideators' also 'influencers' who vote on ideas and develop them further. Using a specially developed algorithm he pays his 'ideators' and 'influencers' 30% of the revenue relative to their impact and contribution. His company currently launches

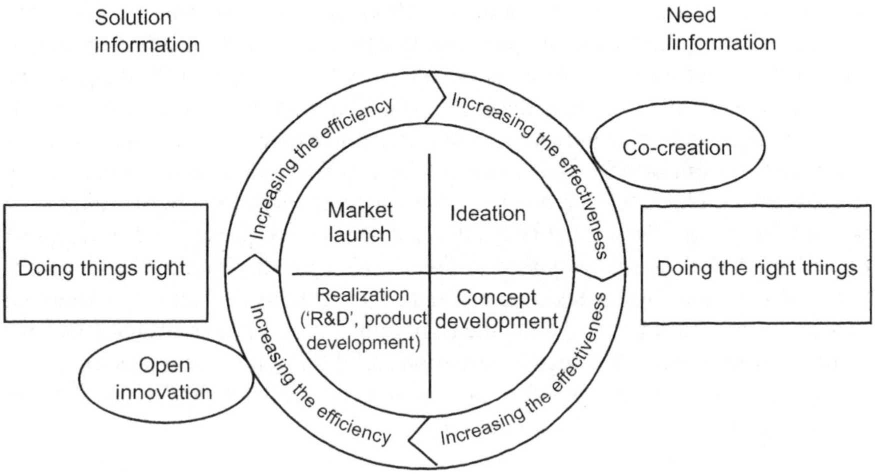

Fig. 4.1 Two key types of information. *Source:* Frank Piller and Dialego AG Aachen (Dialego, 2012).

one new product per day. I believe that the future of innovation is all about speed as Ben's inspiring company shows. Accelerating the innovation, leveraging assets from 'unobvious others'.

4.3 Key capabilities in open innovation

The importance of combining insights and benefits with need-driven innovation and doing the right things are examined in the following subsections.

4.3.1 Combining insights and benefits with need-driven innovation

Academics usually complain that they need a data cloud with enough real life data before they are able to draw any lessons. The upside is that academics help us understand complex systems and realities with many variables by modelling complex systems into easy to understand models. The model by Frank Piller (Dialego, 2012) captures the two basic ingredients for any corporate strategy (Fig. 4.1):

- increasing effectiveness: doing the right things;
- increasing efficiency: doing things right.

OI is mostly related to efficiency, doing things right, accelerating the innovation cycle. Understanding what is required by the market and customers is related to doing the right things.

4.3.2 Doing the right things

In a letter to the Amazon shareholders in 2009, Jeffrey Bezos allowed a peek into his thinking about growing his company. Although most companies would probably see it as common sense and best practice to take the key capabilities of their current organization as a starting point for growth and innovation strategies, he challenged that view. He stated that '…working backwards from customers needs often demands that we acquire new competencies and exercise new muscles, never mind how uncomfortable and awkward-feeling those first steps might be….' (Bezos, 2009). I have left out the preamble where he shows that he is thinking far beyond the 'you need to listen to what the customer wants' paradigm. Like Apple's Steve Jobs, he argues that you need to develop a deep understanding and intuition of the benefits that customers want and only asking them will not help. Although I will not come back to the 'Doing the right things', one of the key challenges for companies is related to this need information part. Many companies lack the vision and entrepreneurship needed for growth and unfortunately these are the key ingredients in successful adoption of OI.

4.3.3 Doing things right: the need-driven innovation approach

After companies understand their customer and market needs there are roughly two ways to solve the identified 'problems': internally or externally. The traditional way companies solve challenges is by taking an idea-driven experimental approach. The R&D community works on platforms, generating a multitude of ideas and they usually devise experiments to test which ideas are best. Doing experiments is the way forward; fail often to succeed sooner. The problem with that approach is that all costs for experiments are absorbed internally and the process is sequential and can take hundreds or thousands of iterations. Companies can profit from parallel processing today as many innovations happen simultaneously across industries and across the world. To connect to these parallel worlds with solutions, companies must learn to innovate by stating clearly the problem that they are trying to solve.

Albert Einstein once commented that when he had only one hour to save the world, he would take fifty-five minutes to think about the problem and five minutes to think about the solution. OI requires a so-called 'needs driven' innovation approach. By taking this needs-driven innovation approach, companies have the opportunity to tap into unobvious solutions and ideas from adjacent technologies and industries. When companies communicate problems to their trusted networks and beyond to global networks as provided by the likes of NineSigma, they tap into what I call cognitive surplus. An interesting example of cognitive surplus is the acclaimed three billion hours per week people are playing games worldwide (McGonical, 2010). McGonical explains that the average young gamer today in an economy with a strong gaming culture will have spent an average of 10 000 h gaming by the age of 21. That is about the same amount of time an American child spends in school between 5th grade and high school graduation assuming full attendance. This abundance in 'problem solving training' , 'collaborative playing' and cognitive surplus is providing an unparalleled resource. This resource has already been tapped by companies through, e.g., the Netflix Challenge, solving their problem through a grand challenge approach (Piller and Ihl, 2009). This Grand Challenge approach mobilized 27 000 registrants competing for the $1 million prize. Similar grand challenges more directed to mobilizing professionals or business to business oriented crowds are used by General Electric in, e.g., the Healthymagination Challenge on Breast Cancer Diagnostics (Resnick, 2011).

For companies to effectively use a challenge or needs-driven approach, five key capabilities are needed from my perspective in a virtuous circle of steps that need to be repeated. First, companies have to develop need portfolios. To develop these need portfolios or innovation shopping lists, companies have to review and rethink their innovation project portfolio. After the definition of the needs, the company has to transform these needs into challenges by a mix of art and science. Well-formulated challenges are easy to understand and difficult to answer. From the challenges the company needs to define what challenges can be broadcasted and shared with the outside network for identification of solutions. In the next step, the identification and acquisition of solutions requires companies to define a review structure to review different solutions efficiently. From the search and identification of solution an ecosystem arises of players in different technology solution domains, effectively leveraging different ecosystems that require development and management capacity as well as development of absorptive capacity to effectively integrate external solutions. Finally the circle is finalized with perhaps the most important and difficult capability related to detecting signals of change. This capability deals with managing the boundaries, i.e. answering which knowledge, capabilities and capacity do I need from a strategic perspective inside the company and which outside the company boundaries? Learning about changes requires a continuous monitoring of key areas as input for (open) innovation strategy.

There are three steps necessary for companies to start a needs-driven innovation approach. The first step is to start communicating innovation challenges; this requires the development of innovation-needs portfolios and transformation of those needs into challenges. Communicating challenges (needs) starts with formulating the problems clearly before developing good solutions. Starting every product research and development project with a well defined statement of the problem and sending that out in the form of well defined briefs to relevant contacts and connections helps them to understand what the options are for accelerating the product creation process and reducing the time to market. It is interesting to observe how very few companies actually first carry out a search before beginning the research. The second step is to develop and manage innovation ecosystems on three levels: the internal company network, the trusted network of known partners and the global open-innovation network. The third step is to organize capabilities to absorb the results and manage the relationships and integration of the innovation acceleration options that are identified in the three ecosystems.

In addition to development of ecosystems and relevant contacts, another important consideration is whether sharing of confidential information in the form of your challenge or problem is an option or not. Arguably the most difficult part is related to accepting and integrating knowledge and solutions from the outside. Often the R&D community feels disqualified, or as one R&D manager at a large chemical corporation stated: 'if we need open innovation then we have failed as R&D'. This perspective is often referred to as the 'not invented here syndrome' and is related to many cultural aspects. Alan Taub, CTO of General Motors explained in a private conversation that for him the cultural shift was the most important objective related to the OI program. He explained that he wanted to realize 25% of the R&D budget invested with external partners and take out any threshold people in his organization might have to make sure that they use the best partners and not just the usual suspects. As an example of one of the actions he took to achieve this he provided free training and made the use of OI tools like broadcasted challenges 'free of charge' and part of his budget.

4.4 From idea-driven innovation to need-driven innovation

Developing problems or needs obviously starts with good insights as discussed in Section 4.3.2 ('doing the right things'). The innovation agenda or program of an organization is usually a good starting point to filter out challenges and good OI projects. Over the years we have developed different questionnaires to qualify projects that are good OI projects, including:

- What has been done to solve the problem?
- What would be the value if the problem was solved?
- What type of solution is required?
- What is the priority of solving the problem?
- What is the budget for spending on outside solutions?
- What happens if the problem is not solved?

The more mature organizations have dedicated scouts that actively engage with the innovation community to identify needs. Best in class examples are from companies such as Kraft who have developed long lists of qualified needs. These needs have a budget and a solid business case justifying the company's willingness to pay for the solution or the development of a solution.

4.5 Case study: melt-proof chocolate bars from Kraft

The problem that Kraft identified was initially assessed by the program management team at NineSigma as what we refer to as the 'holy grail': everyone in the industry is looking for this so called 'unobtainium'. Even the US military attempted to solve the problem of melting chocolate at higher temperatures. Besides white 'bloom' discoloration that appears on chocolate if it is exposed to direct sunlight or heat, the experience of trying to poke melted chocolate out of the wrapper is a known phenomenon for chocolate lovers alike.

The challenge that we defined was 'Kraft Foods seeks novel materials or approaches to packaging that can protect single serve chocolate bars from medium term exposure to warm ambient conditions. Products frequently experience multiple cycles of exposure to controlled and uncontrolled climates. The product may be shipped in controlled and uncontrolled conditions, sold by the retailer in an air-conditioned setting and consumed sometimes later having been carried on the person or in a handbag. Although current technology can provide a solution to the problem, the resultant packaging is both cost prohibitive and excessively bulky.'

A well-defined brief contains, in addition to the challenge, criteria for a good solution, possible approaches and important: approaches not of interest.

The possible approaches were defined as:

- novel insulating materials;
- phase change materials;
- thin film approaches that can store energy and repurpose it;
- active packaging technologies that are triggered by temperature or light; and
- novel cooling or heat-absorbing technologies.

The approaches not of interest were defined as:

- chocolate formula or recipe changes;
- chocolate-processing modifications;
- battery or electrically powered devices;
- common insulating materials, such as foamed polymers, paper-based expanded structures and expanded polystyrene; and
- solutions that cannot protect product through more than one cycle of heating and cooling.

After the brief with the challenge is defined and agreed, a broadcast and instant network relevant to the challenge is developed of around 10 000 contacts in academia, business and research institutes. Although the brief is directly sent to individual experts whom we suspect might have knowledge or contacts to solve this particular challenge, the reason for any of these organizations to respond is that it is a business opportunity to collaborate on a development contract, to licence out intellectual property or win a supplier agreement. The response to this particular challenge resulted in over 30 different proposals from around the world addressing the problem and suggesting an approach or solution. The next phase after processing the results is the assessment and acquisition phase, screening the proposals and organizations that responded and selecting those with whom Kraft will start contract negotiations. The starting point is a technology map that explains the different technologies proposed and that provides an assessment of the maturity of the solution. The long list of 30+ proposals is reduced to 5 or 10 candidates. Additional questions and telephone interviews are organized with those 10 top candidates and sometimes site visits and demonstrations on site are scheduled, usually after signing a non-disclosure agreement. Finally, for one or more candidates, contract negotiations are started and the acquisition finalized.

As the whole process up until reporting of the proposals is strictly non-confidential, the first step after making a first selection of top candidates involves the development of non-confidentiality agreements with these top candidates. This enables the searching company to get more insights into the capabilities of the partner, background of the solution and proposed business model. This business model can be acquisition of intellectual property (IP), licence of the IP, a research contract or any business transaction that is mutually beneficial. Depending on the seeking company and solution provider, the approach to IP can be a difficult sticking point. In some instances this requires mediation from our side. One observation is that not many global companies understand the dynamics of doing business with smaller businesses. The usual modus operandi is to use the standard '80 page contract' and consequent visit of five people to an overwhelmed 20-person small/medium-sized enterprise (SME). Moderation of this conversation and mediation to help each partner feel and behave like a partner is crucial in all cases.

4.6 Conclusions

Much effort and attention from companies is invested in testing and learning from the so called problem broadcasting approaches as described in the Kraft example. In the end, crowdsourcing for technical problems also enables important change within a company. Usually, the observation is that in the course of the project the clients learn a lot about potential partners and more importantly about themselves and their own company. This 'learning about oneselves' may be the largest benefit of networking with the world. However, OI is not an automatic success but one that demands thorough preparation and rigid implementation. From our experiences in 12 years of OI projects in different contexts, we can derive a number of success factors to profit from OI and crowdsourcing of technical problems. Some of our most important recommendations are presented in the following subsections.

4.6.1 Clearly defined ownership of the problem

The most important factor, perhaps, is to have a problem or task that is suited to being crowdsourced. Start with the problem and challenge owner. Determine the objective of your OI venture before hiring an intermediary. Intermediaries differ, and not all are equally suited for the same kind of task.

4.6.2 Create a good open innovation environment

You need a dedicated team to make OI a success. This starts by installing a central project competence for your OI initiative. Successful companies have nominated an internal OI champion who is passionate, outwardly focused and capable of co-ordinating different crowdsourcing projects. Equally important, however, is to educate other team members and employees about the objectives and principles of the OI project so that they understand what is expected from them.

4.6.3 Decide about the span of control

OI is about opening up to the periphery of your firm. But you can decide about the control you want to keep during the knowledge transfer process and the exploitation of the results by selecting an appropriate method and intermediary. For example: will potential solution providers learn who you are and see your name and company logo on the challenge? We know that challenges that reveal the challenge owner's identity receive significantly more and better proposals. However, revealing who you are may also inform your competitors. Knowing about these trade-offs is crucial for OI success.

4.6.4 Create realistic timelines and budgets

First of all, be fast. Challenges and also solution proposals will reach competitors sooner or later. So before you start, get the buy-in for the implementation of returned solutions that match your requirements. Reserve a budget for this. Be sure about the human capacities and resources required for the review process of the solution proposals and for making a technology sourcing decision.

4.6.5 Decide about your resource allocation

Consider what's next in the short term: are you seeking an intermediary that also provides support before and after the challenge is broadcasted, e.g. with evaluating the solutions or negotiating the licensing terms? Or do you want to perform these activities in-house. Both options are available, and have their distinctive pros and cons. Also consider what's next in the long term: Think of, e.g., community management; will the intermediary help you to manage a community of potential solvers? Or are you prepared to take these tasks by yourself?

I believe that food and drink companies are somewhat further developed when it comes to consumer and customer insights driven innovation compared with other industries. This should make it easier for them to define needs portfolios. Kraft is an example of a company that is well developed in this aspect as can been seen at their Innovate with Kraft Foods innovation portal (http://www.Kraftbrands.com/innovatewithKraft/). But Kraft is an exception as most food and drink companies have neither the needs portfolio developed, nor the capability to translate these into challenges and the internal organization to acquire, manage and develop the partner ecosystems. The biggest challenge for companies to become successful in using OI lies interestingly enough on the inside of the organization as it requires resources, development of new skills and capabilities as well as a different mind set.

4.7 Future trends

The food and drink industry and, more specifically, the consumer packaged goods companies such as Procter and Gamble and Unilever have been early adaptors of OI practices. The outreach to the open global innovation community is, as well as for new technology, for every possible challenge: new product ideas, packaging innovation and process innovations. Three trends related to the future of OI for food and drink companies are now listed.

4.7.1 Collaborative sustainability

Currently, already more then 50% of our client projects are related to sustainability. This is a cross-industry phenomenon but the change in the approach is that large corporate clients are joining forces to solve industry challenges such as changing the current oil-based feedstock of packaging material to a sustainable bio-based polymer feedstock. None of the single companies is big enough to solve this challenge in isolation. Consolidating demand and needs in this example enables a business case for the development of a new feedstock.

4.7.2 Value chain open innovation

Companies are struggling with absorbing technologies owing to increasing technology complexity, integration challenges and the increasing gap between research and development. Proposed solutions are often not sufficiently mature for companies to evaluate and integrate. As we already witnessed in the high tech industry, we see that food and drink companies are also requiring their existing supply base to absorb innovations into their pipeline or use partners with design and prototype capabilities to develop the concept to a more mature stage.

4.7.3 Innovation portals

Over the past few years, more and more companies have launched their own innovation portals, all with different collaboration models, target audiences varying from consumer to business networks and objectives. None of the companies seem to get value out of these portals and it is expected that a consolidation (aggregation of portals) phase in combination with a more focused approach will lead to a more mature cost-effective way to expand the networks and allow a more IT-supported and process-driven model.

4.8 References

BEZOS, J. P., 2009. Amazon.com Annual Report 2008, Letter to Shareholders.

CHESBROUGH, H., 2011. Open innovation: where we've been and where we're going, *European Innovation Conference*, Billund, Denmark 2011.

DIALEGO, 2012. Benefit driven innovation, *Front end of innovation*, EMEA conference, Zurich, 2012.

GANSKY, L., 2010, *The Mesh, why the future of business is sharing*, Portfolio/Penguin Group.

MCGONICAL, J., 2010. Gaming can make a better world, *TED conference*, Long Beach California.

PILLER, F, AND IHL, C., 2009. Open innovation with customers: foundations, competences and international trends, Technology and Innovation Management Group, University RWTH Aachen.

PILLER, F. AND WIELENS, R., 2011. Open innovation in the machinery industry: how a German consortium of SMEs in the drivetrain industry benefited from crowdsourcing of technical problems. *BILAT-USA symposium on transatlantic EU-US cooperation on innovation and technology transfer*, Vienna, 22 March 2011.

RESNICK, D., 2011. Behold: the grand challenge, www.NineSigma.com.

5

The impact of open innovation on innovation performance: the case of Spanish agri-food firms

C. Bayona-Sáez, T. García-Marco and M. Sanchez-García, Universidad Pública de Navarra, Spain and C. Cruz-Càzares, University of Barcelona, Spain

DOI: 10.1533/9780857097248.1.74

Abstract: This chapter examines the level of innovation and the degree of openness of Spanish agri-food firms and how the firms' openness affects their innovation performance. With a sample of Spanish agri-food firms and using three dimensions of openness: breadth and depth of information sources, breadth of co-operation agreements and external R&D expenditures, we find a positive effect for breadth and co-operation agreement breadth for achieving radical innovations, but they do not affect incremental innovations. Incremental innovations seem not to be affected by the firms' openness, rather, they depend on firms' internal capability. External information depth and external R&D do not enhance any type of innovation. For the rest of Spanish firms we observed that the level of openness affects both incremental and radical innovation achievement.

Key words: Spanish agri-food firms, open innovation, innovation performance

5.1 Introduction: the agri-food sector and innovation

The agri-food sector in the world and particularly in Europe is facing a situation in which the higher level of product competitiveness from emerging countries, essentially owing to lower labour costs, combines with the greater market penetration capacity on the part of the products from other advanced countries, based chiefly on a more efficient production and marketing structure (Capitanio *et al.*, 2009). Additionally the sector structure is characterized by a large number of micro, small, and medium sized enterprises (SMEs) (Kühne *et al.*, 2010). In Spain the agri-food sector plays an important role in the economy, contributing 8.7% of its GDP and more than 13.5% of total employment (www.fiab.es). Its annual turnover is €81 369 million, representing 16% of all industry. This fact makes it in the largest industrial sector in the Spanish economy and the fifth largest in Europe. In relation to the number of firms, the agri-food sector comprises a total of 30 261 firms, 96% of which are small to medium-sized enterprises (SMEs) that employ 445 475 workers, representing 17% of industrial employment.

Although these agri-food firms have overcome many problems and now resemble industrial enterprises (Serrano and Alonso, 2008), this sector has a number of special characteristics that affect its behavior in the market, with a high proportion of SMEs and a significant dependence on food chain strategies, especially those dominated by the food retail sector. Thus, in a comprehensive study of the European retail sector, Dobson *et al.* (2001) argue that the increased power of retailers may decrease prices but it also reduces product variety and innovation efforts by agri-food firms (Weiss and Wittkopp, 2005). Therefore agri-food firms exhibit marked characteristics that make them more vulnerable to a global, changing and unstable environment characterized, among other factors, by the dependence of the production process on uncontrollable factors, perishable production or raw material and the existence of government policies that may limit profitable production processes (Fayos *et al.*, 2009).

In this context, the agri-food sector's innovativeness is considered one of the most important factors for a firm to challenge major competitors both on national and international markets (Rama, 1996, 2008; Grunert *et al.*, 1997; Traill and Meulenberg, 2002; Capitanio *et al.*, 2009). However, despite this growing importance of the activities in innovation in the agri-food sector, data have revealed a low R&D intensity in general in the sector (Garcia Martinez and Burns, 1999; Capitanio *et al.*, 2009). In the view of these authors, agri-food firms are mainly process-innovation oriented (Archibugi and Pianta, 1996) and use new technologies developed by upstream industries. Moreover, the innovation processes in the agri-food sector are supply-driven (Garcia Martinez and Burns, 1999). That is particularly clear if we compare in-house R&D expenditure by production sector with that by destination sector. Among the other characteristics that define the agri-food sector's behavior is the fact that innovations in the agri-food industry are incremental rather than radical (Galizzi and Venturini, 1996; Grunert *et al.*, 1997). These incremental innovations are good instruments to market development if these actions obtain good results. However, it is necessary to assure the maximum success of these organizations' decisions, because the firms could lose relevant opportunities for growth, turning them into a waste of limited organizational resources. The prevalence of incremental innovations is related to constraints from demand and conservative consumer behaviour (Capitanio *et al.*, 2009).

From a more strategic perspective it has been noted that agri-food firms are driven to innovate and differentiate their products by the need to maintain their competitive advantage over increasingly powerful retail chains that rely on using their own brands (Garcia Martinez and Briz, 2000). Furthermore, related to technological autonomy, Garcia Martinez and Burns (1999) point out that most of the agri-food sector exhibits a higher level of technological autonomy in products than in processes, and that there is a restricted degree of technological autonomy in some agri-food subsectors. Moreover, following these authors, large firms are technologically less dependent in terms of product technology than SMEs, whereas, in terms of process technology medium-sized firms show a higher level of technological autonomy. By the same argument, there is a small nucleus of persistent patentors who contribute around 80% of the total number of patents granted to the multinational agri-food sector (Alfranca *et al.*, 2004), and these are not particularly large companies.

On the other hand, recent studies show a significant growth in market-oriented agri-food sector towards innovation. The recent empirical work on innovation in the agri-food industry by Batterink *et al.* (2006) indicates that successful innovating agri-food processing companies have a strong market orientation (Fortuin and Omta, 2009). This effort for a greater market orientation is reflected in the move from a supplier-dominated industry to another situation where industry is able to generate new innovation, including two-way 'learning by doing' and 'learning by using' (Rama and Alfranca, 2003). Moreover, this favorable situation for the growth of innovation is reinforced by the significant impact on the agri-food sector of the biotechnology revolution, pressures arising from globalization for firms to maintain better process control and exploit economies of scale, the need to ensure agri-food safety, nutritional quality and provide a new generation of functional foods, and consumers' demand for convenience, variety, and quality. These factors all increase the demands for research in the industry (Traill and Meulenberg, 2002). Therefore, newly available technologies and the high level of international business of the agri-food sector can lead to it having a new technological status as a high R&D intensity industry when traditionally it has been classified as low R&D intensity sector (Filippaios *et al.*, 2009).

As already observed, agri-food firms are not free of the need to achieve innovations in order to create a competitive advantage so as to successfully compete in the market. Garcia Martinez and Burns (1999) mentioned that most of these firms have the capability to internally develop product innovations, that is, the traditional strategy of closed innovation. However, is this strategy the correct one for the agri-food firms?

Chesbrough *et al.*, (2006) coined the term open innovation, which is a newer approach to conceiving innovation strategy. He mentions that firms should open their frontiers and look for knowledge beyond their boundaries. According to this approach, firms should also look for external channels to commercialize the innovations and/or inventions that are not in line with the firms' core activity. In this sense, this co-operation interest is studied by Acosta *et al.* (2011), when they indicate that the agri-food industry is not considered to rely on R&D alone but learns and interacts with various actors. Open innovation has received a lot of attention in business management studies, but, until now, little attention has been paid to the agri-food sector. Therefore, the aim of this chapter is to fill this gap.

Some empirical studies have shown the positive effect of open innovation, measured as the breadth and depth of external information sources, on firm innovative performance (Laursen and Salter, 2006; Mol and Birkinshaw, 2009). Following on from this, but also adding two more dimensions of openness, the breadth of co-operation agreements and external R&D, we aim to assess whether the firms pursuing open-innovation practices perform better than their counterparts in terms of the achievement of radical and incremental innovations.

In order to achieve our objective, we structure the remainder of the chapter as follows. The innovative profile of Spanish agri-food firms is presented in Section 5.2. Different measurements of open innovation in Spanish agri-food firms are explained in Section 5.3, whereas the fourth section provides the results of the openness on firm innovative behavior. The last part of the study is reserved for the conclusions, contributions and implications and pointing out the limitations.

5.2 How innovative are Spanish agri-food firms?

In this section we aim to describe the main features of the innovative behavior of the agri-food firms. In order to achieve this goal and to understand the effects of the open-innovation practices in the next section, we used the Technological Innovation Panel (PITEC) Spanish survey which is a dataset panel gathering data from 2003 to 2009. It is carried out by the Spanish National Statistics Institute (INE) in collaboration with the Spanish Science and Technology Foundation (FECYT) and the Foundation for Technological Innovation (COTEC). PITEC is designed as a panel survey (available as a data base available on the FECYT site, http://icono.fecyt.es/) and compiles the information provided by the Spanish Community Innovation Survey (CIS), which follows the guidelines of the Oslo Manual (OECD, 1997). It includes individualized firm data on technological innovation activities, such as the main obstacles for achieving innovations, the main technological information sources, innovation and R&D expenditures, qualifications of the R&D personnel, co-operation in R&D activities classified by origin and type of partners, and the effects of the innovation achievements.

PITEC started in 2003 with two subsamples, the first one including firms with 200 or more employees (28.19% of the total sample in 2009) and, the second one of firms with intramural R&D expenditures (70.32% of the total sample in 2009). In 2004 two new samples were included: firms with fewer than 200 employees, external R&D expenditure and no intramural R&D expenditure (3.46%, in 2009) and a representative sample with fewer than 200 employees and no innovation expenditure (7.76%, in 2009) (COTEC, 2010).

The PITEC dataset includes extractive, agri-food, manufacturing and service firms. Firms in the agriculture, ranching, forestry, fishing –CNAE-2009 code 0000–, food, beverages, and tobacco –CNAE-2009 code 0003– are considered to be agri-food firms for the purposes of this research. The total sample of the PITEC dataset embraces 80 372 observations that correspond to 12,817 Spanish firms. From the total sample 7.17% corresponds to agri-food firms: 1052 firms with 5759 observations across the panel. The internal structure of our agri-food sample is as

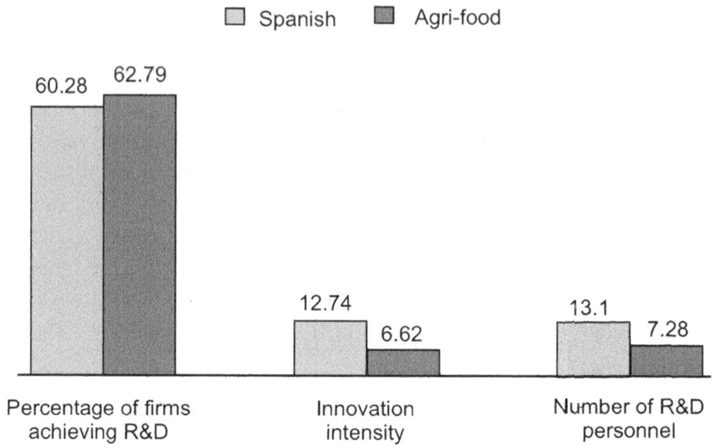

Fig. 5.1 Innovation inputs. *Source:* authors with PITEC dataset.

follows: 15.87% of the sample corresponds to agriculture, ranching, forestry and fishing, the remaining 83.73% corresponds to firms in the food, beverages and tobacco industries.

To measure the innovative character of agri-food and non agri-food firms in Spain, we analyse two perspectives: innovation input and innovation output. The former includes innovation expenditures, R&D employees and innovation intensity and the latter accounts for product innovations, process innovations and patents. If we analyze the agri-food firms' innovativeness and their evolution over the time, we find an important number of innovative firms in both perspectives.

Figure 5.1 shows the three measurements of the innovation inputs dimension. As observed, the first measure accounts for the percentage of firms involved in R&D activities. This graph shows that 60.28% of the Spanish sample as a whole is carrying out these activities and this percentage increases to 62.79% for agri-food firms. Although there are more agri-food firms involved in R&D on average they have lower innovation intensity[1] (6.62%) compared with the Spanish firms (12.74%) in general. Similarly, agri-food firms have on average half of the R&D personnel that Spanish firms have: 7.28 and 13.1, respectively.

In conclusion, we can observe that there are more agri-food firms involved in R&D, but they employ less R&D personnel and they are less innovation intensive than the Spanish firms in general.

However, from the perspective of innovation outputs, when we analyze the percentage of firms achieving process and product innovations (Fig. 5.2), we find that the percentage is higher for agri-food firms than for the total of the Spanish firms. As observed, there are around 10% more agri-food firms achieving process

[1]Measured as: Innovative expenditure/Sales. Innovative expenditure includes internal and external R&D, acquisition of advanced machinery, equipment, hardware and software destined for production of new products or process, acquisition of other external knowledge for innovation, training, introduction of innovations to market and design, other preparations for production and distributions.

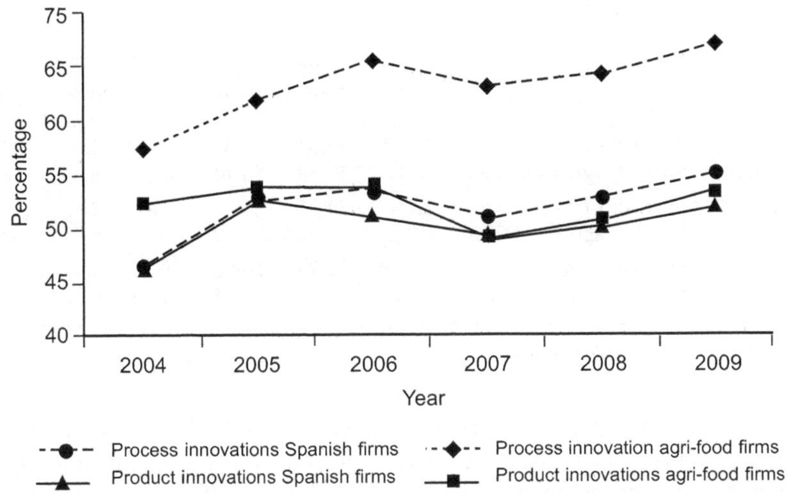

Fig. 5.2 Percentage of innovation firms; product and process. *Source:* authors with PITEC dataset.

innovations than Spanish firms in general; this tendency is constant across the panel and the percentage of firms innovating is above 60%. This fact is in accordance with the idea that agri-food firms are process innovation oriented (Archibugi and Pianta, 1996). In both subsamples we can observe that the number of innovative firms grew from 2004 to 2006 but from 2006 to 2007 a slight reduction was registered in the number of innovative firms. Since 2007, the percentage of firms achieving product and/or process innovations has constantly increased.

In Fig. 5.2, the trends in product innovation development can also be observed. Agri-food firms have registered higher levels of product innovation in relation to the Spanish firm in general though the difference is not as high as in the case of process innovations. The greatest difference is observed in 2004, in 2005, this difference was reduced, and, in 2007, Spanish firms in general slightly outperform agri-food firms. After this year, the difference in the percentage of innovating firms has increased for both samples but with a higher percentage for agri-food firms. In short, we can say that agri-food firms are more involved in achieving product and process innovations than Spanish firms in general, but the main difference between the subsamples is recorded for process innovations. Therefore, it would be worth observing in this database how the agri-food sector also is moving forward with an intense interest in innovation activities in recent years. Furthermore, a tendency can be seen to concentrate important efforts in the innovation process, assuming a change in the patterns of innovation in the sector (Capitanio *et al.*, 2009; Traill and Meulenberg, 2002), and better innovation performance can encourage a greater firm performance.

The PITEC sample allows us to differentiate between two types of product innovations; in goods and services. Patents application is another significant

Table 5.1 Innovative output

		Innovative output (%)					
		2004	2005	2006	2007	2008	2009
Goods innovations	Spanish firms	39.19	43.31	42.6	41.29	42.01	43.23
	Spanish agri-food firms	50.16	50.58	50.74	45.86	48.31	50.12
Service innovations	Spanish firms	22.56	24.11	23	21.6	22.43	23.13
	Spanish agri-food firms	19.16	19.52	18.59	14.21	15.84	15.78
Patent application	Spanish firms	14.86	11.94	10.82	10.29	9.97	9.46
	Spanish agri-food firms	13.08	8.29	7.25	7.43	6.52	5.88

Source: authors with PITEC dataset.

innovation output that we consider in this chapter. In Table 5.1 we present the percentage of firms achieving these innovations and those that applied for a patent. The number of firms completing goods innovations is stable across the panel for both agri-food firms and the rest of the firms. Nevertheless, a higher percentage of goods innovative firms is present among agri-food firms. The percentage of service innovative firms is also stable across the panel but the percentages are much lower than those of the goods innovative firms. Unlike in the case of goods innovation, this time there is a higher percentage of innovative firms for the complete sample compared with the agri-food firms. This result is in accordance with the fact that the complete sample includes enterprises from the service sectors. Finally, the percentage of firms producing patents is very much lower than those producing product, process, goods and service innovations according to the features of the Spanish economy. In this case also the agri-food industry has a lower percentage of firms producing patents compared with the rest of the firms. Interestingly, in this sample a peak of firms patenting can be observed in the year 2004 and afterwards there is a decreasing tendency for both agri-food firms and the rest of the firms. This result confirms the results obtained by Alfranca *et al.*, (2004) and Rama (2008) who detected that a small core of agri-food firms made 80% of innovations in the areas analyzed.

5.3 Measuring open innovation in Spanish agri-food firms

The openness of the innovation activities is not a single best practice but rather firm openness involves a set of different practices that can be empirically differentiable. Following on from other studies (Ebersberger *et al.*, 2011, Laursen and Salter, 2006), in this chapter, we identify three different dimensions of open innovation. The first one stands for the existence of external sources of information for developing the innovation and the importance of each innovation source. The second dimension accounts for the co-operation agreements that a firm has for developing innovations. Finally, the third dimension represents the commitment to the externalization of R&D activities.

In the following subsections, we proceed to describe the different levels of openness related to each of the three main dimensions.

5.3.1 The sources for innovative ideas

As Laursen and Salter (2006) explain, a very important component in the movement toward open-innovation models is a change in the way firms go about searching for new ideas and technologies for innovation. Because of that our first dimension of open innovation involves mechanisms that expose firms to new information and novel ideas from outside sources. We use the concepts defined by Laursen and Salter (2006) named breadth and depth because the organizations that invest in broader and deeper research may have a greater ability to adapt to change and therefore to innovate. The first concept, breadth, refers to external search breadth, which is defined as the number of external sources of information or search channels that firms rely upon in their innovative activities. The second concept, depth, refers to external search depth and it is defined in terms of the extent to which firms draw deeply from the different external sources of information.

Our database contains items that capture the information sources used during the firms' innovation activities. Ten different sources are provided where the respondents rank the importance of these sources on a four-level Likert scale ranging from 'not used' to 'high importance'. These sources could be grouped in three categories: market, institutional and others. The sources contained in the market are clients, suppliers, competitors and private laboratories and consultants. Information sources coming from universities, public research centers and technological centers represent the institutional category. Finally, the other sources include conferences and ferias, journals and scientific papers and industrial associations.

The variable breadth is constructed as a combination of the 10 sources of information for innovation. Each of the sources is coded as a binary variable: 0 being not used and 1 being used. Subsequently, the sources used by a firm are added up so that the variable values range from 0 to 10, where 0 represents no external sources used and 10 stands for the use of all external information sources. It is assumed that firms that use higher numbers of sources are more 'open', with respect to search breadth, than firms that do not.

The second variable, depth, is constructed using the same sources of knowledge as those used in constructing breadth. In this instance, each of the 10 sources is coded with 1 when the firm reports that it uses the source to a high degree and 0 otherwise. The 10 different sources are subsequently added up so that the variable values range, again, from 0 to 10. The depth variable takes the value of 0 when no external knowledge sources are used to a high degree and 10 when all information sources are used to a high degree. Following the same logic as in the former variable, we assume that firms using more information sources to a high degree are more 'open' with respect to search depth than firms that are not.

In Fig.5.3, we show the breadth and depth variables. As in the previous section,

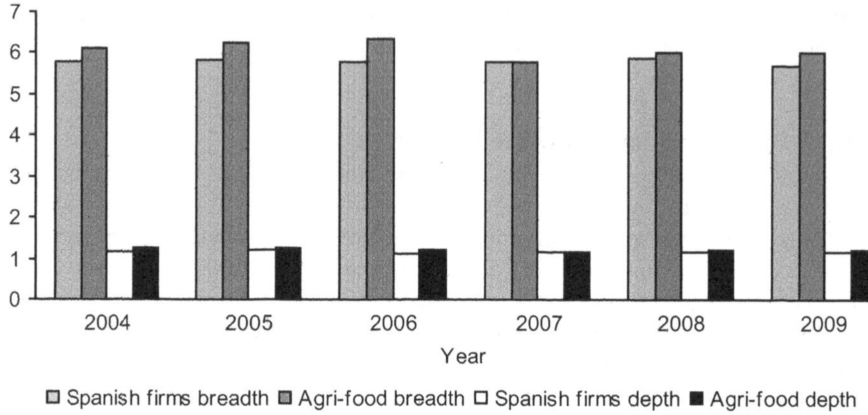

Fig. 5.3 External information search: breadth and depth. *Source:* authors with PITEC dataset.

we compare agri-food firms with the total Spanish sample. The first two columns stand for the breadth of the overall Spanish sample and agri-food firms, respectively. As observed, the agri-food firms have a slightly higher degree of openness over the years under analysis. Across all the years, they have a degree of breadth of approximately six different sources of information to realize innovations. Between 2004 and 2006, they registered a slight increase in the number of information sources but this decreased a little in 2007. For the complete sample, the mean of information sources is below six and although it has been more constant than agri-food firms, almost no difference was detected between the two samples.

When we move our attention to depth interesting results appear. Despite the fact that the agri-food sector and the complete sample seem to gather information from different external sources, the degree to which they use, or consider the sources to be important is very low. As observed in Fig. 5.3, the mean for the depth in agri-food firms is around 1.2 which means that among the six different sources used, only one of them is used extensively or deeply. Again, as in the case of the breadth, the openness of the complete sample is a little bit lower than that of the agri-food firms. For both samples, we can observe that the degree of openness related to depth is quite constant.

5.3.2 Co-operation in innovation

The second dimension to measure openness is the co-operation agreements for developing innovations. The complex nature of the innovation process makes it increasingly necessary for firms to co-operate with other organizations in order to carry through their research and development initiatives. As previously explained, sources of innovation are not to be found exclusively within the company, they are also present in other organizations that are endowed with complementary resources. Therefore, the more intensely the company interacts with these external

Table 5.2 Degree of co-operation in Spanish firms in general and in agri-food firms

Co-operation	Spanish firms	Agri-food firms
Firms co-operating (%)	36.37	40.22
Firms not co-operating (%)	63.63	59.78

Source: authors elaborated with PITEC dataset.

actors, by participating in co-operation agreements, the more it will learn about new opportunities.

Before going into detail about how many agents the agri-food firms usually co-operate with, in Table 5.2 we first present the percentage of firms pursuing co-operation agreements with other agents. As observed, 40.22% of the agri-food firms in our sample are involved in co-operation activities. The overall Spanish sample has a lower percentage of firms engaged in co-operation, 36.37%. This result confirms the necessity of co-operation in the agri-food sector and in the Spanish context in general.

To measure this second dimension, collaboration is captured using the collaboration question in the survey. The particular question asks whether innovation projects have been carried out collaboratively with other actors, and if so with what type of agent. The type of collaborators include:

- other enterprises within the firm's enterprise group;
- suppliers of equipment, materials, components, or software;
- clients or customers;
- competitors or other enterprises in the firm's sector;
- consultants, commercial labs, or private R&D institutes;
- universities or other higher education institutions;
- public research organizations; and
- technological centers.

To emphasize that innovation collaboration is a mutually interactive effort, the questionnaire explicitly states that 'innovation co-operation is active participation with other enterprises or non-commercial institutions in innovation activities'. This excludes pure contracting out of work with no active co-operation.

The variable, co-operation breadth, was composed following the same strategy as for the external information sources, that is, we added up the dichotomous variables that capture the co-operation activities with each of the agents. If a firm co-operates with eight different actors, then it gets a value of eight in the co-operation breadth variable. A firm with no co-operation agreements would obtain a value of zero in the same variables. In Fig. 5.4, we present the co-operation breadth.

As observed, the mean values of this variable are very low showing that on average the Spanish firms as a whole and the agri-food firms co-operate with less than one agent. Recalling the results of Table 5.2, we observed that around 40% of the agri-food firms and the Spanish firms in general co-operate but it seems to be

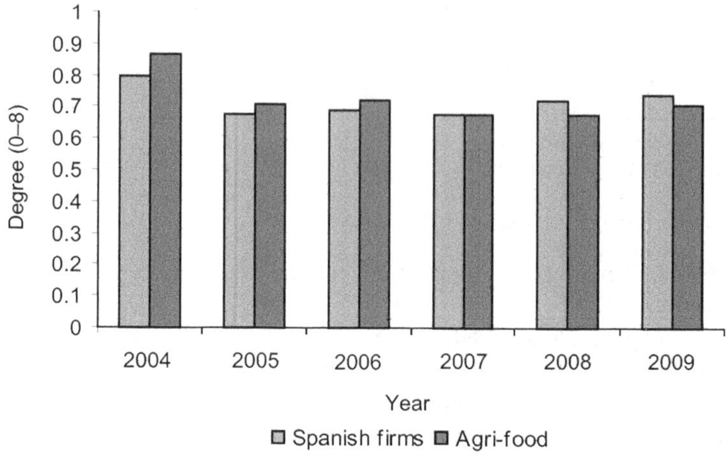

Fig. 5.4 Co-operation breadth. *Source:* authors with PITEC dataset.

that they co-operate exclusively with one or two agents, thus indicating that the Spanish and agri-food firms have not developed the co-operation strategy to its full potential.

This second dimension of the degree of firms' openness seems to capture a really different dimension than that of the breadth of external sources of information. The agri-food firms seem to be open in terms of informal linkages with external agents, but when for formal linkages, such as co-operation agreements, they seem to be very closed.

Unfortunately, owing to a lack of data, we are not able to capture the co-operation depth, that is, the degree in which firms are committed to collaborative activities.

5.3.3 Internal/external innovation expenditure

The third dimension of open innovation is what we will call 'internal/external innovation expenditure'. We include a measure for the percentage of innovation expenditures that correspond to the development of internal and/or external R&D activities. Internal R&D expenditures include current and capital expenditures. External R&D expenditures represent the service outsourcing of R&D and do not include the purchase of machinery and equipment and training. This variable measures the degree of openness in R&D activities, one of the most relevant innovation activities. Through it, we can see the importance that firms give to the acquisition of knowledge as against the internal R&D. However, it is necessary to note that open-innovation concepts are not employed primarily as a rationale for cost reduction or outsourcing of the R&D function, because internal R&D, based in the importance of absorptive capacity, is maintained or even increased.

In Table 5.3, we show the percentage of the innovation expenditures dedicated to internal and external R&D activities. Surprisingly, both Spanish firms in general

Table 5.3 Innovation expenditures in internal R&D and external R&D

	Spanish firms		Agri-food firms	
	Internal R&D	External R&D	Internal R&D	External R&D
2004	42.48	10.38	49.32	10
2005	47.34	7.28	49.31	7.61
2006	42.85	7.65	43.85	8.26
2007	39.88	7.59	40.17	8.28
2008	38.36	7.22	38.5	8.89
2009	35.93	6.89	34.56	8.46

Source: authors with PITEC dataset.

and agri-food firms have recorded a decrease in the percentage of R&D investment over innovation expenditures since 2004. In 2004, agri-food firms dedicated nearly 50% of their innovation expenditures to internal R&D and by 2009 they invested around 15% less. This same tendency is observed for the Spanish firms in general indicating that the firms concentrate their efforts in less uncertain innovative activities.

We also can assume that in terms of this dimension of open innovation, agri-food firms and Spanish firms are rather similar as they invest nearly 30% more in internal R&D than in external R&D, that is, they have more trust in their internal capabilities than in the capabilities of other firms, technological centers or institutions beyond the firm's boundaries. Although this might be the opposite of open-innovation practices, the high reliance on internal R&D could be explained by the absorptive capacity theory (Cohen and Levinthal, 1990). It stresses that a firm needs to develop internal R&D activities to gain the capability to monitor and integrate the external knowledge and without these capabilities the firm could not make the most of the knowledge beyond its boundaries.

In conclusion, with these three open-innovation dimensions we can observe that agri-food firms are slightly more open than Spanish firms in general. This greater openness of agri-food firms can be explained in terms set out by Chesbrough *et al.*, (2006), Dahlander and Gann (2010) and Acosta *et al.* (2011), indicating that network and open innovation are well adapted to this economic sector. The main reasons are focused on the definition of the sector as a food chain and the need for collaboration by the dominant structural dimension.

5.4 The effect of openness on the innovative performance of firms

In this section, we aim to analyze the effect of the three dimensions of firm openness on the innovative performance. With the information available in our sample and following Laursen and Salter (2006), we use two proxies aimed at reflecting various types of innovative performance by firms. First, we use a variable measuring the fraction of the firm's turnover relating to new-to-the-market products. In addition, we include a variable expressing the fraction of the firm's turnover pertaining to products new to the firm. The percentage of sales owing to new products to the market is considered to be radical innovation whereas the percentage of sales owing to products new to the firm is considered as incremental innovations. As observed in Tables 5.5 and 5.6 we include in the model the external information breadth and depth, the breadth of the number of co-operation agreements (breadth_coop) as well as the percentage of innovation expenditures dedicated to internal and external R&D activities (internal R&D and external R&D). We also consider it necessary to control for the importance given to the internal sources of information, which in a certain way represent how closed innovation could be. This is captured by the internal information variable which is measured as the importance given to this source for achieving innovations and is ranked from 1 to 4, where 1 is irrelevant and 4 is very high. Additionally, we controlled for the innovation expenditure, relative to total sales (innovation expenditure), and the firm size, measured by the natural logarithm of number of employees. In Table 5.4 we present the correlation table of the variables used in the model.

Table 5.5 shows the results of the use of Tobit models of radical and incremental innovations for the Spanish firms. In order to increase the reliability of our results and to capture the changes over time we calculated three models for the years 2007, 2008 and 2009. We start the interpretation of our results with the radical innovations and the effect of the informal openness. The breadth variable, which represents the openness based on the number of external sources of information shows a positive and significant effect on firm innovativeness for the three years of analysis. The depth variable, representing how strongly firms rely on external information sources, has a positive impact on firm innovativeness for the years 2007 and 2008, but it loses its significance at the year 2009.

The formal openness represented by the number of co-operation agreements shown in Table 5.5 also indicates that there is a positive effect on the firm's innovativeness during the three years of analysis. From the size of the coefficient, we could assume that this dimension of openness has a stronger effect than the informal openness. Regarding the third dimension of the openness, we are able to observe that the commitment to external R&D activities is not significantly positive for increasing the firm's innovative performance. This indicates that the mere R&D purchase does not improve the percentage of sales owing to new products to the market. Although this result is not the expected one, previous literature has shown that for certain high-tech industries, externalizing the R&D activities does not increase the firm's performance (Cruz-Cázares *et al.*, 2010).

When we turn our attention to the importance given to the internal information and internal R&D activities, we observe that both have a positive impact on the innovative performance and clearly the internal information has a greater impact than the achievement of internal R&D. This behaviour is constant for the year of analysis. All these results indicate that for Spanish radical innovators, it is important to have a combined strategy; both to sustain an active commitment to the internal R&D activity and to become an open innovator.

Table 5.4 Correlations of variables employed in the analysis of the effect of openness on firm behaviour

Variables	Mean	Standard deviation	1	2	3	4	5	6	7	8	9
1. Radical innov.	8.33	21.169	1								
2. Incremental innov.	12.501	26.251	0.042*	1							
3. Breadth	5.803	3.394	0.114*	0.042*	1						
4. Depth	1.173	1.529	0.095*	0.032*	0.427*	1					
5. Breadth_coop	0.64	1.176	0.110*	0.001	0.329*	0.290*	1				
6. Internal info.	3.322	0.933	0.106*	0.038*	0.320*	0.189*	0.173*	1			
7. Internal R&D	42.498	43.721	0.207*	0.184*	0.252*	0.127*	0.111*	0.249*	1		
8. External R&D	7.721	19.841	0.046*	0.053*	0.078*	0.059*	0.103*	–0.007	–0.075*	1	
9. Innov. exp.	0.08	0.255	0.186*	0.109*	0.143*	0.162*	0.180*	0.087*	0.246*	0.059*	1
10. Size	4.238	1.6894	–0.112*	–0.085*	0.079*	0.002	0.128*	0.054*	–0.154*	0.003	–0.237*

Note: *p-value < 0.05.

Table 5.5 Effect of openness in firm innovative performance for Spanish firms in general

Variables	Radical innovations			Incremental innovations		
	2009	2008	2007	2009	2008	2007
Breadth	1.2366***	1.4406***	2.0029***	1.0025***	1.2167***	1.2623***
	(0.269)	(0.256)	(0.2623)	(0.255)	(0.249)	(0.239)
Depth	0.6892	1.3462***	1.2787**	1.119**	1.1651**	0.1801
	(0.508)	(0.488)	(0.504)	(0.495)	(0.493)	(0.482)
Breadth_coop	4.1871***	4.3651***	4.436***	0.9213	1.6127***	1.7461***
	(0.606)	(0.595)	(0.629)	(0.125)	(0.604)	(0.605)
Internal info.	2.9874***	5.9547***	6.8766***	−0.0007	3.0292***	2.5063***
	(0.832)	(0.851)	(0.901)	(0.772)	(0.805)	(0.801)
Internal R&D	0.1381***	0.1415***	0.1399***	0.0287	0.0303	0.0056
	(0.020)	(0.019)	(0.020)	(0.019)	(0.018)	(0.018)
External R&D	0.0629*	0.0354	−0.0187	−0.0422	−0.0069	−0.0878***
	(0.036)	(0.035)	(0.0364)	(0.034)	(0.034)	(0.033)
Innov. exp.	11.669***	14.482***	14.268***	3.1474	−2.436	3.2755
	(2.784)	(2.830)	(2.797)	(2.8502)	(3.054)	(2.811)
Size	−2.473***	−1.1956***	−1.778***	−0.3009	−1.434***	−0.6122
	(0.499)	(0.4916)	(0.513)	(0.4759)	(0.479)	(0.474)
Constant	−25.171	−76.276***	−42.202***	−9.689	−39.36	−32.571
	(25.226)	(27.991)	(24.822)	(27.071)	(25.366)	(24.739)
X2	690.4***	857.59***	856.641***	234.33***	317.47***	302.79***
Log-likelihood	−17091.81	−17771	−17830.624	−20857.06	−21381.875	−21668.362
r2	0.0198	0.0236	0.0235	0.0056	0.0074	0.0069

Note: *p-value < 0.1; **p-value < 0.05; ***p-value < 0.01; standard errors in parenthesis.

For the control variables, we observe that the expenditure on innovations exerts a high impact on the firm's innovative performance and that there is a negative effect of firm size on the innovative performance, that is, small firms tend to achieve more radical innovations than large firms.

In Table 5.5, we move to the effect of open-innovation practices on the incremental innovations and results show that the breadth of the information sources does have a positive and significant effect on the achievement of new products by the firm across the three years of analysis. As for the depth of the information sources it can be observed that in 2007 it did not have a significant effect but later, in 2008 and 2009, the effect becomes significant.

The formal dimension of openness, breadth of co-operation agreements, also have a positive and significant effect on incremental innovations, but this only occurred in 2007 and 2008. However, the externalization of R&D activities is not significant in 2008 and 2009, being significant but negative in 2007.

With regard to the importance to the internal information, results show that this information source has a greater impact on the development of incremental innovations, but only for the years 2007 and 2008. Internal R&D activities seem not to increase the achievement of incremental innovations; rather they seem to be aimed to the accomplishment of radical innovations.

Finally, for the control variables, the innovation expenditures have a significant effect on firm innovativeness only for the year 2007 and the negative effect of firm size can only be observed in 2008.

In general, the open and closed innovative practices are less important for achieving incremental than radical innovations. This could be seen to be a predictable result because being a radical innovator should require more effort to get an innovative result only for the firm.

In Table 5.6, the estimates of the Tobit model for the agri-food firms are presented. As in Table 5.5, there are three models for the radical innovation dependent variable and three models for the incremental innovation dependent variable.

Results show that the breadth of the information sources has a positive effect on the achievement of radical innovations for agri-food firms. This effect is constant for the three years of analysis. Although the number of sources does exert an effect, the depth of these sources of information does not have a significant effect on the achievement of the radical innovations. The second dimension of the openness, collaboration agreements, has a stronger and significant effect on firm innovativeness than the informal breadth dimension. This effect is also constant across the three years of analysis. In line with results obtained for the Spanish firms, the third dimension of the openness, approximated by the externalization of R&D activities, does not produce any positive effect on the achievement of radical innovations.

In the same way as in Spanish firms in general, the importance given to internal information has a positive and significant effect on the innovative performance of firms. Also, its effect appears to be greater than that of the external breadth and depth information sources. However, contrary to the case of the Spanish firms as

Table 5.6 Effect of openness in firm innovative performance for agri-food firms

Variables	Radical innovations			Incremental innovations		
	2009	2008	2007	2009	2008	2007
Breadth	1.5424**	1.4237**	1.8386***	−1.0209	1.2221	1.3336*
	(0.693)	(0.679)	(0.577)	(0.832)	(0.8196)	(0.701)
Depth	−0.8304	−0.403	−0.8895	−0.5100	−0.6966	−0.007
	(1.332)	(1.323)	(1.092)	(1.663)	(1.6567)	(1.386)
Breadth_coop	4.6830***	4.6495**	3.777***	−0.9571	−1.6406	1.3903
	(1.695)	(1.792)	(1.441)	(2.160)	(2.303)	(1.890)
Internal	3.5507	8.8410***	9.6282***	5.3578**	9.3698***	1.0872
	(2.294)	(2.516)	(2.139)	(2.694)	(2.954)	(2.389)
Internal R&D	0.0697	0.0206	0.024	0.1255*	0.0958	−0.0107
	(0.185)	(0.052)	(0.043)	(0.064)	(0.063)	(0.052)
External R&D	0.1025	−0.0588	−0.0362	−0.1719	−0.1264	0.0381
	(0.086)	(0.093)	(0.081)	(0.113)	(0.113)	(0.094)
Innov. exp.	29.937	21.7371	23.199***	33.3211	16.272	6.2766
	(18.491)	(14.297)	(6.816)	(33.594)	(18.886)	(9.703)
Size	0.7093	1.7257	0.798	2.9858	1.4266	0.2642
	(1.491)	(1.481)	(1.211)	(1.817)	(1.8133)	(1.4934)
Constant	−47.857***	−66.222***	−67.969***	−49.748***	−72.999***	−41.078***
	(9.657)	(10.771)	(9.792)	(11.367)	(12.807)	(10.631)
X2	66.57***	56.52***	82.59***	35.11***	42.36***	22.93**
Log-likelihood	−1481.063	−1500.038	−1537.111	−1955.9666	−1963.8123	−1992.001
r2	0.0220	0.0185	0.0262	0.0089	0.0107	0.0057

Note: *p-value < 0.1; **p-value < 0.05; ***p-value < 0.01; standard errors in parenthesis.

a whole, the development of internal R&D activities seems not to significantly increase the percentage of sales owing to radical innovations for agri-food firms.

Regarding the control variables, the expenditures dedicated to innovation activities produced a positive effect on radical innovations for the years 2007 and 2008, but the effect is absent in 2009. The negative effect of firm size observed for Spanish firms in general is not present for the agri-food firms.

These results may show that for agri-food firms, as well as for the rest of the companies, open-innovation practices could be beneficial for companies, but they need a culture of innovation in the organization in order to have a positive effect on its results.

Regarding the models standing for incremental innovations, we can observe that the breadth, depth of information sources and the breadth of co-operation agreements lose a lot of significance compared with their effects on the achievements of radical innovations. The single effect of external information breadth can barely be observed in 2007; the depth variable is not significant in any model, neither are the co-operation agreements. The third dimension of the openness, external R&D, does not increase the probability of increasing the sales owing to new products for the firm.

Corresponding internal variables, the importance given to the internal information, is significant only for the years 2008 and 2009. As for the internal R&D activities, no effect is observed in Table 5.6.

Finally, the expenditure dedicated to innovation activities has positive effects exclusively for the year 2009, the same is true for the firm size.

When we compare the results obtained in the overall Spanish sample and agri-food sample, we observe that the different innovative activities of the firms are more important for radical than incremental innovations. This fact could be interesting for the firm managers. They can consider the different impact of closed or open innovative activities on performance and, in consequence, set the desired commitment to one or other activity. The agri-food firms show more significant variables in the context of radical innovation than incremental actuations. This result could indicate differences in the strategies employed for the firms when they decide to use openness. Thus the firm can search for more alliances when the objectives are radical results and less in the incremental contexts. Finally, it is possible that the agri-food companies have selected the open activities mainly for radical aims.

5.5 Conclusions

In recent years, we have been seeing a growing interest in knowing the degree of implementation of open-innovation practices in companies. Studies have focused on several countries and several sectors. This study concentrates on a sample of Spanish firms with special emphasis on the agri-food sector. The agri-food industry is transitioning from a traditionally poorly innovative sector to a situation where innovation plays a key role. Innovation options, including actions related to open innovation and collaboration, fit very well into this sector characterized by a chain structure (food chain) and a predominance of SMEs.

This higher innovator effort has been observed from the input and output perspective. From a database of Spanish firms we find that there are more agri-food firms involved in R&D, but they employ fewer R&D personnel and they are less innovation intensive than the rest of Spanish firms. However, they are more efficient in transforming their limited innovation resources because the number of firms achieving process innovation is much higher for agri-food firms than Spanish firms. Nevertheless, we could not observe significantly different behaviour between the agri-food firms and Spanish firms with regard to product innovations.

The openness of the innovation activities is not a single best practice and in this chapter we have identified three different dimensions of open innovation. When we analyze mechanisms that expose firms to new information and novel ideas from outside sources as a measure of openness, we find that the agri-food sector has a slightly higher degree of openness than the rest of the sample Spanish firms. Besides, in this study the agri-food firms seem to be open in terms of informal linkages with external agents, but, when it comes to formal linkages, such as co-operation agreements, they seem to be very closed, although they do not have a significantly different pattern of behaviour from the rest of Spanish firms. Finally, the agri-food firms spend a little more on external R&D than the rest of Spanish firms. In conclusion, in this study we are able to observe that agri-food firms have a strong commitment to open-innovation practices, contrary to the traditional view of it as a low-innovation industry. The way agri-food firms follow the three dimensions considered as open-innovation practice, i.e. external information breadth and depth, collaboration agreements and external R&D, fits with the industry structure characterized by a high concentration of SMEs and the food-chain structure.

This slightly greater degree of openness of agri-food firms has an effect only on the radical innovations, not incremental innovations unlike the rest of Spanish firms where the effect of openness is both for incremental and radical innovation. The growth complexity of the market and new technologies forced firms to use collaboration and the agri-food sector is ready for that (Weaver, 2008). These firms' decisions need to be based on a thorough analysis in order to choose whether radical or incremental innovation is the most appropriate for them.

This chapter has certain limitations. The main limitation is that, as discussed in the literature review and as shown in the description of the sample, agri-food firms are oriented towards process innovations, but, owing to a lack of data, we had to analyse the effects of the open-innovation practices on the percentage of sales resulting from new products. It would have been very interesting to study the effect of the open innovation on the main innovation outcomes of the agri-food firms but further research in this regard is needed in this field.

5.6 References

ACOSTA, M., CORONADO, D. AND TORIBIO, M.R. (2011). The use of scientific knowledge by Spanish Agrifood firms. *Food Policy*, **36**(4), 507–516.

ALFRANCA, O., RAMA, R. AND VON TUNZELMANN, N. (2004). Innovation spells in the multinational agri-food sector. *Technovation*, **24**, 599–614.

ARCHIBUGI, D. AND PIANTA, M. (1996). Measuring technological change through patents and innovation surveys. *Technovation*, **16**(9), 451–468.

BATTERINK, M.H., WUBBEN, B.F.M. AND OMTA, S.W.F. (2006). Factors related with innovate output in the Dutch agrifood industry. *Journal of Chain and Network Science*, **6**(1), 31–44.

CAPITANIO, F., COPPOLA, A. AND PASCUCCI, S. (2009). Indications for drivers of innovation in the food sector. *British Food Journal*, **111**(8), 820–838.

CHESBROUGH, H., VANHAVERBEKE, W. AND WEST, J. (2006). *Open innovation: researching a new paradigm*. Oxford: Oxford University Press.

COHEN, W.M. AND LEVINTHAL, D.A. (1990). Absorptive capacity: a new perspective on learning and innovation. *Administrative Science Quarterly*, **35**, 128–152.

COTEC (2010). Panel de innovación tecnológica PITEC. Revisión de 2010. Ed: Fundación COTEC para la Innovación tecnológica.

CRUZ-CÁZARES, C., BAYONA-SÁEZ, C. AND GARCÍA-MARCO, T. (2010). R&D strategies and firm innovative performance: a panel data analysis. *International Journal of Innovation Management*, **14**(6), 1013–1045.

DAHLANDER, L. AND GANN, D.L. (2010). How open is innovation? *Research Policy*, **39**(6), 699–709.

DOBSON, P.W., CLARKE, R., DAVIES, S. AND WATERSON, M. (2001). Buyer power and its impact on competition in the food retail distribution sector of European Union. *Journal of Industry, Competition and Trade*, **1**(3), 247–281.

EBERSBERGER, B., HERSTAD, S., IVERSEN, E., SOM, O. AND KIRNER, E. (2011). *Open innovation in Europe. PRO INNO Europe: INNO-Grips II report*, Brussels: European Commission, DG Enterprise and Industry

FAYOS, T. CALDERÓN, H. AND MIR, J. (2009). Eficiencia, competitividad y políticas de promoción empresarial. Estudio del sector agroalimentario en España. *Economía Industrial*, **373**, 111–122.

FILIPPAIOS, F. PAPANASTASSIOU, M., PEARCE, R. AND RAMA, R. (2009). New forms of organisation and R&D internationalisation among the world's 100 largest food and beverage multinationals. *Research Policy*, **34**, 1032–1043.

FORTUIN, F.T.J.M. AND OMTA, S.W.F. (2009). Innovations drivers and barriers in food processing. *British Food Journal*, **111**(8), 839–851.

GALIZZI, G. AND VENTURINI, L. (1996). Product innovation in the food industry: Nature, characteristics and determinants. In G. Galizzi, and L. Venturini (Eds.), *Economics of innovation: the case of the food industry* (pp. 133–156). Heidelberg: Physica-Verlag.

GARCIA MARTINEZ, M. AND BURNS, J. (1999). Sources of technological development in the Spanish food and drink industry. A 'supplier dominated' industry? *Agribusiness*, **15**(4), 431–448.

GARCIA MARTINEZ, M. AND BRIZ, J. (2000). Innovation in the Spanish food and drink industry. *International Food and Agribusiness Management Review*, **3**(2), 155–176.

GRUNERT, K., HARMSER, H., MEULENBERG, M., KUIPER, E., OTTOWITZ, T., DECLERCK, F., TRAILL, B. AND GÖRANSSON, G. (1997). A framework for analysing innovation in the food sector. In *Product and process innovation in the food industry*, Traill, B. and Grunert , K.G. (eds). Blackie Academic and Professional, London, pp. 1–37.

KÜHNE, B., VANHONACKER, F., GELLYNCK, X. AND VERBEKE, W. (2010). Innovation in traditional food products in Europe: Do sector innovation activities match consumers'acceptance? *Food Quality and Preference*, **21**(6), 629–638.

LAURSEN, K. AND SALTER, A. (2006). Open for innovation: the role of openness in explaining innovation performance among U.K. manufacturing firms. *Strategic Management Journal*, **27**, 131–150.

MOL, M.J. AND BIRKINSHAW J. (2009). The sources of management innovation: when firms introduce new management practices. *Journal of Business Research*, **62**, 1269–1280, DOI: 10.1016/jbusres.2009.01.001.

OECD (1997). *Proposed guidelines for collecting and interpreting technological innovation data: Oslo manual*. OECD Publishing, 2nd (revised) edition.

RAMA, R. (1996). Empirical study on sources of innovation in international food and beverages industry. *Agribusiness*, **12**(2), 123–134.

RAMA, R. (2008). *Handbook of innovation in the food and drink industry*. New York, Haworth Press.

RAMA, R. AND ALFRANCA, O. (2003). Competitive behaviour, design and technical innovation in food and beverage multinationals. *International Journal of Biotechnology*, **5**(3–4), 222–248.

SERRANO, A. AND ALONSO, R. (2008). *Economía de la Empresa Agroalimentaria*. Ediciones Mundi-Prensa, 3ª edición.

TRAILL, B. AND MEULENBERG, M. (2002). Innovation in the food industry. *Agribusiness*, **18**(1), 1–21.

WEAVER, R.D. (2008). Collaborative pull innovation: origins and adoption in the new economy. *Agribusiness*, **24**(3), 388–402.

WEISS, C.R. AND WITTKOPP, A. (2005). Retailer concentration and product innovation in food manufacturing. *European Review of Agricultural Economics*, **32**(2), 219–244.

Part II

Partners and networks for open innovation

6

Partnering with public research centres and private technical and scientific service providers for innovation: the case of Italian rice company, Riso Scotti

V. Lazzarotti and R. Manzini, Cattaneo University – LIUC, Italy

DOI: 10.1533/9780857097248.2.97

Abstract: This chapter focuses on collaborative innovation with a specific set of partners whose unique business activity is developing technology and know-how, such as universities, research centres, technical and scientific services. After discussing the specific characteristics of these partners and the potential benefits and issues in collaborating with them, the experience of the Italian rice company, Riso Scotti, in its relationships with these partners is described. Some general suggestions are given concerning the management and improvements of relationships with universities, research centres and technical and scientific services, by integrating the contributions from other studies with the practical experience described in the Riso Scotti case study.

Key words: open innovation, collaboration, universities, research centres, service providers.

6.1 Introduction

As innovation is increasingly open and distributed, a market for technology is continuously growing (Arora *et al.*, 2001), where private companies and public organizations make business by exchanging technology and know-how. Among these organizations, there are various typologies of actors (Fig. 6.1):

1. Knowledge intensive business service (KIBS) organisations, i.e. (Katsoulacos and Tsounis, 2000; Muller and Zenker, 2001; Windrum and Tomlinson, 1999) private or public companies that, through their professional scientific and/or

technical competence base, offer services embedding a high intellectual value. Such services include banking and finance, marketing, human resource management, and also R&D, design, engineering and prototyping;

2. Technical and scientific service (TSS) providers, which are a subset of KIBS providers, i.e. those KIBS companies that offer services based upon their technical and scientific competence, and are capable of supporting the innovation process of the client firm (Chiesa *et al.*, 2008). TSSs include contract research and development, product design, engineering, testing and prototyping, technology consulting;
3. Public research centres and universities, which are mainly focused on basic research and are primarily funded (but not exclusively) with public funds.

Industrial economics studies have emphasized that KIBS support, and TSS organizations in particular, is relevant for the development and performance of national and regional economies (Hipp and Grupp, 2005; Mansfield and Lee, 1996; Strambach, 2008; Windrum and Tomlinson, 1999). This support becomes even more critical when it is targeted to small companies (Freeman, 1991; Malecki and Tootle, 1996; Rothwell and Dodgson, 1991). Other studies reveal that TSS have a positive impact on the client firm's innovative and new product development performance (Chiesa *et al.*, 2008; Katsoulacos and Tsounis, 2000), and that TSS organizations play a fundamental knowledge-brokering role, that can foster the birth and growth of technology-intensive industries.

However, all these potential benefits are frequently hindered by the difficulties

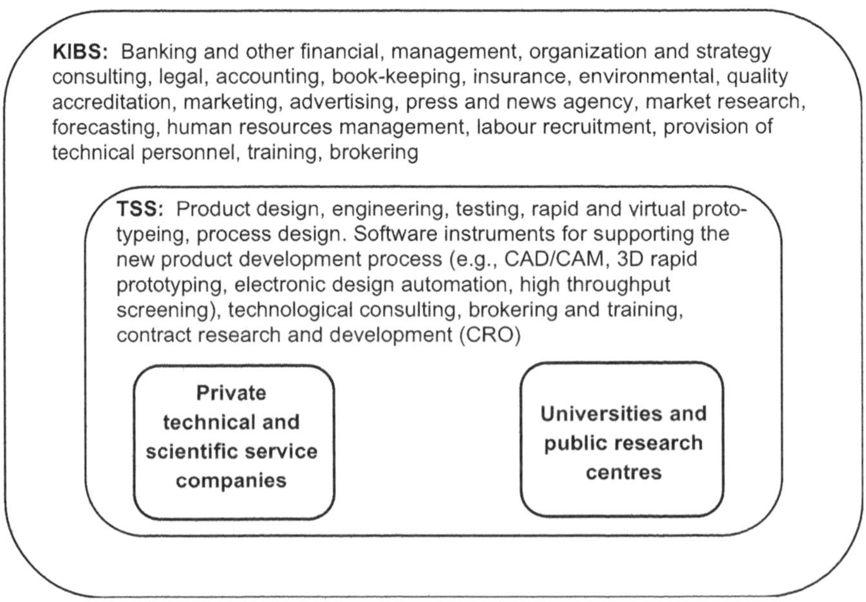

Fig. 6.1 Actors in the technology market (adapted from Chiesa *et al.*, 2008).

arising when collaborations with KIBS are actually implemented, such as, for example, the difficulties inherent in the transfer of tacit and complex knowledge (Jorde and Teece, 1990) and the lack of absorptive capacity in the recipient firm (Cohen and Levinthal, 1990). Studies on these aspects are still limited (Chiesa *et al.*, 2008), therefore, in order to analyse these aspects in more depth, this chapter focuses exclusively on two types of KIBS: private TSS and public research centres (particularly universities), in an attempt to provide evidence regarding the most relevant problems and to put forward some of the solutions found in previous studies. The Riso Scotti case study shows how the problems (and relative solutions) highlighted by previous studies are actually faced by companies and offers some suggestions for further research.

6.2 The role of private technical and scientific service (TSS) providers: advantages and limitations

According to the most recent studies on the topic (see, for example, Chiesa *et al.*, 2008; Lazzarotti and Pizzurno, 2010) companies that access TSS private organizations to support their innovation processes may find several advantages. The most relevant is specialization; in fact, each TSS company is characterized by the specific type of activity (or stage within the innovation process) it is able to support or by the specific competences available inside (i.e. the specific field of knowledge). Specialization, in turn, ensures that the specific activity or the specific field of research is supported in an efficient and effective manner. As a matter of fact, specialization allows TSS companies to achieve the critical mass that is necessary to use the most efficient technology and/or to have available various technologies, tailored for the needs of different innovator clients. Furthermore, the focus on a limited set of competences allows them to stay up-to-date in the studied field and to achieve considerable depth in production knowledge. Another advantage of private TSS companies is that they are strongly focused on time and costs, because they work only and always with a definite order of a specific client, with a precise contract. All this represents a strong trigger to respecting time, costs and quality of the output in coherence with the client's expectations. For the same reason, settlement of intellectual property (IP) problems, even if it is complex as in all collaboration contexts, becomes less difficult as it has to be always defined when signing the contract.

Working with TSS private organizations, however, poses some challenges. The most relevant is that quite often managerial competences are very limited inside these companies. In particular, marketing and communication capabilities are very low, because all managers are usually totally concerned with technology and scientific know-how, and are far less interested in other managerial issues. This means, first of all, that it is sometimes very difficult for a company searching for technological support to identify the potential partner offering the desired technology or competences. Furthermore, the lack of communication capabilities sometimes make the collaboration difficult, especially when some unexpected events (that are very frequent in R&D activities) occur and need to be faced in a co-ordinated way between the innovator company and the technology service provider. The lack of managerial competences also concerns the economic and financial areas; this makes it very difficult to define in a convincing and satisfactory way the value of the specific service that the client can actually purchase. An additional challenge concerns the fact that, at the moment, the majority of TSS companies are SMEs. This may limit their capacity to stay up-to-date with technology and know-how and to achieve the necessary critical mass to be efficient in their work. Furthermore, because of their small size, TSS companies are sometimes perceived as unreliable partners.

6.3 The role of universities and public research centres: advantages and limitations

Studies of university–industry collaborations highlight several advantages that industry can achieve by collaborating with universities (see, for example, Bonaccorsi and Piccaluga, 1994; George *et al.*, 2002; Hall *et al.*, 2003; Meyer-Krahmer and Schmoch, 1998; Rohrbeck and Arnold, 2006) and that the so-called 'third mission' of universities is becoming more and more relevant, especially in a context of increasingly open innovation. The mentioned advantages include primarily unique access to excellent and unique research skills, availability of up-to-date technological knowledge, usage of specialized laboratories, risk sharing and cost reduction for basic research, availability of an excellent, timely and efficient recruiting channel, and stabilizing of long term research projects. Other relevant indirect benefits can be identified in the access to a wide, international and open network of competences (because the academic world is strongly interconnected).

The same studies agree that university–industry collaborations are very difficult in practice and underline the main barriers (Laukkanen, 2003; Rappert *et al.*, 1999; Rohrbeck and Arnold, 2006). These difficulties can be grouped into three main sets: cultural barriers, institutional barriers and operational barriers. Cultural barriers mainly refer to the fact that industry and university have completely different goals in knowledge production (profits for industry, validation and reputation for universities), and, as a consequence, have conflicting secrecy policy (universities need to disclose their research results, whereas industry needs to keep them secret in order to create a temporary monopoly). Cultural differences concern also the language and the time horizon of research activities (industry always refers to a window of market opportunity, universities conceive research results in a continuous, long-term perspective). Institutional barriers relate to the different nature style of work (universities are used to working on complex and wide-ranging problems, in a long-term perspective, and in a less formal and hierarchical environment, whereas industry tends to focus on more limited and specific problems, with a short–medium term perspective), the diverging perception of what the 'product' of R&D is (industry positively considers results when they have a clear market application, university always considers new knowledge as a result). Finally, operational barriers refer to the completely different organization and incentive systems (universities are more bureaucratic and have limited incentives explicitly correlated to results), the lack of project management competences of universities (that frequently lose the focus on the required time and costs), and the lack of knowledge about the partner and his processes.

In conclusion, it can be argued that, potentially, TSS companies, universities and public research centres can significantly improve the efficiency and effectiveness of companies' innovation process, but several difficulties may arise that prevent companies from actually achieving such benefits. Riso Scotti represent an example of a company which started to open its innovation process with the aim of exploiting knowledge produced by universities and research centres, but then decided to strongly limit and focus the access to such external sources of knowledge because of the negative experience with some of these actors.

6.4 Riso Scotti case study

Founded in 1860 and incorporating several high-tech companies, specializing in the cultivation, research, testing, production and transformation of rice, for over 150 years Riso Scotti has produced and commercialized best-quality rice. Riso Scotti has successfully combined the rice-growing traditional competences with innovative technologies, able to ensure high-quality standards and an enlarged product range. Nowadays, Riso Scotti products include not only rice, but also pasta, snacks, rice-based drinks, bakery products, oil and ready meals: the 40% of sales in Italy is not made up by rice, but by the diversification of rice-based products. The company is in fact the market leader in Italy for pasta made from rice (80% of market share), an exceptional product which has enabled consumers to eat gluten-free pasta.

Riso Scotti has created its cutting-edge technological centre in Italy: 'Bivio Vela', near Pavia, whose strong innovation lies in its systemic process with which traditional work is handled. The complete production control guarantees high-quality standards and full optimization of raw materials: an example of an integrated cycle, which utilizes all resources and minimizes waste and, thus, any environmental impact.

Internationalization is crucial for Riso Scotti: its products are distributed in 60 countries throughout the world, 23 of which are outside of Europe, over five continents, exploiting a solid network of partnerships with capable local operators. Eastern Europe is particularly relevant and, in 2005 Riso Scotti started the 'Danube Project' in Romania, an innovative agro-industrial project through which Riso Scotti exports not only its products, but also the Italian culture to rice manufacturing (in just five years Riso Scotti has bought 10 000 acres of land and built two factories that have achieved a turnover of €22 million). The internationalization process moved beyond Europe, to Panjab, India, where, with the participation of a local rice mill, Riso Scotti produces, manufactures and exports to Europe the Basmati rice, an ideal side dish of increasing demand for the European consumer. In this case, Italian specialists collaborate with Indian employees in order to guarantee and safeguard the product quality. New initiatives are ongoing also in the United States. The company is highly internationalized also in terms of sales: it exports about 30% of sales in 54 countries worldwide, including 23 non-European countries, being able to take advantage from strong partnerships with local skilled workers.

6.4.1 Innovation

Riso Scotti has internalized the innovation process very recently: until two years ago innovations were mainly developed outside the company and then brought inside to be branded as 'Riso Scotti'. The revolutionary decision to introduce an internal R&D unit has completely changed the nature and relevance of innovation in the company, and implied for a cultural, mindset change, which is still in progress. Together with internationalization, innovation is perceived as the most relevant source of growth and competitive advantage for the future of the company. At the same time, this radical change is perceived as very risky, because the Riso Scotti brand is widely known as a high quality brand and, hence, new products and new markets should be exploited maintaining the same level of quality.

At the moment, the R&D department is very small and the R&D budget is a small percentage of turnover, but R&D is not the only 'locus' of innovation: innovation activities are widespread within the whole company, from marketing through production and logistics, to packaging and commercialization. All employees from all departments within the company are aware of the relevance of innovation for the growth of the company and, hence, give their contribution in terms of new ideas generation and experimentation, suggestions for improvements of existing products, introduction of process innovations. In other words, it can be said that Riso Scotti is adopting its characteristic 'lean' and pragmatic approach even in innovation activities. Within R&D, research represents 10% of the total R&D expenses, development is 90%.

6.4.2 Product innovation

Usually, new product ideas come from the marketing department, which, by analyzing market data and competitors' behaviour, identifies new customer needs and opportunities for introducing new or improved products. On average, around 10% of ideas proposed by the marketing unit are approved for further development and usually 2–3 projects are identified to be brought to the market. Then, the R&D department carries out the necessary research activities and very rapidly the new product moves into the experimentation phase, with a pilot production. This phase mainly concerns two aspects: first, the actual adaptability of available technologies for developing and manufacturing the new product and the identification of the necessary technical modifications; second, the tasting of the new product, which is analyzed by means of internal taster or, in some case, through a marketing research conducted by external agencies. The results of the experimentation are discussed with the marketing unit for final tuning and, after that, if technological issues have been successfully solved, the product is ready to be produced and commercialized. The commercialization of new products in Italy does not create any problem, because distribution channels are consolidated and the Riso Scotti brand is well known and appreciated. For foreign markets, on the contrary, Riso Scotti is still growing and then it is necessary to deal with local importers and taste new products with local customers. The failure rate of new products is very high, around 70%. The total innovation time is between 2 and 4 months, as a result of the time required by the market. Product innovation at the moment mainly concerns ready meals, easy-to-cook products and healthy food.

6.4.3 Process innovation

Technological competences available within the company are really excellent and this allows Riso Scotti to work internally for the introduction of process improvements and to implement the necessary modifications when new products are to be introduced in production. Hence, process innovation in Riso Scotti is conceived more as a continuous improvement process, rather than an episodic event.

6.4.4 The Riso Scotti approach to open innovation

Innovation at Riso Scotti is mainly an internal process. However, given the limited resources internally available for R&D, the ongoing shortening of product lifecycles and the rapid change in customer needs force Riso Scotti to exploit some external support.

In the research phase, (to which, as already specified above, only 10% of the R&D budget is dedicated) Riso Scotti carries out internally the literature searches concerning the specific fields in which Riso Scotti is interested for developing new products. This exploratory phase is conducted internally because it is considered critical for identifying new potentially interesting areas for innovation and for understanding the remaining potential of technologies currently used by the company (i.e., positioning the current Scotti technologies in the S-curve). In other words, literature searching is used internally as a tool for technology intelligence (Lichtenthaler, 2005). After the internal technology intelligence process, the most relevant and promising scientific and technological areas are identified where new research activities could be conducted. Given the limited internal R&D resources, Riso Scotti searches for an external partner to conduct such activities. A supposed 'optimal' partner for this type of research could be a university, because usually it involves mainly basic and some applied research. Riso Scotti have contacts and relations in several universities in Italy, and have attempted to collaborate with some of them. However, several problems emerged in these collaborations. First of all, research teams in universities demonstrate lack of the necessary focus on time and costs, critical factors for companies, which know very well that losing the right window of opportunity means sure market failure and that respecting the target cost imposed by competition means going to market quickly. As a consequence, after some collaboration that ended with longer time and higher costs than expected, Riso

Scotti managers decided to search for a different type of partner. An analysis was conducted to find private research and service centres, with the required competences and technologies for conducting those research and experimental activities considered relevant for future innovations. Even if this analysis required a strong effort and is always ongoing, in order to improve the quality of set of potential partners, Riso Scotti has developed a consolidated, long-term oriented network of TSS companies, each specialized in a specific type of technology or competence to support research and (some) experimentation on recipes, ingredients and treatments.

In summary, it can be argued that the major part of research activities is conducted with external partners, mainly private TSS organizations, whereas, at the moment, collaborations with public research centres and universities have been limited to a few cases, involving researchers that have demonstrated a more entrepreneurial attitude in their relationships with industry, with which Riso Scotti has established a long-term relationship.

Concerning development activities, from experimentation to commercialization, innovation is mainly internal and Riso Scotti is reluctant to open the process. However, some collaborations with suppliers and technology specialists are necessary to complete the innovation process. In more detail, the main partners are:

- ingredients suppliers;
- machinery suppliers; and
- packaging suppliers (such as, for example, Tetrapack).

With all these suppliers, the approach adopted by Riso Scotti is based upon the identification of excellent companies, and the creation of long-term collaborations with them. By this approach, mutual trust with partners is guaranteed, opportunistic behaviour is avoided and coherent long-term collaborations are ensured. Relationships are usually based upon continuous, informal contacts and common work, which represent the background on which collaboration contracts are signed concerning specific innovations. Obviously, excellent suppliers/partners are more expensive, but they ensure the quality of the final result. Furthermore, given that the IP strategy of Riso Scotti is still embryonic and not highly sophisticated, this approach represents a guarantee against spill over. Furthermore, even if Riso Scotti has some patents (concerning mainly rice, such as, for example, the so called 'gemma di riso'), its excellent know-how as well as the main recipes are mainly protected with trade secrets and trust.

Concerning packaging, in some instances, collaborations are set up for the graphic and aesthetic aspects.

Other partners with which Riso Scotti sometimes collaborate are expert consultants in the food industry and become necessary particularly when new national and international norms and regulations must be studied and introduced into the company. Concerning these regulatory aspects, Riso Scotti will probably increase the number of its relationships with institutional bodies and associations, either at the national or international level, because being continuously up-to-date with regulations is becoming increasingly complex and, at the same time, absolutely essentail to remain competitive.

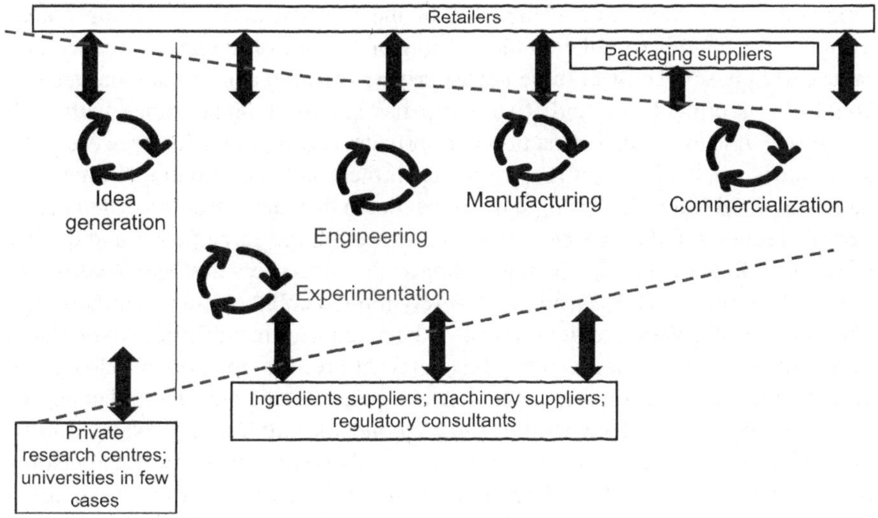

Fig. 6.2 The open innovation process in Riso Scotti.

The collaboration with retailers should also be mentioned, because this is based upon long-term, informal relations, by which Riso Scotti continuously monitor customer evolution and the positioning of the Riso Scotti brand, essential information for screening and selecting innovation projects.

The open-innovation model adopted by Riso Scotti is depicted in Fig. 6.2.

6.5 Conclusions and managerial implications

Riso Scotti is an example of a company that is beginning to open its innovation process in a very cautious and careful way: only the exploratory research/idea generation phase is actually open to a wide set of TSS, and, in terms of governance, the network of collaborations is always strictly guided and controlled by Riso Scotti, who determines the research areas to be investigated and the experiments to be conducted. The other phases of the innovation process, from development up to commercialization, are open only to a very limited set of (vertical) partners, with whom Riso Scotti has a consolidated, long-term relationship.

This case study suggests some conclusions concerning three aspects. First, previously observed problems of university–industry and TSS–industry collaborations have been verified. Second, the experience of possible solutions and approaches have been adopted adopted to exploit such collaborations. Third, some considerations have been drawn concerning the potential benefits that companies operating in the food industry may find in collaborating with TSS and universities.

With reference to published conclusions on collaborations between TSS companies and university, Riso Scotti is a sort of paradigmatic case. For Riso Scotti, the need to exploit external sources of technology and know-how is very high. The limited size of the company and, in particular, the limited resources and competences available for R&D is stimulating the opening of the innovation process, especially in its early phases. To meet these needs, private research centres and universities could play a critical role and Riso Scotti has already collaborations with both categories of actors. The interaction with private TSS organizations proved to be satisfactory, even if finding the appropriate partners and consolidating the network has been a hard work. Such companies have shown that they are able to satisfy Riso Scotti's requests in the expected time and respecting the agreed costs and quality of results. On the contrary, the interaction with universities has been so difficult that, at the moment, Riso Scotti has severely limited collaborations with this type of partner. All the main problems that have been reported in published studies have been experienced by Riso Scotti. Most relevant are: the operational barriers, in particular the lack of project management competences and the consequent lack of focus on time, costs and quality of results; second, the cultural barriers, in terms of differences in languages, time horizons and fundamental goals; third, the institutional barriers, particularly with reference to the style of work, with a low attention to hierarchy and respect of time. Another quite predictable observation concerns the problem of IP and secrecy: because Riso Scotti protects its innovations mainly by trade secrets, with a very limited use of patents, this reinforces the problems related to appropriability and intellectual property, with both private TSS companies and universities. Such problems, in turn, further limit the attitude and inclination towards open innovation. In summary, it can be stated that the difficulties arising when Riso Scotti attempts to collaborate with public and private research centres probably prevent the company from actually exploiting the full potential of external sources of knowledge and technology.

In spite of these difficulties some recommendations can be made that may help in improving the efficiency and effectiveness of the problematic collaborations depicted so far. These recommendations can be summarized by one essential point: establishing long-term relationships with TSS companies and universities, which will allow better reciprocal knowledge concerning managerial and organizational issues, greater trust, better understanding of mutual interests and identification of routines and procedures to help problem solving. All these aspects, in turn, represent the bases for reducing potential conflicts relating to IP rights. Creating and consolidating a network of relationships is not an easy task, requiring a continuous effort to find the appropriate partners, nurture the relationship, and identify the conditions for a mutual long-term interest in the collaboration.

Finally, some conclusions can be drawn from the Riso Scotti case study, and these can be generalized to be appropriate for a large part of the food and drink industry. Limited R&D investments, the focus on incremental rather than radical innovation and on new product development rather than on research, and the demand-pull prevalent nature of innovation are typical in the industry (Sarkar and Costa, 2008; Garcia Martinez and Briz, 2000; Lagnevik *et al.*, 2003). For these reasons, private TSS organizations and universities can play a critical role, providing food and drink companies with excellent and specialized technology and research expertise that they cannot develop inside. Also the increasing risks and costs of innovation, owing to the high failure rate of new products (which is around 70% in the industry) push in the same direction, because collaborations with public and private research centres means a reduction and sharing of both the risks and costs. However, in order to exploit these potential benefits, companies have to be strongly proactive. The case study presented in this chapter suggests two ways to be proactive: first, by creating a long-term, consolidated and dynamic network of relationships; and second, by defining an internal IP policy and strategy that can properly support the opening of innovation processes, at least in the early phases of the innovation process, i.e. those concerning exploratory research and idea generation.

6.6 References

ARORA A., FOSFURI A. AND GAMBARDELLA A. (2001) Markets for technology and their implications for corporate strategy, *Industrial and Corporate Change*, **10**(2), 419–451.

BONACCORSI A. AND PICCALUGA A. (1994) A theoretical frame work for the evaluation of university–industry relationships, *R&D Management*, **24**(3), 229.

CHIESA V., FRATTINI F. AND MANZINI R. (2008) Managing and organising technical and scientific service firms: a taxonomy and an empirical study, *International Journal of Services, Technology and Management*, **10**(2), 211–234.

COHEN W.M. AND LEVINTHAL D.A. (1990) Absorptive capacity: a new perspective on learning and innovation, *Administrative Science Quarterly*, **35**(1), 128–152.

FREEMAN C. (1991) Networks of innovators: a synthesis of research issues, *Research Policy*, **20**(5), 499–514.

GARCIA MARTINEZ M. AND BRIZ J. (2000) Innovation in the Spanish food and drink industry, *International Food and Agribusiness Management Review*, **3**, 155–176.

GEORGE G., ZAHRA S.A. AND WOOD D.R. (2002) The effects of business university alliances on innovative output and financial performance: a study of publicly traded biotechnology companies, *Journal of Business Venturing*, **17**(6), 577–609.

HALL B.H., LINK A.N. AND SCOTT J.J. (2003) Universities as research partners, *The Review of Economics and Statistics*, **85**(2), 485–491.

HIPP C. AND GRUPP H. (2005) Innovation in the service sector: the demand for service-specific innovation measurement concepts and typologies, *Research Policy*, **34**(4), 517–535.

JORDE T.M. AND TEECE D. (1990) Innovation and cooperation: implications for competition and antitrust, *Journal of Economic Perspectives*, **4**(3), 75–96.

KATSOULACOS Y. AND TSOUNIS N. (2000) Knowledge intensive business services and productivity growth: the Greek evidence, in Boden M. and Miles I. (eds.) *Services and the knowledge based economy*, London: Continuum.

LAGNEVIK M., SJÖHOLM I., LAREKE A. AND ÖSTBERG J. (2003) *The dynamics of innovation clusters: a study of the food industry*. London: Edward Elgar Publishing Company.

LAUKKANEN M. (2003) Exploting academic entrepreneurship: drivers and tensions of university-based business, *Journal of Small Business and Enterprise Development*, **10**(4), 372–382.

LAZZAROTTI V. AND PIZZURNO E. (2010) What is the place of technical and scientific service companies (TSS) in the process of developing new products? Insights on their managerial and organizational features, *International Journal of Engineering, Science and Technology*, **2**(9), 39–53.

LICHTENTHALER E. (2005) The choice of technology intelligence methods in multinationals: a contingency approach, *International Journal of Technology Management*, **32**(3–4), 388–406.

MALECKI E. AND TOOTLE D.M. (1996) The role of networks in small firm competitiveness, *International Journal of Technology Management*, **11**(1–2), 43–57.

MANSFIELD E. AND LEE J.Y. (1996) The modern university: contributor to industrial innovation and recipient of industrial R&D support, *Research Policy*, **25**(7), 1047–1058.

MEYER-KRAHMER F. AND SCHMOCH U. (1998) Science-based technologies: university–industry interactions in four fields, *Research Policy*, **27**(8), 835–851.

MULLER E. AND ZENKER A. (2001) Business services as actors of knowledge transformation: the role of KIBS in regional and national innovation systems, *Research Policy*, **30**(9), 1501–1516.

RAPPERT B., WEBSTER A. AND CHARLES D. (1999) Making sense of diversity and reluctance: academic–industrial relations and intellectual property, *Research Policy*, **28**(8), 873–890.

ROHRBECK R. AND ARNOLD H.M. (2006) Making university–industry collaboration work – a case study on the Deutsche Telekom Laboratories contrasted with findings in literature, *International Society for Professional Innovation Management Conference, Networks for Innovation*, Athens, Greece, 2006.

ROTHWELL R. AND DODGSON M. (1991) External linkages and innovation in small medium sized enterprises, *R&D Management*, **21**(2), 125–138.

SARKAR S. AND COSTA A. (2008) Dynamics of open innovation in the food industry, *Trends in Food Science & Technology*, **19**(11), 574–580.

STRAMBACH S. (2008) Knowledge-intensive business services (KIBS) as drivers of multilevel knowledge dynamics, *International Journal of Services, Technology and Management*, **10**(2), 152–174.

WINDRUM P. AND TOMLINSON M. (1999) Knowledge intensive services and international competitiveness: a four country comparison, *Technology Analysis and Strategic Management*, II, 3.

7

Consumers as part of food and beverage industry innovation

S. E. Kemp, Consultant, UK

DOI: 10.1533/9780857097248.2.109

Abstract: Most new food and beverage products fail commercially. The chance of successful innovation can be improved by making consumers part of the innovation process. Consumer-driven food and beverage innovation, which designs products to meet consumer needs, can be achieved through a company culture focused on the consumer and by applying appropriate consumer input throughout the innovation process. Co-creation offers a new way to innovate, in which prosumers work in mutually beneficial collaboration with companies to develop products, often through social networking.

Key words: co-creation, company culture, crowdsourcing, innovation, prosumerism.

7.1 Introduction

Successful innovation is vital for company survival and growth, yet most innovation in the food and beverage industry fails commercially. It is estimated that 75% (Buisson, 1995) to 90% (Traill and Grunert, 1997) of new products fail in their first year, wasting a considerable amount of resources, and resulting in missed opportunities for many companies. In a survey in which food, beverage and consumer product companies were asked the cause of new products failure, the top two reasons given were 'consumer found no new or unique value proposition' (71%) and 'the product failed to meet consumer needs' (61%) (Grocery Manufacturers Association, 2009). Successful innovation must result in a unique and superior product. It must also, at a minimum, meet the needs and expectations of consumers, and ideally, it should exceed them. Success can be improved and the risk of failure reduced by making consumers part of the innovation process from the outset in order to design products that meet their needs, wants and likes. Building consumer input into new product development (NPD) projects to a high standard can double the success rate of such projects and give a 70% higher market share than NPD projects that do this poorly (Cooper *et al.*, 1998).

Really new innovation, as opposed to derivative innovation, such as line extensions, represents the highest risk but the greatest reward. In the US, there has been a 42% decrease in time-to-market from 41.7 to 24 months in the last 10 years (Cooper, 2011). Sales derived from new products have dropped from 32.6 to 28% of business sales, although R&D spending has remained constant. Businesses are focusing their resources on simpler, faster, cheaper, lower-risk products that are derivative. It is estimated that only 2% of total product launches are radically new, whilst 77% represent nil or an incremental level of novelty only (Costa and Jongen, 2006). Developing a really new product is more challenging, and requires better consumer research, particularly in the early stages, but this is where companies can gain a real competitive advantage.

As with any other product category, it is important that food and beverage innovation is driven by the consumer (Costa and Jongen, 2006; Jaeger and MacFie, 2010; Lundhal, 2011; MacFie, 2007; Moskowitz *et al.*, 2006). Food and beverages, however, differ from many other product categories in that sensory properties (taste, aroma, flavour, texture, appearance and sound) are key benefits that must be liked and preferred by consumers (Tuorila, 2007). Brand drives awareness and product trial, but product liking drives repeat purchase. If consumers do not like a food product, they will not buy the product again regardless of branding. It is dangerous to rely on brand alone to sell such products. To improve success and reduce risks, it is important to understand consumers and make them part of the innovation process in order to create offers they want to purchase, use and enjoy.

The aim of this chapter is to give an overview of consumer-driven innovation. First, the importance of consumer-driven innovation is discussed and the factors that impact consumers when they choose a food or beverage product are examined. Three methods of consumer-driven innovation are described: consumer-focused company culture, consumer-driven innovation processes, and co-creation. At a basic level, making the consumer central needs to be part of the company culture and it should be second nature for all individuals to consider the consumer in their daily project work. On a more sophisticated level, companies can carry out a systematic programme of research with consumers to guide and lead innovation at every stage of the innovation process, so that new food and beverage products are designed from the outset with the consumer in mind. Against this background of innovation using traditional consumer-driven NPD, a paradigm shift is occurring through the use of on-line social networking that is enabling companies to carry out co-innovation directly in partnership with consumers through a dialogue that enables rapid innovation; a process termed 'prosumerism'. A summary, future trends, sources of further information and references are given at the end of the chapter.

The focus of this chapter is on research and development (R&D) activities in innovation, rather than marketing activities, although it is equally important to consider the consumer in marketing. It is beyond the scope of this chapter to cover topics related to consumers as customers, such as shopper behaviour. In the limited space available, it is not possible to cover consumer research techniques in any detail. Readers are directed to other texts for further information. This chapter is aimed at the food and beverage industry, but the majority of the principles outlined can also be applied across the fast-moving consumer goods (FMCG) sector to the cosmetic, personal care and household product categories.

7.2 Understanding food and beverage consumers and their world

The objective of consumer-driven innovation is to create the right product to fulfil consumer needs and expectations. It is vital for the success of a company to gain a competitive advantage by understanding and delivering what consumers want in a manner that delights them. To do this, it is necessary to understand consumers, the world they live in and how they interact with products in that world. However, consumers live in an increasingly complex, fluid environment in which they are presented with an ever increasing number of choices.

The market is changing from a localised supply economy to a globalised demand economy, with many new trends, such as the economic downturn. The way consumers communicate and get information continues to change as the internet, mobile phones and social networking continue to develop. Consumers now have a global outlook, rich access to information, and the ability to network at speed. Consumers themselves continue to change, becoming increasingly more sophisticated and knowledgeable, with evolving expectation and behaviour, such as increased product repertoire and different shopping behaviour through on-line ordering and delivery; increased impatience and an expectation that response will be rapid; and an expectation that they will have a say, which has been greatly facilitated by digital communication, whether it be voting on a reality television show, sending feedback to companies, starting their own web-based community or putting their own thoughts and ideas on-line. Today's consumers are willing and able to voice their opinions, good or bad, rapidly and on a global scale. This rapid evolution makes understanding and delivering to the consumer ever more challenging, yet companies must do so in order to survive, ideally on an ongoing, proactive basis to enable a faster response.

There are many models that describe consumer behaviour during food and beverage choice and consumption. General texts include Frewer *et al.* (2001), Marshall (1995), Meiselman and MacFie (1996), and Shepherd and Raats (2010). Meiselman (2007) gives an overview and categorises the field into three types of variables: people (consumer), food (product), and environment. Tables 7.1–7.3 list variables in each category. Although much is known about each of the variables alone, less is known about their interaction. Some key points related to each category are discussed below.

Table 7.1 Consumer-related variables influencing food and beverage choice and consumption

Variable	Description
Demographics	Demographics can be used to assess the characteristics of a population such as age, gender, ethnicity and income, giving generalities about a population. It may be possible to identify a typical consumer, but not to identify aspects at an individual level.
Psychographics	Psychographics is segmentation based on psychological aspects such as personality, values, attitudes, motivation (e.g., anxieties, emotions), interests, lifestyles, life stage, personality, beliefs (e.g., ethics, risk (Saba, 2007)), knowledge, past experience, learning, memory, expectations (Cardello, 2007), perceptions, cognition, decision making and neophobia. It is worth noting that consumers can have certain attitudes and beliefs about innovation and the technologies used in innovation, related to the degree of risk they are willing to take (Siegrist, 2007). For example, irradiation is a safe and effective way of sterilising and cooking food, but is rejected by consumers in some countries because of the perceived risks. Consumer attitudes depend on the information given and the manner in which it is given. Consumers may be more accepting of novel foods and technologies if it can be shown there are tangible benefits (Frewer *et al.*, 2003). Media and interests groups may generate misinformation and exaggerate risks. Genetically modified foods are an interesting case. Foods of genetically modified origin (GMO) are stigmatised for the European consumer. The use of GMO foods in Europe is highly restricted. In the US, however, GMO has not been an issue for consumers owing to the manner in which it has been presented to them and GMO ingredients are commonly used in food. Consumers are more comfortable with the familiar and have a tendency to reject novel foods, known as food neophobia (Pliner and Savvy, 2006). This can make it more difficult to introduce very new food innovations into the market and particular strategies may be necessary, such as special testing methodologies and trial before purchase. Foods with more complex sensory properties, such as wine, may also be rejected initially and consumers liking increases as they learn to appreciate them. On the other hand, boredom may set in when a food is eaten routinely, and again, special techniques can be used to assess and reduce this propensity (Köster and Mojet, 2007).
Behaviour	Behavioural variables, such as habits (loyalty), shopping behaviour, product usage and usage rates. (Kardes *et al.*, 2010) are important for understanding product purchase and use. An ideal scenario is to create a purchase habit, so that consumers no longer go through the entire decision making process each time they buy a particular product, but habitually buy the same product each time automatically. If a product is really new, consumers may need to change their behaviour to use product. Behaviour, however, can be difficult to change, and this can be an advantage or disadvantage. In innovation, a consumer is often thought of as someone who buys and uses one particular product. However, consumers buy an entire inventory of products and recognising purchase patterns can create opportunities. For example, in the US, it was recognised that fathers are often sent out in the late evening to 24 h stores to buy diapers, at which time they often make an

Table 7.1 Continued

Variable	Description
	impulse purchase of beer. Putting beers near diapers increased beer sales. Consumers' product choice can change depending on the situation. Consumers themselves change over time. This may be a gradual evolution or a rapid change in response to an event. Even the new product introduction itself can change consumer behaviour. However, it is now possible to track consumer purchase behaviour in great detail using web-based tracking and loyalty card data.
Physiographics	Genetics, health, nutrition, physical condition (related to age, gender and lifestage), immune system, sensory acuity, intestinal physiology, dentistry, need state (hunger or thirst), eating disorders and psychobiology (Yeomans, 2007). Sensory-based segmentation is of particular importance in food and beverages. Not all consumers like the same flavours and textures. It is possible to segment consumers based on sensory preferences in a particular product category and design products to cover those segments, for example, by using different flavour varieties. Brand loyal consumers may become very sensitive to changes in the sensory properties of their favourite products and care may be needed during reformulation to prevent alienation of core users.

Table 7.2 Product-related variables influencing food and beverage choice and consumption

Variables	Description
Intrinsic product factors	Examples include ingredients, sensory properties, nutritional content (e.g., reduced calorie, low fat, functional foods, low GI) and origin (natural, organic)
Extrinsic product factors	Examples include claims, brand, label, packaging, sustainability and price

7.2.1 Consumers

Not all consumers are the same. It is important to define a target consumer: the consumer at whom the product is aimed. A product that has been designed to try and suit everyone is likely to lead to a mediocre product that is no-one's favourite, and is easily beaten by targeted competition. Indeed, increasingly small consumer segments are being targeted through customisation and personalisation. Research may be needed to determine consumer segments and hence the target consumer. Consumer segments can be based on many aspects, such as demographics, psychographics, behaviour and physiographics (Table 7.1). The person who buys the product may not be the person who pays for it or uses it. For example, mothers purchase food for their children. It may be necessary to investigate the buyer, payer and user separately.

7.2.2 Product

At a minimum, food and beverages need to be of a certain quality and be safe to meet consumer expectations. Sensory quality is of particular importance (Tuorila, 2007). Ideally, food has also to be of an acceptable nutritional standard. To give increased competitive advantage and distinctiveness, a food product needs to have additional benefits, such as performing a particular function, appealing to the emotions, conveying status and lifestyle, or coming from a particular origin. The product is sold as part of a whole marketing mix that serves to re-enforce these benefits including packaging, name, price, advertising and branding. Table 7.2 gives examples of intrinsic and extrinsic product factors.

Table 7.3 Environmental variables influencing food and beverage choice and consumption

Variables	Description
General environment	Examples include political and economic, technological
Trends and events	Examples include economic downturn, terrorism, natural disasters and food scares (e.g., contamination and health scares)
Social influences	Examples include culture (Rozin, 2007), religion, taboos, trust in industry and government Current models of food choice and consumption are largely based on Western patterns. Eating behaviour varies considerably around the world (Walker, 2002). Consumers from different regions and countries may differ considerably in their preferences, particularly with respect to food. Food is particularly strongly linked to culture. Rozin (2007) notes that when people move from one country to another their food repertoire is the last part of their cultural heritage to be relinquished. If a product is to be launched globally or across a wide geographic area, it needs to be tested with consumers from different regions to either find a product that appeals to them all or to determine if local variations are needed. Cultural and religious taboos need to be taken into account, such as eating in public, or avoiding particular foods or food combinations
Personal influences	Examples include family, peer group, social class, time available and immediate physical surroundings
Eating situation	Examples include context (part of a dish/meal, snack or experience), occasion (social, seasonal, etc.), place (home, restaurant, etc.) Traditionally, eating context and occasion are important. Some food products are eaten alone, whereas some are always eaten as part of a dish or part of a meal. Some food products may be eaten infrequently, for example as a treat, whereas others are eaten daily. Food products may be designed for particular occasions, such as for a present, for a particular time of year, such as Christmas, or for an event, such as a birthday or wedding. Increasingly, innovation is moving from selling a product to selling an experience.

7.2.3 Environment

Consumers live in a complex and ever-changing world, as mentioned above, which is influenced by a variety of environmental factors (Table 7.3). Increasingly, innovation is moving from selling a product to selling an experience, which for maximum impact is emotional, interactive and often has a particular context associated with it (Schifferstein and Hekkert, 2007). Examples include a chocolate bar that provides not only an amazing taste but hedonistic pleasure in the form of a small piece of indulgent time for yourself; a pizza meal shared with the family in a fun, friendly restaurant; a healthy dish eaten with the family round the table at home that helps mothers look after the happiness and wellbeing of their families. The trend is for the product and experience to become more personalised and customised, as it becomes easier to reach individual consumers. As globalisation continues, it remains important to understand different cultures to be able to innovate for global or regional markets.

7.3 Consumer-centric company culture for innovation

A key factor for successful innovation, and indeed a successful business, is to have a company culture where the consumer mind-set is foremost, and pervades the company's strategy and activities. This may be called 'voice of the consumer', 'connecting with the consumer' and 'putting the consumer at the heart'. The key to success is a mind-set of putting the consumer at the centre of everything the company does. This can be done on two levels: the level of the corporation and the level of the individual employee.

7.3.1 Consumer-centric corporate culture

A primary factor necessary to achieve a consumer-centric culture is the support of senior management. It is not unknown for some to view consumer data as an enemy that stifles gut-feel decisions and kills off pet projects with a poor score. Senior management need to lead by example in their passion for truly wanting to deliver to the consumer.

Companies need to have the right structure to make the most of consumer insights that are gained, so they can be adequately injected into innovation. Traditionally, the consumer has been the territory of marketing. Market and marketing research tends to reside in marketing departments and sensory and consumer research in R&D. These departments often work in silos at different points of the process. There is a need to integrate departments that have responsibility for investigating the consumer, so that the entire product experience can be explored through the eyes of the consumer.

Market, consumer and sensory research departments have evolved from being service providers who run tests, to consultants with the expertise to help guide innovation projects. Companies need to capitalise on, consolidate and integrate this expertise. Each innovation project should have consumer experts on the team from the initial planning stages, who can ensure the consumer is represented in the process and is uppermost in the minds of the innovation team by employing the

Table 7.4 Framework and approaches for incorporating consumer mind-set into company cultures

Elements of of the framework	• Blueprint approach to describing consumers on whom the company wants to focus • Databasing qualitative and quantitative findings • Documenting the value set for the person and product (value diagramming) • Strategies that connect technologies to product features, marketing and consumer • Ability to value the impact for consumers/customers • Good communication around consumer issues
Activities	• Desk research, such as business publications and books, to identify current cultural interests and influential thinking • Trend-spotting to identify trends and meta-trends • Census, segmentation and usage and attitudes (U&A) data to understand the target consumer • Syndicated research, e.g., from scanners, diary panels for metrics about products in a category, such as product sales • Semisyndicated research, i.e., regular studies with a specific purpose that allow for some customisation • Custom research at all stages of the NPD process • Secondary research services, such as reports on particular topics that give simplified analysis and interpretation of data

Source: Moskowitz *et al.* (2006).

best and most appropriate tools for the job. The principles of open innovation should be used, so that suppliers of market, consumer and sensory research, as well as relevant academics are incorporated into the innovation process as appropriate.

Moskowitz *et al.* (2006) describe an organisational structure, framework and approaches for incorporating the consumer-mind set into company cultures, using multiple methodologies and disciplines (Table 7.4).

7.3.2 Consumer-conscious employees

A bottom-up, as well as a top-down approach can be taken to bring the consumer into the organisation. Individuals in a company need a passion for considering and delivering to the consumer. R&D staff are stereotypically thought to focus on technologies and products. In fact, it is vitally important that innovators understand consumers, and their interaction with products in their world, to develop preferred products. Technology-based staff need to be as passionate as marketers in considering how to deliver to the consumer.

Major companies have programmes to draw their employees into the world of the consumer and to enable them to see things through consumers' eyes that include approaches such as inductions, regular consumer updates, regular trips to watch consumers whilst they shop or eat in restaurants, in-home visits, overviews of consumer trends, and circulation of relevant articles. Even the smallest food company can take some of the simple and cheap steps above to educate their employees and start to create a consumer-centric culture from the bottom up.

Furthermore, there are many simple steps that everyone can personally take to become more consumer orientated. Watch consumers as they move around you, live their lives, get their information, cook, eat and drink. Attempt to immerse yourself in the consumer's world. We are all consumers of food and beverages. Don't blindly use products: use them consciously. Think about how you buy them, how you use them, how they taste, and how they perform. Watch how other people are buying and using products. Try walking a mile in your consumers' shoes.

7.4 Consumer-driven innovation process

The following subsections give a detailed analysis of the consumer-driven innovation process with an example of a project.

7.4.1 Overview of the consumer-driven innovation process

Over time, the consumer has become more a focus of the innovation process. Earlier, supply-chain, thinking dominated innovation and products were pushed onto the consumer with little thought of what the consumer wanted, as illustrated by Henry Ford's approach that consumers could have any colour they wanted as long as it was black. The focus was on the product. Value chain thinking now dominates, so that delivering what the consumer wants is recognised by most as vital for the success of all parties in the chain. Consumers are recognised as part of the value chain and now 'pull' products through it. Indeed, consumers themselves have recognised this, as they become increasingly empowered and proactive.

The focus and aim of consumer-led food product innovation is to identify and fulfil consumer needs, rather than develop a product per se (Costa and Jongen, 2006). Care is taken by major companies to understand the consumer from the outset and products are designed to meet their needs. It is commonplace to involve external value chain partners in consumer and market research for open innovation, and even to get them to pay for a portion of the research if they want their component, such as a fragrance or flavour, included in a test. Elsewhere in the value chain, supermarkets are able to identify consumer preferences and co-ordinate new product development via innovation networks by exploiting information gathered directly from their customers at point-of-sale and from data mining (Cox and Mowatt, 2004).

In major companies, consumer input is sought at every stage of the innovation process, including planning, often using a systematic sequence of standard techniques at particular points in the process (Moskowitz *et al.*, 2006). Costa and Jongen (2006) suggest there are three obstacles to the implementation of consumer-led food innovation strategies in NPD: lack of concrete guidelines for the effective implementation in everyday industry practices; the sequential nature of consumer-led NPD in contrast with the reality experienced by R&D practitioners; and the lack of intra- and interorganisational co-ordination and integration of R&D and Marketing's research activities and knowledge.

Traditional consumer research strategies for innovation are becoming perceived as slow and inflexible. Increasingly sophisticated techniques, which are faster and more flexible, are being used to define and refine innovation (Beckley *et al.*, 2007). As innovation moves from delivering a product to delivering an experience, Lundahl (2011) proposed a rapid food product innovation framework based around emotional research.

7.4.2 Consumer input at stages of the consumer-driven innovation process

Many schemes have been put forward for managing and carrying out innovation and these are becoming increasingly more sophisticated, with most elements such as product, packaging and marketing development running in parallel and with iterative steps. One of the most widely used is the stage gate process introduced by Cooper (1986), which involves a series of stages with a decision point or gate after each stage. Urban and Hauser (1993) have put forward a process for consumer-led NPD that is similar. Consumer input at each stage is now discussed.

Opportunity identification

It is difficult to ask consumers what new products they want, as they don't know or can't articulate ideas. Listening only to current consumers of a product may lead to missed innovation opportunities. One of the aims of consumer research at this stage is to identify consumers' latent unmet needs; needs they are not aware they have, and identify gaps in the market that might be an opportunity to make an offering. Companies who invest more and use a greater range of consumer research techniques are more successful (van Kleef and van Trijp, 2007). A proactive approach is required to continually gather consumer information, indentify consumers' unmet needs and how to satisfy them, and stay up to date with the changing market and consumer. This helps companies stay ahead of the competition by identifying breakthrough innovations and getting to market more quickly.

Concept

A concept is a translation of an idea into an explanation of how the product provides a solution to consumers' problems and meets consumers' unmet needs. It provides the blueprint for the development process. It is important to test concepts with consumers in order to identify, focus on and optimise the strongest concepts. Consumers may aid in concept selection and design using traditional concept testing or by systematically testing elements or attributes using experimental design and conjoint analysis (Moskowitz *et al.*, 2005 and 2006; Moskowitz, 2007).

Business case

Once the best concepts have been identified, a decision needs to be taken on which concept to take forward based on expected business performance. It is important that the target consumer is considered as part of this strategy.

Development

Delivery to the consumer needs to be central to the development process and testing with consumers should be carried out as an iterative process throughout. Consumer-relevant product, packaging and marketing are developed in parallel. Ideally, product features, such as physicochemical and sensory properties, should be manipulated to design products with optimal consumer liking, cued benefits, and coherent communication. Emotional benefits and the 'product experience' are becoming increasingly important in food and beverage innovation and there are consumer-research techniques to determine emotions linked to product attributes (Lundahl, 2011; Schifferstein and Hekkert, 2007, Thompson, 2007). Claims should be designed into the product and substantiated by consumer testing (ASTM, 2007).

The quality of food and beverages was traditionally defined by experts, such as tea tasters, cheese graders and wine tasters, who judged or graded products according to defined scales or their individual preferences. Likewise, quality assurance (QA) and quality control (QC) parameters were originally driven by the manufacturing team with input from experts and product specialists. Quality is now also driven by consumers (Kilcast, 2010; Munoz *et al.*, 1992). QA and QC parameters, including non-microbiological shelf-life parameters (Hough, 2010), are determined by how much deviation in key attributes is acceptable to consumers.

Validation before launch

Product testing prior to launch serves to ensure the product is optimal to meet consumer needs, expectations and preference. Any failure of the product after this point will be in public, and could damage the brand and the company reputation, as well as wasting considerable resources. At this point, the marketing, launch and business plans are finalised and a decision to launch is taken.

Launch

Launch involves following the business plan to roll out the product locally or nationally with appropriate advertising and promotions. Again, the consumer needs to be considered in that provision needs to be made for receiving consumer feedback and consumer sales contact/service.

Post-launch evaluation

After launch, checks need to be made to ensure that the product performed according to business plan, but also whether it delivered to consumers as anticipated in terms of expectations, quality, liking, usage, etc. At this stage, consumer testing on initial and repeat purchase is carried out. Consumer feedback and complaints are also assessed. Improvements are then made to the product based on learnings. Even when a product has been successfully launched in the market, maintenance is still needed to ensure it continues to perform in the continually changing world of the consumer and market throughout its lifecycle. As well as tracking sales volume during introduction, tracking of consumer awareness, trial and repeat purchase is necessary. For example, if trial and awareness are low, advertising may need to be enhanced. If repeat purchase is low, the product may need to be improved. Tracking during the growth phase allows unexpected success to be capitalised upon and possible failure to be averted. In the maturity stage, consumer research to find opportunities to revitalise the product may be undertaken. In the decline stage, research may show ways to slow or prevent decline (Pope, 1993).

7.4.3 Tools for consumer-driven innovation

Most major companies have a systemised approach so that particular techniques in qualitative and quantitative market research (Beall, 2010; Birn and Forsythe, 2002; Hague *et al.*, 2004; Mariampolski, 2012; McQuarrie, 2011; Xu, 2005), marketing research (Bradley, 2010; Pope, 1993; Wilson, 2006), consumer research (Jaeger and MacFie, 2010; MacFie, 2007; Moskowitz *et al.*, 2006; Van Trijp and Steenkamp,

Table 7.5 Test design considerations when testing experimental foods and beverages

Consideration	Explanation
Types of experimental food and beverages	• Novel ingredients • Ingredients from novel sources, e.g., GMO ingredients • Ingredients made using a novel process • Products and ingredients from outside the country in which testing is to be carried out, which may not have been approved for use in the test country • Novel processes and technology applied to food production and manufacture
Toxicological safety	Toxicological safety of novel ingredients and their levels of use must be assessed before any consumption, even for informal tasting of trial samples by employees. Clearance may be required from an ethical committee. The UK Advisory Committee on Novel Foods and Processes (ACNFP, 2000) has produced guidelines on the conduct of taste trials involving novel foods or foods produced by novel processes. Similarly, highly novel processes, such as irradiation, may need to be approved.
Food safety and hygiene standards	Many experimental and development food and beverage samples produced during the innovation process are made in a laboratory, especially during the early stages of NPD. It is important that all food and beverage samples to be eaten, even if only by employees in an informal trial, are prepared using good food safety and hygiene standards. In particular, experimental samples should be approved as microbiologically fit for human consumption.
Protection of intellectual property	It is important to consider implications for protection of IP before running research with consumers. Using a new product, ingredient or technology in the public domain, such as research with consumers, may invalidate subsequent patent claims, even if consumers have signed confidentiality agreements. Publishing results before being granted a patent can establish 'prior' art, which can be desirable or undesirable depending on the circumstances.

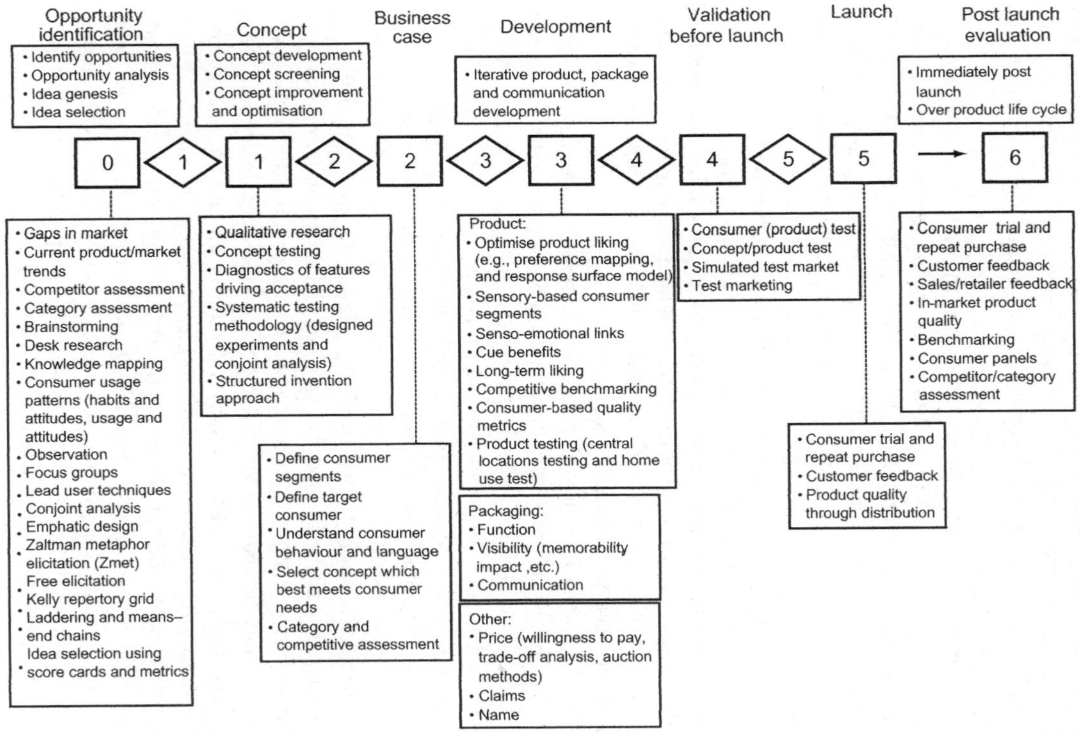

Fig. 7.1 Application of selected consumer research techniques in the innovation process. Note: For simplicity the innovation process is shown in a simple, linear fashion with well-defined stages and application of selected techniques. In reality, the process is iterative within and between stages. Some stages may overlap. For example, early product development might begin during concept development. Product, packaging and advertising are developed in parallel. (Techniques to test communication and advertising are not included). Some techniques may provide input to multiple stages. (Sources include Moskowitz *et al.* (2006), MacFie (2007), Pope (1993), Stone and Sidel (2004), van Kleef *et al.*, 2005 and van Trijp and Steenkamp (2005).)

2005) and sensory research (Gelinas, 1996; Kemp *et al.*, 2009; Lawless and Heymann, 2010; Meilgaard at al., 2006; Resurreccion, 1998; Stone and Sidel, 2007; Stone and Sidel, 2004) are routinely applied at appropriate stages in the innovation process. Figure 7.1 illustrates the application of selected techniques in the consumer-driven innovation process. In addition, consumer groups may be used to integrate the consumer voice into all aspects of an innovation project, such as consumer design panels, insight panels, and experience panels, which comprise not company employees but people who work for the company giving advice and guidance on innovation projects.

Research into consumers should be carried out using the same standards as other types of research, such as setting a hypothesis, using appropriate tests and analyses, using safe and ethical practices (ESOMAR, 2008; IFST, 2010; MRS, 2006, 2010), avoiding bias, and considering test limitations during interpretation. Special design considerations are necessary when testing novel food and beverage innovations (Table 7.5). Testing is typically done with the target consumer. In general, the smaller the target consumer segment, the more difficult and expensive it is to test with them. Results must be actionable and it is good practice to set an action standard before undertaking the test. Many large companies and research agencies have databases of results and norms against which to compare new results and attempt to predict whether they are successful. In order to create databases, tests or test elements need to be standardised.

Poor consumer research in innovation can have major consequences for a company. In 1985, The Coca Cola Company launched New Coke but were forced to re-introduce the original formula less than three months later. There are many reasons why New Coke failed, but inappropriate consumer research played a part (Prendergrast, 2000). The company placed an overreliance on taste testing (Schindler, 1992), an approach which failed to assess liking when drinking a typical quantity as part of the marketing mix. Taste testing did reveal that a small percentage of participants strongly preferred the original formulation, a finding which was also echoed in focus groups, but was not picked up on by executives.

There are times when consumer research should be used cautiously. Very novel food and beverages may be rated as less liked because they are unfamiliar. This applies to very new innovations and foods taken from one culture and introduced to another. In certain situations, testing with typical consumers may lead to a mediocre product. These include avant-garde, trend-setting products, where lead-users may be used in research, and high quality, speciality products, as 'high quality' may mean different things to typical consumers and developers. There are situations when set patterns of behaviour involving product purchase and use may be difficult to change and these need special consideration.

7.4.4 Example of a consumer-driven innovation project

Company X, a snack food company, has a consumer-centric culture. It proactively and continuously seeks out new ideas generated from a variety of sources, including identifying gaps in the market by analysing the market overview;

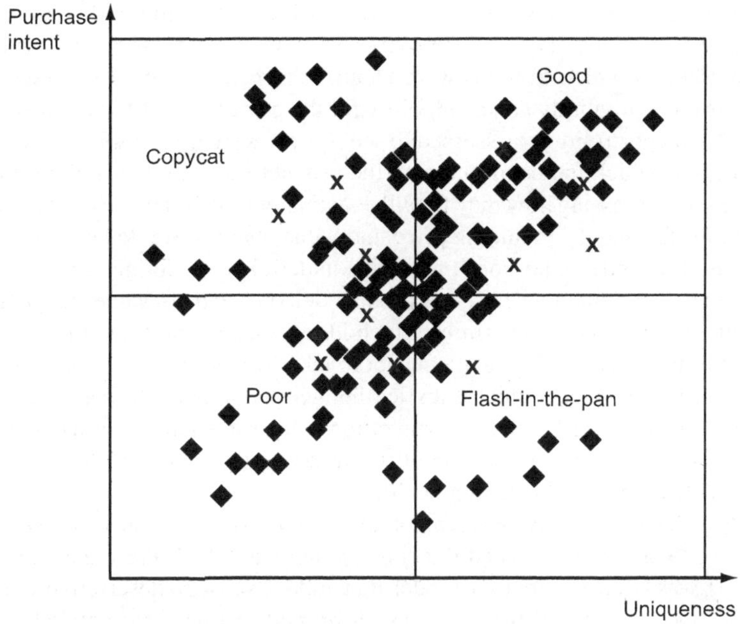

Fig. 7.2 Illustration of concept classification based on liking and purchase intent scores (test concepts = X; database = ◆).

assessing current trends; assessing competitors; brainstorming; desk research; and consumer feedback. They noted that on-the-go men had an unmet need for a convenient, easily-accessible, filling, savoury snack. They also noted consumer trends for healthy products.

The company carried out a knowledge mapping and also looked for starting points within the company, including exploiting a new extrusion technology. The company investigated consumers' needs, wants, and desired benefits, using qualitative consumer research techniques. Laddering and means–end chain-theory techniques within the savoury snack food category were used to link product attributes to benefits and, in turn, to consumer values, and these were used to develop communication and product ideas that were harmonious. Several ideas were selected using a score card of metrics and taken forward to build into concepts.

Concepts for ideas were developed. Qualitative consumer testing was used to probe and optimise concepts. Quantitative concept testing was carried out with consumers on metrics, such as purchase intent, uniqueness, fit, and expectations, to understand the rewards, risks and strategic fit of the concept to the brand and the business. The key measures, purchase intent and uniqueness, were plotted against each other (Fig. 7.2), and compared with historical databases to select the best concepts. Diagnostic tests were used to identify features that drive acceptance, including sensory attributes, ingredient information, functionality, packaging, size, and possible claims.

Business and marketing elements were considered to build a strategy for the new innovation. Category and competitive assessment were carried out and gaps in the market on which to focus were identified. Technologies were assessed to determine their suitability and availability for the project. The fit with the existing product/brand portfolio was assessed to see if there were opportunities in existing product lines, which included mapping the product lines, understanding product life cycles and assessing how they fit with existing brands and the overall company strategy. Profit models, including price and value chain, were developed.

Research to understand consumers was undertaken to define consumer segments, consumer behaviour for good product delivery and consumer language for use in advertising, and to determine which ideas/concepts best met the consumer needs and market gaps. The final concept to take forward was chosen: a convenient, savoury, snack bar, which satisfied hunger and was healthy and tasty, that could be sold and stored without refrigeration in the same manner as a confectionery bar. The target consumer was identified as active men aged 18–30 of moderate income with an interest in staying healthy.

Product design and development began with a protocept that was tested with consumers to ensure it matched the concept and was liked. Preference mapping (Fig. 7.3) was used to create a model that linked sensory-descriptive data and consumer-liking data and the key sensory properties that were noticed by, and were important to, consumers in driving liking were determined as flavour, brown colour of coating, texture of the coating upon first bite and cohesiveness. Consumer segments based on sensory preferences for different flavours were also identified. Designed experiments and predictive modelling using response-surface methodology were carried out to optimise the sensory properties driving liking by identifying the product and process variables that affect these sensory properties and systematically varying them to design varieties of the product with optimal liking for the two largest segments: chicken and beef. Healthy formulation was used by keeping salt and fat relatively low, and including 'healthy' ingredients, such as vegetables. Key benefits of healthy and tasty were cued to consumers using product features and communication indentified from earlier research. Emotional consumer profiling of the product was incorporated so that product features and communication worked harmoniously to deliver an emotional context of satisfaction. As a claim was going to be made regarding tasting as good as refrigerated meat-based snacks, parity testing for hedonic equality was run with consumers according to ASTM guidelines (ASTM, 2007) to substantiate the claim.

Product performance, function and delivery against consumer needs were assessed with consumers as development progressed. As the direction for the product became more certain, product testing with consumers was scaled up from cheaper and quicker small scale central location testing (CLT), to larger scale product testing, including home use testing (HUT). Different versions of the product were tested against each other to select the best. The product was also benchmarked against competitor products to check performance.

Once the two final prototypes were selected, a process for production was chosen taking into account feasibility, capacity and costs. The process was then

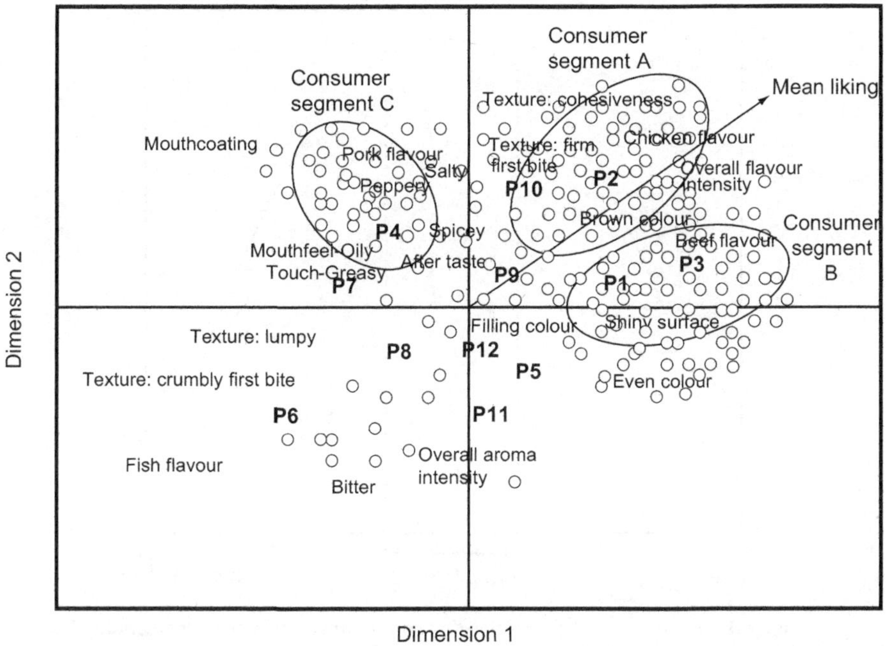

Fig. 7.3 Illustration of preference mapping of savoury snack bars showing individual consumer preferences (open circles), consumer segments and sensory properties for products (P).

scaled up from the lab to pilot plant to full-scale production. The product was checked using sensory descriptive analysis, discrimination testing and consumer testing to ensure the products were the same as the prototypes.

Consumer-relevant quality standards were determined. The key consumer attributes identified from preference mapping were varied in a systematic way and liking measured, so that the levels of tolerance were determined, as illustrated for colour of coating in Fig. 7.4. Key consumer attributes were also assessed in shelf-life testing to identify the time at which quality became unacceptable to consumers to help set a 'best before' date. For example, the coating of the product became softer over time until it reached a softness which consumers perceived as a deterioration in quality.

Packaging was developed in parallel with the product. Elements of the packaging were tested with consumers using visibility and functionality tests to ensure the package delivered as it should. In this case, convenience of opening, ability to eat from the packet, resealability and disposability were important elements in testing, as well as colour, name, graphics, panel elements, instructions, attractiveness and fit to concept, and technical tests, such as strength, barrier properties, and product/pack interactions. Branding and imagery was also tested with consumers in parallel with product development. Several techniques were used to set price, including economic modelling and market research with consumers, including trade-off analysis.

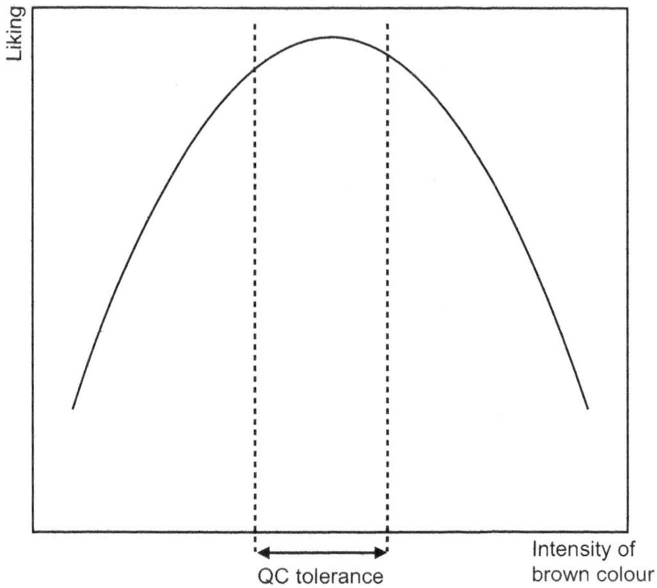

Fig. 7.4 Illustration of consumer-driven tolerance for a quality attribute.

Product testing on the final product was undertaken as a validation before launch. The test was run with the target consumer in-home under normal usage conditions. Key measures included liking and purchase intention. The test was run both blind and branded. Blind testing determined product liking, which gave a guide to repeat purchase. Branded testing determined initial take-up and effectiveness of communication. Comparing the two test results gave an idea of how well the product and communication performed relative to each other, and was balanced in this case.

A simulated test market (STM) was carried out with consumers before launch to predict volume, market share and cannibalisation of existing products, and to assess the complete marketing mix of product, name, package and advertising in finished form to ensure the product performed. During the STM, consumers were exposed to advertising, had the opportunity to purchase the product in a simulated shopping environment, used the product over a period of time, had a follow-up interview, and the opportunity to repurchase the product. Modelling was based on purchase intent and share of preference, measured using a trial from a competitive set to estimate trial. Repeat purchase was estimated using attitudinal scales and normative databases held by the research agency. Estimates of trial rate, repeat rate and purchase frequency were made to give an estimate of potential volume. Diagnostics were included to determine the elements driving predicted sales volume and give guidance on how to improve the mix to improve the sale forecast. The company was confident enough in the results not to go to the time, expense and security risk of a full test market and made the decision to launch based on the results of the STM.

The product, with two variants of chicken and beef, was launched following the business plan. A website and helpline were set up to generate customer interest and receive queries and feedback. After launch, checks were made on sales and product performance. Consumer testing on awareness and trial indicated that the advertising was performing as expected. However, repeat purchase results and consumer feedback indicated that consumers were finding the product inconvenient and messy to consume, as the coating was more crumbly following distribution than expected. Adjustments to the product formulation were undertaken to make the coating less crumbly.

Continuous tracking of sales and consumer perception was carried out over time. Early success was capitalised upon by adding additional flavour varieties based on consumer preferences. Several years later, sales reached a plateau and consumer research indicated that consumers had begun to perceive the product as old fashioned. The core product was revamped and relaunched incorporating current flavour trends, such as chicken tikka, and the line was extended to include a vegetarian option intended to appeal more to young women.

7.5 Consumers as co-creators

Traditionally, consumers have been thought of as passive recipients of new products. However, they have long been actively involved in the co-creation of value from products, through activities such as customisation and assembly (Toffler, 1980). This process is known as 'prosumerism'. The terms prosumerism, prosumer and prosumption are derived by combining the words (non-corporate) producer or professional and consumer. Kozinets *et al.* (2008) give an overview of collective innovation, propose a model of developmental progression of individual to collective creativity and provide a topology of online creative consumer communities in two dimensions, collective innovation orientation and concentration, and four segments: crowds, hives, swarms and mobs.

The advent of the web and social networking is revolutionising the innovation process as consumers are able to have a direct dialogue with companies regarding innovation. As Traitlor and Saguy (2009) aptly state: 'Open social networks will enable organisations to search, find, and implement just about any potential solution to an opportunity from just about any source, and become the best and fastest in translating, adapting, and executing such solutions. It means that we may have to define the field of innovation in a new way'.

On-line trends in prosumerism have been evident for some time. The growth of the wiki and so-called 'citizen scientist', in which ordinary people are actively involved in scientific activities, such as posting inventions on line or participating in organised experimental studies, such as Bird Watch UK in which people are asked to submit observational survey data on bird species. The resurgence of customisation and personalisation is an obvious market trend fuelled by the internet. Prosumers have also started their own user groups, notably in the computer software and gaming industry where open sourcing and user-generated content is common place, offering solutions to product problems or further product developments.

Consumers are now becoming directly involved in the innovation process, and consumer and company roles are merging. Co-creation via digital communications is resulting in a win–win for both consumers and companies. Consumers have often found it necessary to do things in a non-optimal or inconvenient manner, or find ways to compensate when a product does not perform as they would like. Historically, there has been little opportunity for consumers to ask for different solutions, but they now have opportunity to do so. Advantages for companies are:

- co-creation of value;
- intimate access to consumers;
- immediate feedback;
- less costly than using traditional market research;
- saving resources in the early stages of innovation by focusing more quickly on consumer-viable and economically viable ideas;
- increasing consumer involvement and awareness, thereby increasing the probability that consumers purchase the finished product;
- positive word-of-mouth from consumers;
- development of stronger relationships with consumers; and
- creating consumer communities for future co-creation and promotional activities.

Companies need to remember that co-creation involves genuine dialogue with consumers, a degree of access to the product and some transparency, which may also carry an element of risk. Companies are still experimenting with co-creation and need to change their culture and procedures to have commitment and defined follow-up processes to bring co-created ideas to market.

There are a range of approaches that companies have adopted to tap into prosumerism, many based on crowdsourcing that enables consumers to submit and refine ideas. Approaches used for co-creation with consumers are:

- Co-creation communities formed on an ad-hoc or permanent basis, e.g., P&G Vocal Point.
- Forums that allow for consumer interaction in the innovation and development process, soliciting ideas for new and improved products, contributions to the product, e.g., computer applications add-ons, or new or unusual uses for the product.
- Innovation as an experience that creates interest and excitement using interactivity, networking, emotions and personalisation, to offer an experience that the consumer would like to repeat.
- Competitions that ask for inventions or new ideas.
- Competitions that ask for consumers to use the product in new or creative ways. An example is recipe-based competitions, where consumers are asked to send in recipes using the product, often accompanied by photos or video clips.
- Crowdsourcing ingredients thereby involving consumers at an early stage in the value chain.

- Feedback and complaint forums to identify problems and potential solutions.
- Product/brand interest forums that discuss the product/brand and generate awareness. These may be useful for tapping into brand-loyal users.
- User toolkits for innovation that allow consumers to develop a custom product via iterative trial and error (von Hippel, 2001).
- Blog mining.
- Organised consumer–producer interaction, such as the classification of user-producer interactions based on pharmaceutical and food innovation developed by Moors *et al.* (2010) that offers the potential to optimise the use of user–producer interactions in the innovation process.

Crowdsourcing has been applied to various aspects of food and beverage innovation. There are several notable examples of crowdsourcing applied to development, including the two examples given below and the case study described in Section 7.5.1. In 2008, Walkers ran a crowdsourced competition 'Do us a flavour' in which UK consumers were asked to submit innovative flavours for crisps (potato chips). Six finalists were selected from 1.2 million entries: builder's breakfast, cajun squirrel, chocolate and chilli, crispy duck and hoi sin, fish and chips and onion bhaji. Crisps with these flavours were put on sale and consumers were asked to vote for their favourite flavour. The winning flavor, builder's breakfast, was then added to the product range for a period of time. The winner received £50 000 and 1% of the retail sales of that flavour. The competition reinvigorated the Walkers brand by encouraging consumers to try the whole range of six flavours and then capitalised on the consumer preference displayed during voting by marketing the most preferred flavour. As a result, year-on-year sales rose by 14% and brand equity by 6%. (Gottlieb, 2010).

In 2009, Glacéau Vitaminwater crowdsourced a new flavour for a flavoured water with health benefits using 'Flavor Creator', a Facebook app (http://www.facebook.com/vitaminwater). In a three-month, three-step programme, the 'Flavor Creator' consumer community of more than 1 million voted for their favourite flavour, and played games and answered quizzes to give input on aspects such as the beverage's functional benefits, label design and product name. The winner received $5000. The new product was launched in 2010: 'Connect' flavoured with black cherry and lime with added caffeine. The program is an example of co-creation in which company experts and the consumer community worked together in a controlled way to develop a product. The company was able to get feedback about on-trend flavours, produce a product that was more likely to sell well, and create brand and product interest.

Crowdsourcing is being used to source food ingredients. Giapo Gelato in New Zealand produces an all-natural line of healthy gelato and sorbets. It is using crowdsourcing to source organic fruits for its 'Giapo Certified Organic' line from ordinary consumers with surpluses. Consumers must guarantee that no herbicides or pesticides have been used to grow the fruit and allow random testing to ensure compliance. The price of the fruit supplied is then calculated based on current market prices, and suppliers are given free Giapo Gelato products in return. Details can be obtained from http://www.giapo.com/blog/ingredient/.

Crowdsourcing is also being used for menu development. For example, in 2011, MacDonalds in Germany invited consumers to submit ideas for burgers. Five finalists, selected from more than 116,000 entries, received votes on Facebook and the winner of breaded chicken, herb sauce and cucumber was chosen. Each finalist burger was put on sale for one week of a five week period over the summer, with the winners appearing in the TV advert for their burger. The campaign created interest and encouraged consumers to visit MacDonalds to try all five finalist burgers. (BurgerBusiness.com, 2011).

There may be profound implications for traditional marketing with the launch of crowdsourced non-branded products (Kozinets *et al.*, 2008). Free beer is an initiative from a community committed to open-sourcing, who open-sourced a community-made recipe for home-made beer with the goal of spreading the idea of do-it-yourself non-branded beer. (More details can be found at http://freebeer.org/blog/). The recipe was released under a Creative Commons licence, which means anyone can use the recipe to brew their own beer or create a derivative. A prime objective is to create an open business and question the prevailing system of copyright and trade. As a result, breweries have opened around the world manufacturing untrademarked, non-branded beer.

Crowdsourcing approaches have also been used in food and beverage advertising. Interactive and customisable adverts have long been available. There are many promotions and competitions that allow consumers to develop their own adverts, which provide the company with a wealth of ideas and finished adverts. For example, Dorritos runs annual competitions in several countries to submit a short advert (for an example see www.crashthesuperbowl.com) (Kozinets *et al.*, 2008). The winning advert is used as part of a television advertising campaign and the winner receives prize money.

Crowdsourcing offers huge potential for innovation, but as it starts to grow rapidly, there is likely to be competition for co-creators as more projects are launched. At the moment, co-creators are willing to offer their commitment and ideas for free or relatively low reward, often with little commitment from the company to take the innovation forward. In the future, co-creators are likely to become more selective and demanding, and companies will need to understand motivations and increase the value of participation for the co-consumer by increasing the tangible rewards (money and product-related prizes) and non-tangible rewards (recognition, visibility, co-creation experience and related social interaction, innovation outcomes, such as seeing the product brought to market). Of particular importance is being seen to act on co-creator input. This signals that the company is not seeking to exploit co-creators or use them flippantly, such as by crowdwashing, which involves stripping co-creators of ideas or asking for ideas for promotional reasons without the intention of using them. Companies need to recognise and value co-creator commitment and contribution, in order to attract future co-creators and maintain a good reputation and relationship with their consumers. (Hück, 2011).

In summary, the trend for prosumerism via social networking is set to revolutionise innovation and forward-thinking companies are already gaining competitive advantage through this route.

7.5.1 Case study: Pepsico's Mountain Dew® 'DEWmocracy™'

In 2007 and 2008, Mountain Dew® used the DEWmocracy™ campaign to co-create a new flavour variant Mountain Dew® Voltage™, with raspberry, citrus and ginseng, in which consumers were asked to elect product elements. Part of the process involved playing a story-based interactive on-line game.

In 2009, the campaign was continued and extended with DEWmocracy™ 2, in which co-creators were asked to select the flavour, colour, name, package design and advertising for three variants, one of which would join the Mountain Dew® range permanently (see www.dewmocracy.com).

In June and July, 2009, DEW Labs trucks drove around the US with seven new Dew flavours, which were tasted by more than 200 000 consumers. At the same time, kits of the seven flavours and video cameras were sent to 50 DEW 'fanatics' who were asked to taste, debate and vote on the flavours. The top three flavours: Punch of Tropical DEW, Lime Blasted DEW and Smooth Citrus DEW, were selected and sent to the DEW Labs community of 4000 passionate DEW fans, who each chose their favourite flavour and joined one of the three 'Flavour Nations' to collaborate in building other elements of the product.

Colours were then chosen via a live web event called 'Shoot Your Shade'. Nine colours selected by DEW Labs members were voted on by DEW fans and votes were illustrated by pelting paintballs for a limited time at a Flavour Nations representative wearing a lab coat. The three final colours selected were Red Cloud, Deep Green and White Flash.

Flavour Nations members selected nine new names, put them on Twitter and asked Dew fans to follow their favourite. On 15 October 2009, the names with the greatest number of followers were selected for each flavour.

A competition was held to choose the pack design. Entrants were asked to design a basic Mountain Dew® label and DEW fans voted for the top 10. The Flavour Nations picked the final three designs and the designers worked with the Mountain Dew brand team and DEW Lab to create the finished designs. The winners received $10 000 and an Apple MacBook Pro®, whilst the top ten designers also won MacBook Pros.

Dew fans selected the advertising by voting for their favourite 12 s videos submitted by advertising agencies and individuals. Six finalists were selected, from which the DEW Labs community selected three winners to work with them and the DEW brand team to create TV ads for each flavour.

The three final products; Mountain Dew® Typhoon™ with Punch of Tropical flavour and coloured Red Cloud; Mountain Dew® Distortion™ with Blast of Lime flavour and coloured Deep Green and Mountain Dew® White Out™ with Smooth Citrus favour and coloured White Flash were put on sale. Products were promoted in the '2010 Dewmocracy campaign trail' via the web, roadshows and events around the US and almost two million votes were cast by the closing date in mid-June 2010. The winning variant, Mountain Dew® White Out™, was launched as a permanent member of the Mountain Dew® product range in October 2010.

This campaign has created a long-lasting relationship with consumers, established core user groups willing to co-innovate, and created high awareness, as well as delivering new, preferred and highly-publicised products.

7.6 Conclusion

New innovation in the food and beverage sector is largely commercially unsuccessful, with fewer than 25% of new products still on the shelves after their first year. Not only does this waste immediate resources, but affects the company business plan for growth and profit and indeed may threaten its very survival. Successful innovation is vital to gain and sustain a competitive advantage and deliver to shareholders.

Today's market environment is challenging. Consumers live in an increasingly complex world that is changing from local to global, supply to demand, passive to interactive, and insular to connected. Consumer outlook is changing as they become more connected, informed, sophisticated, discerning and impatient, fuelled by the internet and social networking. They are empowered as never before and expect to be able to voice their opinion, whether to a company or their peers, and get what they want, the way they want it, quickly and at ever better value driven by the global economic downturn. There are an ever increasing number of product choices that can be customised and purchased in a number of ways.

Increasing the chance of successful innovation in such an environment is difficult but can be achieved by identifying consumer needs and designing products that meet them. Companies that seek to understand the consumer in a proactive and ongoing manner, have a chance to respond more rapidly than competition. Companies that seek to gain consumer input throughout the innovation process to drive NPD, against a backdrop of a consumer-centric corporate culture, have a better chance of success. Consumer input may well have the highest payoff when used at the front end of innovation in identifying and refining opportunities, i.e., unmet consumer needs, offering opportunities that are more likely to succeed, minimising missed opportunities and short-cutting development. Working with consumers as co-creators is a new way to undertake innovation that can create mutual value, whilst generating awareness at relatively low cost.

Food and beverage consumption and choice is complex and influenced by many interactive variables. Consumer research is still a relatively new field, only taking off in the 1980s and 1990s, and methods and their application continue to be developed. Sensory quality is key and it is now common place to link sensory attributes to ingredients, benefits, values, and emotional elements of the brand to design products to meet the sensory quality preferences of sensory-based consumer segments. Innovation is moving towards delivering not just a product, but a product within the context of an emotional experience, so understanding consumer behaviour and motivations in particular contexts is important.

Some companies have fully embraced a consumer-driven approach to innovation, with a consumer-centric culture, consumer-led innovation and experimentation in co-creation. Some companies continue to use traditional consumer-based methods. However, many companies have failed to recognise the benefits of including consumers as part of the innovation process. The food and beverage industry, in general, lags behind other sectors in its R&D spend and is traditionally slow to adopt new approaches.

A paradigm shift is occurring in the market and a paradigm shift is needed in innovation to meet this challenge. Making consumers part of food and beverage innovation, through a consumer-centric corporate culture, a consumer-led innovation process and by harnessing the power of prosumerism in co-creation, offers a route to success.

7.7 Future trends

The role of consumers in innovation will continue to grow as companies recognise that their survival depends on gaining competitive advantage by rapidly launching new products with a better chance of success. New company models will emerge that involve a single, fully integrated function including all those working on consumer understanding, so that all innovation team members share a common goal to identify and satisfy consumer needs.

There is potential for innovation to undergo a paradigm shift with the advent of prosumer co-creation through social networking, resulting in a win–win for companies and consumers. Companies will need to adapt their organisation and processes in order to capitalise on this trend, and learn how to work with consumers in a genuinely co-operative manner to attract good co-creators whilst avoid being seen as exploitative.

Innovation is moving towards creating experiences rather than products and more effort will be put into developing methods to investigate and deliver emotional benefits. As a consequence, more reliance may be put on measuring overall satisfaction, rather than liking. Sensory properties are being recognised as part of the brand and attempts are already being made to protect intellect property (IP) in this area through patenting and trademarking, although with limited success.

As described earlier, the consumer and their world will continue to change rapidly and successful companies will need to pick up on trends early to stay ahead of their competition. On one hand, globalisation is increasing and innovation will increasingly design products for delivery to consumers on a global scale. On the other hand, customisation will increase. This may go beyond traditional consumer segmentations to the level of the individual. For example, innovation for sensory-based segmentation is currently possible, but customisation has the potential to reach the level of the individual through the use of genomics and tastomics. Similarly, individual tailoring of products with nutritional and health benefits is a possibility.

Consumer understanding will improve as company consumer research becomes more proactive and takes a broader approach. Underpinning this will be progress on better scientific understanding of the complex nature of food and beverage choice and consumption. This will open the door to resolving many issues such as whether there are general rules for food choice, holistic understanding of consumers (rather than product-specific understanding), whether the Western model of food choice and consumption fits other cultures, and how to introduce new technologies to consumers. The way we eat, including ingredients, foods, meals and eating occasions, has changed considerably over the last 50 years and more (Kipple and Ornelas, 2000; Walker, 2002) and is likely to continue to change. Understanding the factors affecting change may enable future eating patterns to be predicted.

Method development for improved consumer-led innovation will continue. Traditional methods are being refined to give better, more rapid predictions of product success, for example, by make them faster, more flexible and more realistic; e.g., usage in the appropriate context, over the appropriate time period, and over multiple uses. Sophisticated, multidisciplinary techniques continue to be developed. Computerisation and the internet continue to facilitate and speed up research, e.g., rapid data collection; rapid feedback; continued sophistication of modelling with large, complex data sets; better trendspotting, databasing and datamining; and objective analysis of qualitative data and consumer-written text using text analytic software. The framework and guidelines for applying methods for consumer-led innovation will continue to be refined to make it structured yet flexible.

Brand new consumer research methods are also being developed. A notable example is neuroimaging. Brain activity in response to stimuli can be measured using brain imaging and inferences made about cognitive processing. A seminal study by McClure *et al.* (2004) showed that Coca Cola and Pepsi activated an area of the brain associated with reward, but prior presentation of a visual image of a can of Coca Cola changed subsequent brain activation. This has led to the new field of neuromarketing, which seeks to understand the brain mechanisms involved in consumer behaviour with the aim of influencing purchase behaviour through stimulation of the appropriate area of the brain.

7.8 Sources of further information and advice

Consumer research for consumer-led food and beverage innovation is a diverse and multidisciplinary field and information may be gained from books, journals, societies and organisations in diverse areas such as marketing, psychology, sociology, perception, innovation, food science and technology, and nutrition. Only key sources are listed here.

Key books include *Consumer-led food product design* (MacFie, 2007) and *Sensory and consumer research in food product design and development* (Moskowitz *et al.*, 2006).

Key journals include *Advances in Consumer Research, Food Quality and Preference, (International) Journal of Research in Marketing, Journal of Consumer Behaviour, Journal of Consumer Marketing, Journal of Consumer Psychology, Journal of Consumer Research, Journal of Consumer Satisfaction, Dissatisfaction and Complaining Behaviour, Journal of Consumer Studies, Journal of Marketing Research* and *Journal of Sensory Studies*.

Major professional bodies include the Market Research Society (MRS), the Marketing Research Association (MRA), the European Society for Opinion and Market Research (ESOMAR), the European Sensory Science Society (E3S); the Society of Sensory Professionals (SSP), the Institute for Food Science and Technology Professional Food Sensory Group (PFSG), the Institute of Food Technology Sensory and Consumer Sciences Division (IFT SCSD), the Association for Consumer Research (ACR), the Society for Consumer Psychology (SCP) and the Sensometrics Society.

Organisations that set standards in market, consumer and sensory research include the International Standards Organisation (ISO), the American Standards for Testing and Materials (ASTM) (including Committee E18 on Sensory Evaluation) and the British Standard Institute (BSI) (including Committee AW/012 Sensory Analysis).

The European Sensory Network (ESN) is an international network of leading research institutions and industrial partners at the cutting edge of sensory and consumer sciences. ESN members and partners share their knowledge and expertise to explore the value and applications of sensory and consumer sciences in food and non food industrial practices.

The societies listed above hold regular conferences. In addition, the Pangborn Sensory Science Symposium is held bi-annually.

7.9 References

ACNFP (2000), Guidelines on the conduct of taste trials involving novel foods or foods produced by novel processes. The UK Advisory Committee on Novel Foods and Processes. Available from: http://www.acnfp.gov.uk/acnfpapers/inforelatass/guidetastehuman/guidetaste (Accessed 23 April 2012).

ASTM (2007), *ASTM E1958 – 07e1 Standard guide for sensory claim substantiation*, Philadelphia, American Standards for Testing and Materials.

BEALL AE (2010), *Strategic market research: a guide to conducting research that drives businesses*, Bloomington, IN, iUniverse.

BECKLEY JH, FOLEY MM, TOPP EJ, HUANG JC AND PRINYAWIWATKUL W (2007), *Accelerating new food product design and development*, Iowa, Blackwell Publishing Professional.

BIRN R AND FORSYTH P (2002), *Market research*, Oxford, Capstone Publishing.

BRADLEY N (2010), *Marketing Research: tools and techniques. (Second edition)*, Oxford, Oxford University Press.

BUISSON PD (1995), 'Developing new products for the consumer', in Marshall D, *Food choice and the consumer*, London, Blackie Academic and Professional.

BURGERBUSINESS.COM (2011), 'McDonald's uses social media for menu development', BurgerBusiness.com. Available from: http://www.burgerbusiness.com/?p=7553 (Accessed 3 January 2012).

CARDELLO AV (2007), 'Measuring consumer expectations to improve food product development', in MacFie H, *Consumer-led food product development*, Cambridge, Woodhead Publishing Limited, 223–261.
COOPER RG (1986), *Winning at new products*, Reading, Mass, Addison-Wesley Publishing Company.
COOPER RG (2011), *Winning at new products: creating value through innovation*, New York, Basic Books.
COOPER RG, EDGETT SJ AND KLEINSCHMIDT EJ (1998), *Portfolio management for new products*, Reading, Mass, Addison-Wesley.
COSTA AIA AND JONGEN WMF (2006), 'New insights into consumer-led food product development', *Trends in Food Science & Technology*, **17**, 457–465.
COX H AND MOWATT S (2004), 'Consumer-driven innovation networks and e-business management systems', *Qualitative Market Research*, **7**(1), 9–19.
ESOMAR (2008), *ICC/ESOMAR International code on market and social research*. Available from: http://www.esomar.org/knowledge-and-standards/codes-and-guidelines.php (Accessed 3 January 2012).
FREWER L, RISVIK E AND SCHIFFERSTEIN H (2001), *Food, people, and society: a European perspective of consumers' food choices*, Berlin Heidelberg New York, Springer-Verlag.
FREWER L, SCHOLDERER J AND LAMBERT N (2003), 'Consumer acceptance of functional foods: issues for the future', *British Food Journal*, **105**, 714–731.
GELINAS A (1996), 'Creative applications: Sensory techniques used in conducting packaging research with consumers', *ASTM Special Technical Publication*, Philadelphia, American Society for Testing & Materials.
GROCERY MANUFACTURERS ASSOCIATION (2009), *CPG innovation & growth: developing the right innovation & product lifecycle management (PLM) strategies for today and tomorrow*. Grocery Manufacturers Association, Accenture, and Information Resources, Inc. (IRI). Available from: http://www.gmaonline.org/resources/research-tools/research-and-reports/ (Accessed 3 January 2012).
GOTTLIEB M (2010), 'Walkers 'Do Us a Flavour' case study', UtalkMarketing. Available from: http://www.utalkmarketing.com/Pages/Article.aspx?ArticleID=18384 (Accessed 3 January 2012).
HAGUE PN, HAGUE N AND MORGAN C-A (2004), *Market research in practice: a guide to the basics*, London and Sterling, VA, Kogan Page.
HOUGH G (2010), *Sensory shelf life estimation of food products*, Boca Raton, CRC Press.
HÜCK S (2011), 'The strategic relevance of ethics, commitment and rewards in co-creation'. Available from: http://openbusinesscouncil.org/2011/08/be-prepared-for-the-coming-%E2%80%9Dwar-for-co-creators%E2%80%9D/ (Accessed 3 January 2012).
IFST (2010), *IFST Guidelines for ethical and professional practices for the sensory analysis of foods*. Available from: http://www.ifst.org/documents/misc/practicesforsensoryanalysis1.pdf (Accessed 31st March 2012).
JAEGER S AND MACFIE H (2010), *Consumer-driven innovation in food and personal care products*, Cambridge, Woodhead Publishing Ltd.
KARDES FR, CRONLEY ML AND CLINE TW (2010), *Consumer behavior: science and practice, international edition*. Mason, OH, South-Western Cengage Learning.
KEMP SE, HOLLOWOOD T AND HORT J (2009), *Sensory evaluation. A practical handbook*, Oxford, Wiley-Blackwell.
KILCAST D (2010), *Sensory analysis for food and beverage quality control: a practical guide*, Cambridge, Woodhead Publishing Ltd.
KIPPLE, KF AND ORNELAS KC (2000), *Cambridge world history of food*, Cambridge, Cambridge University Press.
KOZINETS RV, HEMESGERGER A AND SCHAU HJ (2008), 'The wisdom of consumer crowds. Collective innovation in the age of networked marketing', *Journal of Macromarketing*, **28**(4), 339–354.
KÖSTER EP AND MOJET J (2007), 'Boredom and the reason why some new food products fail',

in MacFie H, *Consumer-led food product development*, Cambridge, Woodhead Publishing Limited, 262–280.

LAWLESS H AND HEYMAN H (2010), *Sensory evaluation of food: principles and practices. (Second edition)*, New York, Springer.

LUNDAHL D (2011), *Breakthrough food product innovation through emotions research*, London, Waltham, San Diego, Academic Press.

MACFIE H (2007), 'Preference mapping and food product development', in MacFie H, *Consumer-led food product development*, Cambridge, Woodhead Publishing Limited, 551–590.

MARIAMPOLSKI H (2012), *Qualitative market research. (Second edition)*, Thousand Oaks, London, New Delhi, Sage Publications, Inc.

MARSHALL D (1995), *Food choice and the consumer*, Glasgow, Blackie Academic and Professional.

MCCLURE SM, LI J, TOMLIN D, CYPERT KS, MONTAGUE LM AND READ MONTAGUE P (2004), 'Neural correlates of behavioural preference for culturally familiar drink', *Neuron*, **44**, 379–387.

MCQUARRIE EF (2011), *The market research toolbox: a concise guide for beginners*, Thousand Oaks, London, New Delhi, Sage Publications, Inc.

MEILGAARD MC, CARR BT AND CIVILLE GV (2006), *Sensory evaluation techniques. (Fourth edition)*, Boca Raton, CRC Press.

MEISELMAN HL (2007), 'Integrating consumer responses to food products', in MacFie H, *Consumer-led food product development*, Cambridge, Woodhead Publishing Limited, 3–33.

MEISELMAN HL AND MACFIE HJH (1996), *Food choice, acceptance and consumption*, London, Blackie Academic and Professional.

MOORS EHM, BOON WPC, NAHUIS R AND VANDEBERG RLJ (2010), 'User-producer interaction in emerging pharmaceutical and food innovations', in Flowers S and Henwood F, *Perspectives on user innovation*, London, Imperial College Press, 161–188.

MOSKOWITZ H (2007), 'Consumer-driven concept development and innovation in food product development', in MacFie H, *Consumer-led food product development*, Cambridge, Woodhead Publishing Limited, 342–382.

MOSKOWITZ H, PORRETTA S AND SILCHER M (2005), *Concept research in food product design and development*, Iowa, Blackwell Publishing Professional.

MOSKOWITZ HR, BECKLEY JH AND RESURRECCION VA (2006), *Sensory and consumer research in food product design and development*, Iowa, Blackwell Publishing Professional.

MRS (2006), *Conducting research with children and young people*, Market Research Society. Available from: http://www.mrs.org.uk/standards/ (Accessed 3 January 2012).

MRS (2010). *Code of conduct*, Market Research Society. Available from: http://www.mrs.org.uk/standards/ (Accessed 3 January 2012).

MUNOZ A, CIVILLE GV AND CARR BT (1992), *Sensory evaluation in quality control*, London, Chapman and Hall.

PLINER P AND SAVVY S-J (2006), 'Food neophobia in humans' in Shepherd R and Raats M, *The psychology of food choice*, 75–92, Oxford, CABI Publishing.

POPE JL (1993), *Practical marketing research*, New York, AMACOM.

PRENDERGRAST M (2000), *For god, country and coca-cola: the definitive history of the great American soft drink and the company that makes it. (Second edition)*, New York, Basic Books.

RESURRECCION AVA (1998), *Consumer sensory testing for product development*, Maryland, Aspen Publishers, Inc.

ROZIN P (2007), 'How does culture affect choice of foods?', in MacFie H, *Consumer-led food product development*, Cambridge, Woodhead Publishing Limited, 66–80.

SABA A (2007), 'How do risk beliefs and ethics affect food choice?', in MacFie H, *Consumer-led food product development*, Cambridge, Woodhead Publishing Limited, 108–121.

SCHIFFERSTEIN HNJ AND HEKKERT P (2007), *Product experience*, San Diego, New York, London, Burlington, Elsevier Science.

SCHINDLER RM (1992), 'The real lesson of New Coke: the value of focus groups for predicting the effects of social influence', *Marketing Research*, **4**, 22–27.

SHEPHERD R AND RAATS M (2010), *The psychology of food choice*, Oxford, CABI Publishing.

SIEGRIST M (2007), 'Consumer attitudes to food innovation and technology', in MacFie H, *Consumer-led food product development*, Cambridge, Woodhead Publishing Limited, 122–138.

STONE H AND SIDEL JL (2004), *Sensory evaluation practices (Third edition)*, New York, Academic Press.

STONE H AND SIDEL JL (2007), 'Sensory research and consumer-led food product development', in MacFie H, *Consumer-led food product development*, Cambridge, Woodhead Publishing Limited, 307–320.

THOMPSON D (2007), 'Sensoemotional optimisation of food products and brands', in MacFie H, *Consumer-led food product development*, Cambridge, Woodhead Publishing Limited, 281–306.

TOFFLER A (1980), *The third wave*, New York, William Collins Sons & Co. Ltd.

TRAILL B AND GRUNERT K (1997), *Product and process innovation in the food industry*, London, Blackie Academic and Professional.

TRAITLOR H AND SAGUY IS (2009), 'Creating successful innovation partneships', *Food Technology*, **63**(3), 22–35.

TUORILA H (2007), 'Sensory perception as a basis of food acceptance and consumption', in MacFie H, *Consumer-led food product development*, Cambridge, Woodhead Publishing Limited, 34–65.

URBAN GL AND HAUSER JR (1993), *Design and marketing of new products. (Second edition)*, Englewood Cliffs, NJ, Prentice-Hall.

VAN KLEEF E AND VAN TRIJP HCM (2007), 'Opportunity identification in new product development and innovation in food product development', in MacFie H, *Consumer-led food product development*, Cambridge, Woodhead Publishing Limited, 321–341.

VAN KLEEF E, VAN TRIJP HCM AND LUNING P (2005), 'Consumer research in the early stages of new product development: a critical review of method and techniques', *Food Quality and Preference*, **16**(3), 181–201.

VAN TRIJP JCM AND STEENKAMP JEBM (2005), 'Consumer-oriented new product development: an update on principles and practices', in Jongen WMF and Meulenberg MTG, *Innovation in food production systems: product quality and consumer acceptance. (Second edition)*, Wageningen, Wageningen Press.

VON HIPPEL E (2001), 'Perspective: User toolkits for innovation', *Journal of Product Innovation Management*, **18**(4), 247–257.

WALKER H (2002), *The meal*, Totnes, UK, Prospect Books.

WILSON A (2006), *Marketing research: an integrated approach. (Second edition)*, Harlow, Financial Times/Prentice Hall.

XU J (2005), *Market research handbook: measurement, approach and practice*, Bloomington, IN, iUniverse.

YEOMANS MR (2007), 'Psychobiological mechanisms in food choice', in MacFie H, *Consumer-led food product development*, Cambridge, Woodhead Publishing Limited, 81–107.

8

Co-creation of value with consumers as an innovation strategy in the food and beverage industry: the case of Molson Coors' 'talking can'

M. Garcia Martinez, Kent Business School, University of Kent, UK

DOI: 10.1533/9780857097248.2.139

Abstract: A new perspective in value creation is examined where personalised consumer experience takes central stage as opposed to a product and firm-centric view, and the ways in which open innovation and collaborative knowledge-creation processes influence the ability of organisations to innovate in new territories are discussed. A case study is presented of Molson Coors Brewing Company UK (MCBC-UK), a leading UK beer company part of the global Molson Coors group, which, in the search for value addition to remain competitive in a declining market, embraced collaborative open innovation. The joint innovation effort between the company and a supplier is leading to more open approaches as consumers are involved in the process. By understanding what consumers' value and engaging in active dialogue and interaction, the company has been able to develop superior value propositions relevant to their target consumer base.

Key words: open innovation, co-creation, value creation, food industry, consumers.

8.1 Introduction

Open-innovation research has largely focused on innovation collaboration through organisational linkages and ecosystems with contributions from across a network of partners ranging from suppliers of raw materials and equipment and research institutes to consumers and customers that create value for the end consumer (West and Lakhani, 2008). More recently, interest has moved to consider co-creation interactions with consumer and unidentified individuals on the internet in the process of product development and value creation (Howe, 2008; von Hippel, 2005; Surowiecki, 2005; Tseng and Piller, 2003). High-quality interactions that enable consumers to co-create unique experiences with companies are the key to unlocking new sources of competitive advantage (Prahalad and Ramaswamy, 2004b, p.7). Moreover, the consumer's direct involvement in value creation helps consumers forge a deep emotional bond with the company that can lead to enhanced consumer loyalty, consumer satisfaction and consumer loyalty perceptions (Bendapudi and Leone, 2003). These developments have been possible owing to advances in Internet, online design tool and new Web 2.0 technologies to support collaboration with consumers at each stage of the innovation process (Morgan and Wang, 2010; Terwiesch and Ulrich, 2009; Zwass, 2010).

In the food and drink industry, true innovation is limited, with the majority of introductions representing range extensions and new flavours (Costa and Jongen, 2006; Sorescu and Spanjol, 2008), there is, therefore, an opportunity for companies in this sector to add value and extricate themselves from commodity sectors where the lowest cost provider holds sway by embracing consumers' ideas as part of the innovation process. By understanding what consumers value and engaging in active dialogue and interaction, food and drink companies can seek to maximise the lifetime value of desirable customer segments (Payne and Frow, 2005). Value creation is a moving target where new expectations develop and old value concepts are commoditised putting continuous pressure on companies to create new value (Geraerdts, 2012). This is a major challenge for the food and drink industry where the lack of true innovation is making it increasingly harder for consumers to perceive the added value of slightly new food products and thus to reward incremental innovative behaviour.

This chapter aims to add to the existing co-creation studies by looking at this new perspective in value creation where personalised consumer experience takes central stage as opposed to a product and firm-centric view (Prahalad and Ramaswamy, 2004b). There are still large gaps in our understanding of how open-innovation models and co-creation are used strategically and of the levers of value creation in business markets (Geraerdts, 2012). Empirical studies on the benefits of open innovation and co-creation have been largely based on high-technology industries (Bianchi *et al.*, 2011; Christensen *et al.*, 2005; Ferrary, 2011; Smals and Smits, 2012; Todtling *et al.*, 2006), raising questions about the underlying openness decision process in more traditional and mature industries (Santamaria *et al.*, 2009).

This chapter examines how openness and collaborative knowledge-creation processes influence an organisation's ability to innovate in new territories. The chapter focuses on the Molson Coors Brewing Company UK (MCBC-UK), a leading UK beer company that is part of the global Molson Coors group. In search of value addition to remain competitive in a declining market MCBC-UK embraced collaborative open innovation. A joint innovation effort between the company and a supplier is leading to more open approaches as consumers are involved in the process. By successfully managing value co-creation and exchange, MCBC-UK seeks to maximise the lifetime value of desirable customer segments.

In Section 8.2, there is an overview of the research into co-creation with consumers. Section 8.3 presents the research design and the data used in the analysis. Section 8.4 reports MCBC-UK's innovation strategy, which is increasingly focused on consumer-oriented value-creation initiatives in order to invigorate a category in decline. Section 8.5 discusses how MCBC-UK moved to adopt an open-innovation approach and describes the processes and support mechanisms that were implemented to access and leverage external knowledge. Having established the model for collaborative innovation, Section 8.6 discusses the process of value creation that MCBC-UK engaged with active consumers as part of their open-innovation strategy with a supplier. Section 8.7 provides a discussion of the benefits of open innovation and co-creation by moving beyond the purely transactional to capture real synergy. Section 8.8 provides a summary and concluding remarks.

8.2 Co-creation of value with consumers

The concept of value creation with consumers has recently been the subject of marketing and innovation management studies as companies increasingly recognise that 'connected, informed and active' consumers are a source of competence they must tap into in order to add value to their products (Prahalad and Krishnan, 2008; Prahalad and Ramaswamy, 2004c; Vargo and Lusch, 2004). Consumers can play important roles in various innovation and value-creation activities; for instance their involvement at the stage of product design assists firms to improve the 'fuzzy front-end' process of identifying consumers' needs and wants (Lusch and Vargo, 2006). In fast-paced or turbulent markets, in particular, co-operating with so-called lead users has been described as an important source of innovation for firms (von Hippel, 2005).

The idea of a customer-centric firm challenges the conventional view of consumers in the innovation process that they are either passive or 'speaking only when spoken to' in the course of market research or concept testing (von Hippel, 1978, p. 40). The focus has shifted to a process of co-creating value through the exchange of knowledge and skills with customers and partners (Vargo and Lusch, 2004) to co-construct unique experiences (Prahalad and Ramaswamy, 2004b). Companies take consumers as a partner or co-producer instead of an external element (Firat *et al.*, 1995). Thus, the firm-centre notion of value in exchange (i.e., making a value proposition for the passive consumer to accept or reject) is being replaced by greater focus on the co-creation of value by the company and the 'connected, empowered and active consumer' (Prahalad and Ramaswamy, 2004b, p.8) who determines value 'uniquely and phenomenologically' (Vargo and Lusch, 2008, p.7) during the consumption.

Superior value propositions are created through personalised interactions that are meaningful and sensitive to a specific consumer (Prahalad and Ramaswamy, 2004a). Thus, critical to the creation of value is the company's ability to engage in active dialogue and interaction with the consumer rather than one way promotion

(Payne *et al.*, 2008). However, the wide spectrum of personal value together with the diversity, and sometimes even the opposite, of personal preference poses a major challenge for companies (Geraerdts, 2012). Prahalad and Ramaswamy (2004c) argue that organisations in the co-creation age have to become increasingly flexible. To satisfy individual demands, firms have to develop the most appropriate product or service from multiple combinations of parts and components, thus requiring greater flexibility. Companies can, however, be misguided by setting up the whole operation for the convenience of provisioning without putting the 'voice of the consumer' at the centre. Hence, looking at customers as individuals and proactively developing products to offer them at the price they are willing to pay, and at schedules that they are willing to wait for, is by no means a straightforward task (Piller and Ihl, 2009). A primary factor necessary to achieve a consumer-centric culture for innovation is to have a culture where the consumer mind-set is foremost, and pervades the company's strategy and activities (Hienerth *et al.*, 2011).

8.3 Research design

To understand the process of value creation with consumers within the framework of an open-innovation corporate strategy, this study follows an exploratory case study design (Yin, 2003) to provide a rich illustration of the phenomenon under analysis (Eisenhardt and Graebner, 2007). The research focuses on the open-innovation transition of MCBC-UK, a leading UK beer company that is part of the global Molson Coors group, and discusses how opening the doors to consumers and embracing their ideas as part of the innovation process has enabled the company to innovate in new territories.

The data for the study was gathered through a series of semistructured interviews with MCBC-UK's open-innovation steering committee during 2010, complemented with extensive documentary and media analysis. Interviews were conducted on site and lasted on average 2 h. All discussions were tape-recorded and fully transcribed. Secondary information sources have been triangulated with data drawn from direct interviews (Yin, 2003). Finally, telephone follow-up interviews were conducted to validate the case study and complete information gaps.

Information gathered during the interviews concerned the company's approach to open innovation, including the drivers for open innovation, the organisational and management structures developed to manage the transition to open innovation; risks and barriers (e.g. culture, internal resistance) and enablers (e.g. communication, rewards). Having established the model for collaborative innovation, the enquiry moved to consider the process of value creation that MCBC-UK engaged with active consumers as part of their open-innovation strategy with a supplier.

8.4 Molson-Coors Brewing Company UK (MCBC-UK): the need for consumer-driven innovation

The need for value creation strategies where the voice of the consumer is central to business and marketing strategies is clearly exemplified in the case of MCBC-UK. The company had to reconsider its value proposition to remain competitive in a declining market where beer sales have persistently fallen in the last decade and the value of beer brands significantly diminished by continuous discounting and promotions in the low-margin off-trade (sales in shops and supermarkets) (Key Note, 2011). Adding to this depressed business environment, past optimisation of production costs and company overheads negatively impacted MCBC-UK's innovative performance resulting in a diminished technical department and depleted product pipelines.

The company has an extensive product portfolio; however, the consumer value of some products and their potential to invigorate a category in decline is questionable. As a result, MCBC-UK has reassessed the value bases and identified products that answer various consumer's needs, while simultaneously phasing out those products that are failing to contribute to the category growth. This evolution from a narrow product focus innovation process towards a holistic, consumer-led value creation framework has led to the development of a series of strategic themes or 'WANTs' (Slowinski, 2004) that will guide MCBC-UK's innovative efforts in coming years. Value creation is the backbone of MCBC-UK innovation strategy to spearhead an eventual recovery in beer sales by developing new products with true added value. Full execution of its innovation strategy will require a fundamental paradigm shift to be achieved only by working collaboratively with a network of partners ranging from suppliers of raw materials and equipment and research institutes to consumers and customers. The challenge for MCBC-UK is to find an innovation partner willing to co-create value and see through the innovation process from idea generation to commercialisation.

Recognising the need for consumer-driven innovation, MCBC-UK has a cross-functional innovation structure with the brand innovation team bringing together the commercial and technical teams (Fig. 8.1). Innovation activities respond to the brand innovation briefs produced by the brands, marketing and sales team setting out the innovation priorities for the next 3 years. WANTs are then internally screened by the cross-functional innovation team in terms of their potential consumer's relevance and technical requirements in order to deliver technical solutions aligned to the strategic WANTs. The newness of the required technological solutions and whether or not the WANT is part of the company's core competencies determines whether innovation projects are developed using internal resources or, conversely, whether knowledge gaps in MCBC-UK technological know-how require the company to seek technologies and ideas from the outside world. The objective is to find the best technology there is, relevant to MCBC-UK's brands and innovation strategy, and to find the best way to bring it in. Sometimes this technology is found internally, sometimes not. A make/buy/partner decision-making protocol (Slowinski and Sagal, 2010) brings rigour to this determination by benchmarking internal developments against external alternatives to ensure the best option is selected.

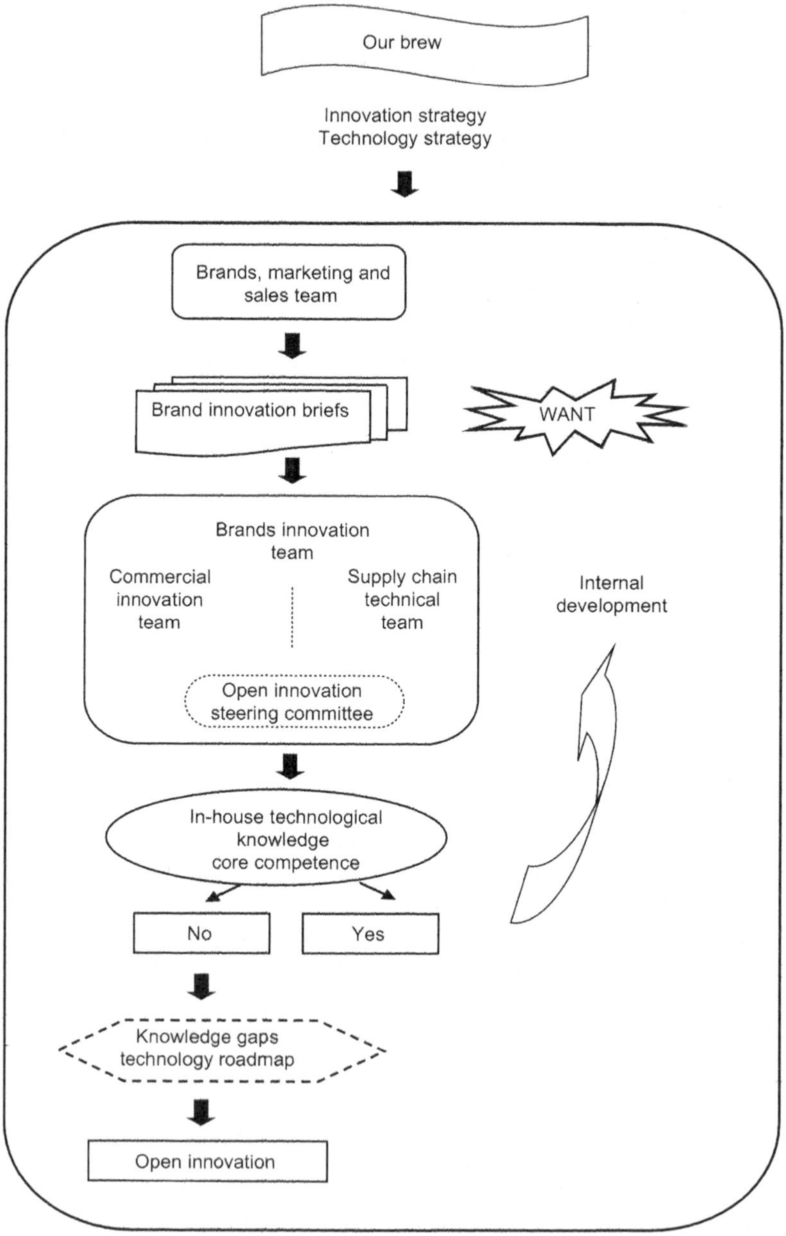

Fig. 8.1 MCBC-UK innovation organisational structure.

8.5 The discover style of open innovation

To kick start the journey towards open innovation, MCBC-UK has established a multidisciplinary open-innovation steering committee within the brands innovation team to scope out what opportunities a collaborative and open approach to innovation would deliver over and above existing partnerships with key suppliers (Fig. 8.2). The process initially involved consulting and benchmarking open-

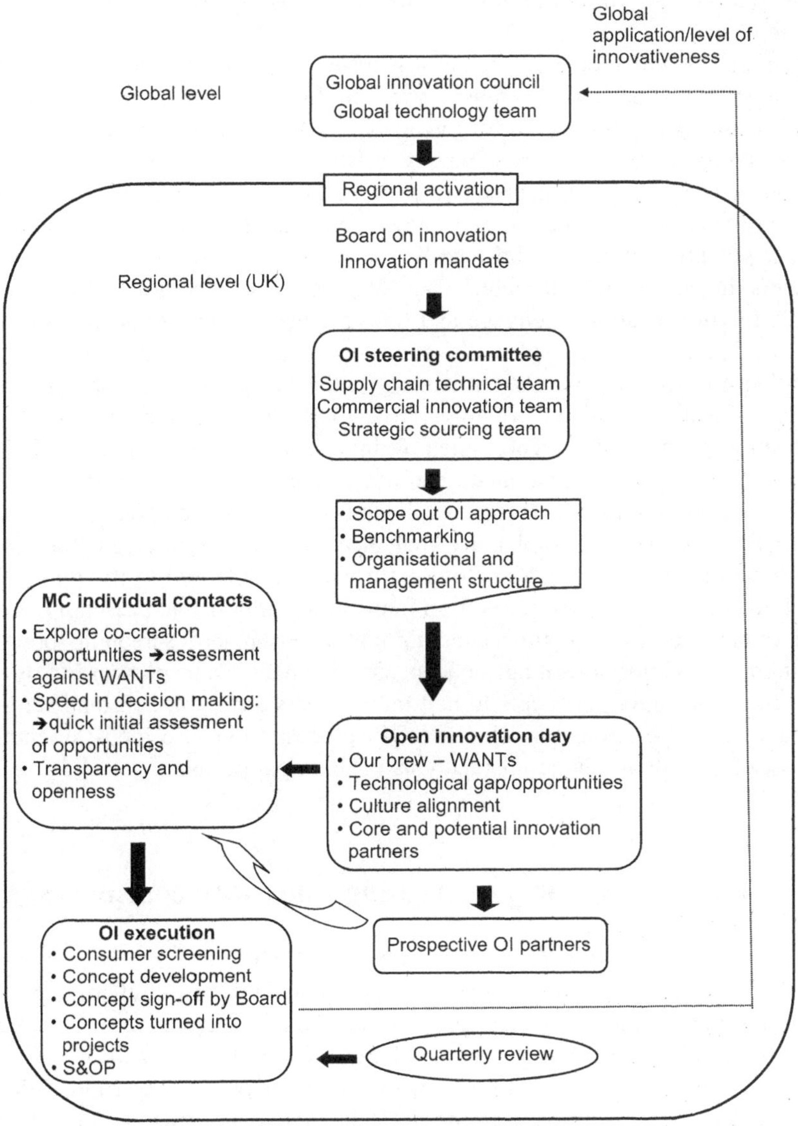

Fig. 8.2 The discover style of open innovation.

innovation models of leading companies in the industry. MCBC-UK has yet to determine the internal organisational structure of open innovation, that is, whether to set up a dedicated resource within the company (i.e., an independent open-innovation business unit) to manage open innovation; or, alternatively, to embed open-innovation activities within existing functions with a steering committee aligning partnerships to WANTs so that open innovation becomes a responsibility and, eventually, the established model of all innovation terms. The outcome of emerging collaborative partnerships would ultimately determine the system of collaboration that best fits MCBC-UK's innovation requirements.

Having implemented an incipient organisational structure for its open-innovation strategy, MCBC-UK has started to look for skill sets outside the organisation to execute its innovation agenda and created a competitive innovation marketplace. In 2010, the company organised an open innovation day for a core body of existing partners where both the innovation strategy and WANTs were presented and business opportunities and gaps in the market discussed as well as potential returns for those willing to engage with MCBC-UK. Interested partners in co-creating particular WANTs were assigned individual people within the organisation to engage and translate their thinking into possible co-creation initiatives. The need to move away from a faceless organisation is key to successful open innovation, particularly in the stages where transparency and speed in decision making are imperative to foster trust and openness with innovation partners. This represented an important cultural change to MCBC-UK, which was more accustomed to ad hoc connections with suppliers and less formalised internal mechanisms to evaluate potential technologies.

MCBC-UK then assessed individual ideas against the strategic themes to ensure alignment with WANTs (long-term view) as opposed to the innovation pipeline (short-term view). Ideas from innovation partners that passed the initial assessment process then proceeded through the standard screening process, including consumer screening, concept development, concept sign-off by the Board on Innovation, concepts turned into projects and projects going through the sales and operations planning (S&OP) process to ensure focus, alignment and synchronisation among all functions of the organisations.

8.6 The 'talking can': co-creating value with consumers

In search of value addition to its premium brands, MCBC-UK in partnership with Ball Packaging Europe, the front-runner in optimising the use of thermochromic inks on steel and aluminium cans, has developed a new thermochromic beverage can. Applied to the company's global flagship Coors Light brand in the 'cold activated can', the entire Rocky Mountains logo changes colour from white to shimmering blue indicating that the beer is at the right drinking temperature, giving the ultimate drinking experience. The thermochromic technique is based on

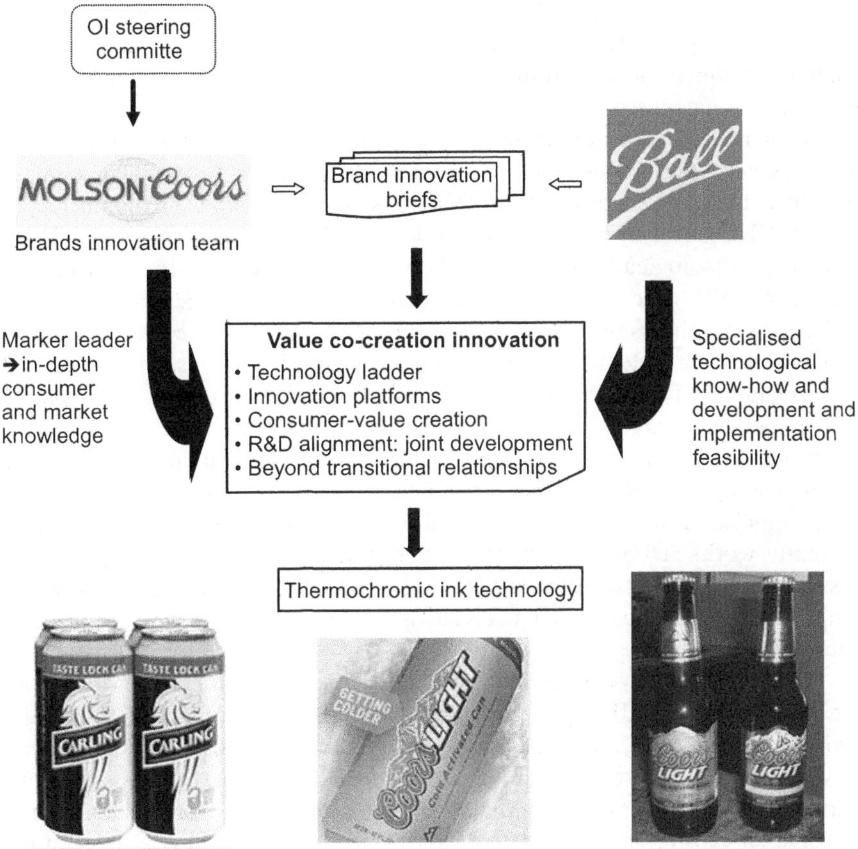

Fig. 8.3 Value creation innovation by open innovation: the talking can.

special pigments in the inks which change colour as the temperature rises or falls. The ink, which begins to change colour at 11 °C turns blue when the beer has reached the ideal drinking temperature (between 6–10 °C). The new generation of thermochromic ink on a large scale will promote the ice cold refreshment positioning of the brand.

This innovative technology has been developed through a process of co-creation of value between the firm, its supplier and active consumers. MCBC-UK recognised that consumers are a source of competence that they must tap into in order to add value to its premium brands ('We need to add the voice of the consumer to our product ideas'). The initial process involved MCBC-UK being open and sharing the WANTs with Ball to discuss the concept of value (i.e., immediate return on a particular product versus longer term, strategic value creation initiatives) and opportunities both for value creation and value extraction (Fig. 8.3). This was a significant step change from the past focus on cost cutting initiatives. Having a point on the horizon (as opposed to short-term targets) and a roadmap to a process that is informed by consumers were key to secure Ball's buy-in to a collaborative knowledge-creation process. Conscious of the significant capital investment needed to innovate in the beer sector and the financial risks involved, both companies decided to follow a 'technology ladder' approach with the sequential development of innovative technologies responding to particular WANTs (i.e., premiumisation) as opposed to particular products in order to maximise return on investment. By creating synergies around core competencies (i.e., MCBC-UK's in-depth consumer and marketing knowledge with Ball's superior technological know-how in beverage cans) a number of packaging-related WANTs were co-developed and screened with consumers.

The co-creating sessions took place in 2009 and involved 40 consumers engaged in active dialogues and development of shared solutions with MCBC-UK and its supplier. Dialogue centred on packaging ideas, how consumers related to the options and features that MCBC-UK had on offer and how those offerings might be of more value to customers. The two-day sessions provided a better understanding of how consumers approach the beer category; what is the consumer mindset in this category; what value consumers are looking for; and what currently works and does not work for consumers. On the second day, creative sessions were organised around the co-developed solutions where consumers took innovation ideas and used them to create their own ideas and then presented them back to each other.

The idea of a thermochromic device was discussed as a solution area to increase the value of MCBC-UK's premium brands where consumers were first asked to provide feedback on various cans on display. The dialogue then moved onto ways to visualise cold where themochromic ink was talked about. Surprisingly, some participants were not aware that MCBC-UK has already incorporated this technology in their cans prompting questions about the value of the device in its current form if consumer were not registering or valuing its presence. Consumer consensus was that the idea of themochromic ink was a valuable one ('it adds excitement and relevance to the category') but in order to create superior consumer value (i.e., consumers getting excited about it) it would require the entire can to change appearance rather than just the logo. This dialogue was particularly powerful to the technology supplier and generated a better understanding of consumer preferences largely lacking in B2B businesses resulting in a more targeted, consumer driven R&D programme. Consumers' value perceptions have made Ball understand the value MCBC-UK place on themochromic ink as a piece of brand architecture and identify a consumer validated business opportunity to be realised by taking the technology to the next level.

The previous firm-centric view on the value of thermochromic ink was largely dictated by the cost of having the technology on the cans, which, in turn, determined the ink coverage and ultimately how much the technology is part of the can's look and feel. Consumers however were failing to perceive this cost-driven value proposition. Although the idea resonates with consumers ('cold you can see'), the financial benefits (or value) would only emerge once the technology becomes an integral part of the offering. This required the company to better align the design with the brand message highlighting the technology's relevance ('emulating the 'cool' perception of the iPhone technology among consumers').

Technologically, the new generation of thermochromic ink on a large scale is an extrapolation and optimisation of the current printing technologies. The technology promotes the ice-cold refreshment positioning of the brand where the entire Rocky Mountains logo changes colour from white to shimmering blue indicating that the beer is at the right drinking temperature. Although printing technology has longevity in the market, future technological developments will require a different rung on the technological ladder by adding functionalities to the can, such as printing a thermometer on the can, for which consumers would be willing to pay a premium price. Active printing technology is getting closer to commercialisation opening collaborative knowledge creating opportunities for MCBC-UK to cross-fertilise with a wider network of suppliers (i.e., of bottles, cans and cardboard).

8.7 Discussion

It is difficult to establish the (economic) benefits of open innovation and co-creation because there are many other consumer touch points (such as merchandising and promotions) affecting product performance, but undoubtedly open innovation has emerged as a true 'win–win' strategy for both companies. It has created excitement ('something to talk about in the market place') and brand awareness, allowing MCBC-UK to escape the commoditisation trap and create consumer value desperately needed to invigorate a declining category. There has been a realignment of Ball's R&D strategy around MCBC-UK's strategic themes, thus securing additional business opportunities for Ball by becoming the 'preferred technological supplier' for value co-creation.

The technology promotes the ice cold refreshment positioning of the brand and generates the consumer pull to eventually drive a price differential for cold beer in a category where retailers constantly want cheap deals and consumers regularly switch in and out of promotions. Developments in printing technology will allow the introduction of tangible functionalities by co-creating experiences of value with consumers for which they would be willing to pay a premium. Co-creating interactions will further give MCBC-UK an indication of the technological leap required to make its products be of more value to consumers and to determine whether thermochromic packaging is regarded as a real benefit to consumers (i.e., core to the offering and thereby worth investing in) or rather part of the augmented product and more subject to changes in consumer's demands.

Open innovation is founded on relationship building and trust and both companies had a long-standing good business relationship before starting the partnership (Fig. 8.4). MCBC-UK had in the past turned to open-innovation intermediaries to locate partners for particular technological problems; however, this time they needed a more profound collaborative arrangement to execute their innovation strategy. Above all, they needed a partner willing to co-create

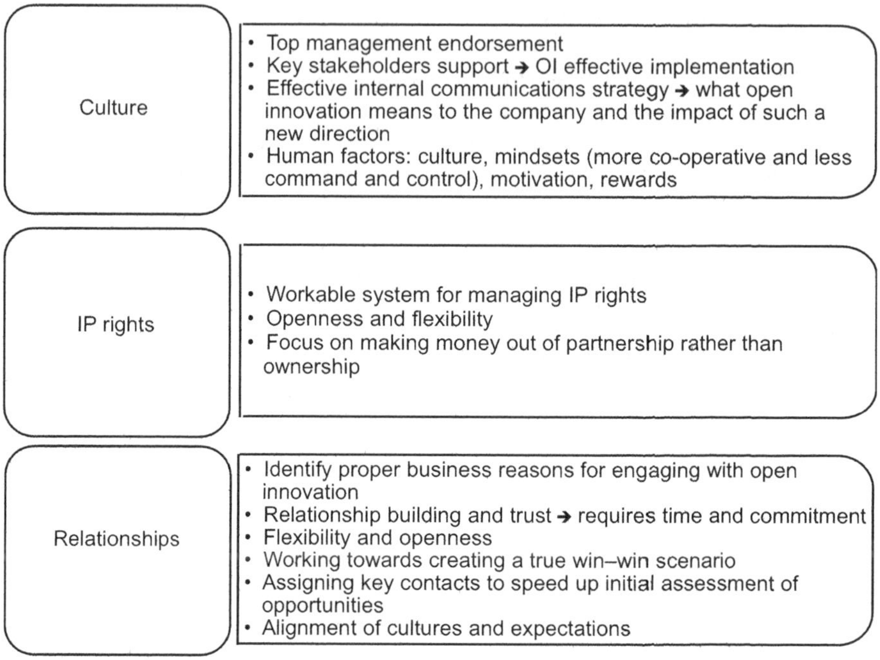

Fig. 8.4 Key factors to successful open innovation.

value and see through the innovation process from idea generation to commercialisation. Although MCBC-UK initiated the process, they were quite flexible and open in the earlier discussions with Ball, recognising that a prescriptive one-sided approach would not have worked. Initial discussions centred on MCBC-UK's WANTs and how they could jointly develop them and work towards creating a true win–win scenario (rather than focusing on their own gains). They also had an open approach to intellectual property (IP) rights with the partnership based on a contract of supply agreement with equal participation in future business earnings. This flexible, open approach is an important strategic change for MCBC-UK from a situation where all business aspects were highly protected, which reinforces the need for IP openness in value co-creation. Overall, MCBC-UK regards its partnership with Ball as a success story and lessons have been learnt that will inform the company's emerging open-innovation strategy. The challenge is to replicate Ball's experience with a wider number of innovation partners working on different/same strategic themes. Creating innovation ecosystems would ensure that technological synergies among innovation partners are maximised, but managing these network structures would be a challenge.

8.8 Conclusion

Openness has influenced MCBC-UK's ability to innovate in new territories by working collaboratively with innovation partners. Although the company is still in the early stages of its open-innovation journey, its partnership with Ball has created the learning experience needed to drive the process further. As the new products enter the market, the company will receive results that it can share with its internal and external stakeholders. Open innovation has brought about 'true' innovation, much needed to invigorate a declining category. Innovation velocity might not have been improved given the time required to build and nurture relationships, but a mutually beneficial innovation agenda has certainly emerged.

The case study highlights a number of enablers to successful open innovation, in particular the need to identify proper business reasons (WANTs) for engaging with external partners. The brand innovation briefs acted as the catalysts to co-creation by setting a technology roadmap to develop the strategic themes. Emerging open-innovation initiatives will progress faster as these reference documents become better understood by internal and external stakeholders. However, it is key that WANTs remain relevant and reflect future and emerging consumer trends.

The extent to which open innovation becomes the prominent innovation model for MCBC-UK, and Molson Coors in general, would be a major strategic decision. How the company strategically scouts and makes connections with potential partners beyond its existing supplier base and develops a peripheral vision to identify potential technological development in unrelated sectors will be a major challenge.

MCBC-UK has been and still is involved in a major restructuring of its innovation structures and processes as it develops its open-innovation model so they are both aligned and reinforced. Arguably this has delayed its earlier experience in open innovation; however, a staged approach supported by results is more likely to ensure full alignment of organisational units, particularly operational ones, with open-innovation initiatives.

8.9 References

BENDAPUDI, N. AND LEONE, R., 2003. Psychological implications of customer participation in co-production. *Journal of Marketing*, **67**, 14–28.

BIANCHI, M., CAVALIERE, A., CHIARONI, D., FRATTINI, F. AND CHIESA, V., 2011. Organisational modes for open innovation in the bio–pharmaceutical industry: an exploratory analysis. *Technovation*, **31**(1), 22–33.

CHRISTENSEN, J. F., OLESEN, M. H. AND KJÆR, J. S., 2005. The industrial dynamics of open innovation: evidence from the transformation of consumer electronics. *Research Policy*, **34**(10), 1533–1549.

COSTA, A. I. A. AND JONGEN, W. M. F., 2006. New insights into consumer-led food product development. *Trends in Food and Science Technology*, **17**, 457–465.

EISENHARDT, K. M. AND GRAEBNER, M. E., 2007. Theory building from case studies: opportunities and challenges. *Academy of Management Journal*, **50**(1), 25–32.

FERRARY, M., 2011. Specialized organizations and ambidextrous clusters in the open innovation paradigm. *European Management Journal*, **29**(3), 181–192.

FIRAT, A. F., DHOLAKIA, N. AND VENKATESH, A., 1995. Marketing in a postmodern world. *European Journal of Marketing*, **29**(1), 40–56.

GERAERDTS, R., 2012. Customer value creation: A journey in the search of excellence. *Industrial Marketing Management*, **41**(1), 11–12.

HIENERTH, C., KEINZ, P. AND LETTL, C., 2011. Exploring the nature and implementation process of user-centric business models. *Long Range Planning*, **44**, 344–374.

HOWE, J., 2008. *Crowdsourcing: Why the power of the crowd is driving the future of business.* Crown Business, New York.

KEY NOTE, 2011. *Breweries & the Beer Market. Market Report 2011.* Richmond Upon Thames: Key Note Ltd.

LUSCH, R. P. AND VARGO, S. L., 2006. *The service dominant logic of marketing: dialog, debate and directions.* M.I. Sharpe, Armonk, NY.

MORGAN, J. AND WANG, R., 2010. Tournaments for ideas. *California Management Review*, **52**(2), 77–97.

PAYNE, A. AND FROW, P., 2005. A strategic framework for customer relationship management. *Journal of Marketing*, **69**, 167–176.

PAYNE, A. F., STORBACKA, K. AND FROW, P., 2008. Managing the co-creation of value. *Journal of the Academy of Marketing Science*, **36**(1), 83–96.

PILLER, F. AND IHL, C., 2009. Open innovation with customers: foundations, competences and international trends. Germany: Technology and Innovation Management Group, RWTH Aachen University, http://www.internationalmonitoring.com/fileadmin/Downloads/Trendstudien/Piller-Ihl_Open_Innovation_with_Customers.pdf.

PRAHALAD, C. K. AND KRISHNAN, M. S., 2008. *The new age of innovation: driving co-created value through global networks.* McGraw-Hill, New York.

PRAHALAD, C. K. AND RAMASWAMY, V., 2004a. Co-creating unique value with customers. *Strategy and Leadership*, **32**(3), 4–9.

PRAHALAD, C. K. AND RAMASWAMY, V., 2004b. Co-creation experiences: the next practice of value creation. *Journal of Interactive Marketing*, **18**(3), 5–14.

PRAHALAD, C. K. AND RAMASWAMY, V., 2004c. *The future of competition: co-creating value with customers.* Harvard Business School Press, Boston, MA.

SANTAMARIA, L., NIETO, M. J. AND BARGE-GIL, A., 2009. Beyond formal R&D: taking advantages of other sources of innovation in low- and medium-technology industries. *Research Policy*, **38**, 507–517.

SLOWINSKI, G., 2004. *Reinventing corporate growth.* Alliance Management Press, Gladstone, NJ.

SLOWINSKI, G. AND SAGAL, M. W., 2010. Good practices in open innovation. *Research Technology Management*, September–October, 38–45.

SMALS, R. G. M. AND SMITS, A. A. J., 2012. Value for value: the dynamics of supplier value in collaborative new product development. *Industrial Marketing Management*, **41**(1), 156–165.

SORESCU, A. B. AND SPANJOL, J., 2008. Innovation's effect on firm value and risk: insights from consumer packaged goods. *Journal of Marketing*, **72**(2), 114–132.

SUROWIECKI, J., 2005. *The wisdom of crowds: why the many are smarter than the few.* First Anchor Books Edition, New York.

TERWIESCH, C. AND ULRICH, K. T., 2009. *Innovation Tournaments.* Harvard Business School Press, Boston, MA.

TODTLING, F., LEHNER, P. AND TRIL, M., 2006. Innovation in knowledge intensive industries: the nature and geography of knowledge links. *European Planning Studies*, **14**(8), 1035–1058.

TSENG, M. M. AND PILLER, F. T., 2003. *The customer centric enterprise: advances in mass customization and personalization.* Springer-Verlag, Berlin.

VARGO, S. L. AND LUSCH, R. F., 2004. Evolving to a new dominant logic for marketing. *Journal of Marketing*, **68**, 1–17.

VARGO, S. L. AND LUSCH, R. F., 2008. Service-dominant logic: continuing the evolution. *Journal of the Academy of Marketing Science*, **36**(1), 1–10.

VON HIPPEL, E., 1978. Successful industrial products from customer ideas. *Journal of Marketing*, **42**(1), 39–49.

VON HIPPEL, E., 2005. *Democratizing innovation.* MIT Press, Cambridge, MA.

WEST, J. AND LAKHANI, K. R., 2008. Getting clear about the role of communities in open innovation. *Industry and Innovation*, **15**(2), 223–231.

YIN, R. K., 2003. *Case study research: design and methods.* Sage Publications, Thousand Oaks, CA.

ZWASS, V., 2010. Co-creation: towards a taxonomy and an integrated research perspective. *International Journal of Electronic Commerce*, **15**(1), 11–48.

9

Collaborative product innovation in the food service industry. Do too many cooks really spoil the broth?

A. I. A. Costa, The Catholic University of Portugal, Portugal

DOI: 10.1533/9780857097248.2.154

Abstract: Collaborative product innovation (CPI) is a form of open innovation highly suited to tap customers for innovative concepts, support technology development and create value networks for new product launch. This chapter presents a case study describing how the CPI initiatives of manufacturing companies and haute cuisine chefs shaped the diffusion path of sous vide technology in the US. The case illustrates how collaborating with expert technology users can create value to companies by generating new product ideas and supporting internal R&D. Importantly, it shows how CPI can accelerate the adoption of novel foods by increasing technology acceptance in target markets.

Key words: foodservice, collaborative product innovation, sous vide cooking, innovation diffusion, haute cuisine chefs.

9.1 Introduction

The human capital inputs of innovation processes, i.e., the individual skills and knowledge employed in research and development (R&D) and commercialization activities (Romer, 1990), can be sourced both inside and outside corporate boundaries. Innovation processes in which human capital inputs are sourced mainly within a company's boundaries have been broadly designated as 'closed innovation', as opposed to innovation processes in which such inputs are, to a large extent, purposively sourced outside the company, a business strategy commonly known as 'open innovation' (Chesbrough, 2003). The latter is about harnessing the inbound and outbound flows of ideas, technology and skills across a company's boundaries (which are channelled through its multiple interorganizational links), with the intent of accelerating internal innovation processes and establishing additional, external paths for the commercialization of their outcomes (Chesbrough, 2003; Simard and West, 2006). The establishment and management of interorganizational relationships with customers, competitors, suppliers, public and private research institutions or even seemingly unrelated businesses, with the aim of acquiring additional knowledge and skills for innovation processes, is increasingly seen as an important way for companies to augment their innovation capability (Gatignon *et al.*, 2002; OECD and Eurostat, 2005).

To date, open innovation has been commonly associated with fast-growing, technology-intensive industries (e.g., information and communication technology and pharmaceuticals). There is, however, increasing evidence that this concept and associated strategies may also prevail in more traditional and mature industries (Huston and Sakkab, 2006), particularly when certain sets of circumstances arise. Among such circumstances is a high dependence on other entities, such as other companies, public research institutions and end-user communities, for the supply, development and/or commercialization of new technologies (Chesbrough and Crowther, 2006; Maula *et al.*, 2006; Vanhaverbeke and Cloodt, 2006). Cross-boundary product innovation management should thus be a widespread practice in food supply chains and networks, mainly owing to the number of actors in different areas involved in food supply and their difficulties to single-handedly meet all the heterogeneous (and often contradictory) requirements of intermediate customers, end-users and legislators (Costa and Jongen, 2006; Grunert *et al.*, 2005; Mikkelsen *et al.*, 2005). Empirical evidence of food manufacturing and foodservice companies engaging in open-innovation strategies (Knudsen, 2007) is, however, scarce. Most importantly, a detailed analysis of such activities, their rationale and market outcome is, with the exception of a few case studies (Huston and Sakkab, 2006; Thomke and von Hippel, 2002; Vanhaverbeke and Cloodt, 2006), equally absent from both academic and practice-oriented literature.

Consequently, the main aims of this chapter are:

- to provide a better understanding of the open-innovation practices taking place in mature industries, namely within the food area, as well as of their main antecedents and consequences;
- to analyse the effects of collaborative product-innovation activities on the innovation capabilities and market outcomes of food manufacturers and foodservice operators.

The chapter is structured as follows. In Section 9.2, a critical review of studies on the prevalence of open-innovation strategies in the food area is presented. In Section 9.3, the main findings of a case study analysing the collaborative product innovation activities of three manufacturers of sous vide meals for the US foodservice industry are presented and discussed. These highlight the impact of the open-innovation strategies employed by these companies on their innovation capabilities and market outcomes. In Section 9.4, the main conclusions drawn from the case study are presented and relevant managerial implications are derived; areas where more empirical research is needed are also highlighted.

9.2 A review of open-innovation practices in the food industry

Practices for open innovation in the food industry are examined in the following subsections, including drivers, collaborative innovation and examples.

9.2.1 Drivers of open innovation in the food industry

The food industry is typically described as a relatively mature and slow-growing area of business, which displays a relatively low level of R&D investment and is quite conservative in the type of innovations it introduces in the market (Costa and Jongen, 2006). This sector perceives its end-customers to be, to a large extent, wary of radically new products and changes in consumption patterns. Such perceived wariness, together with the necessary stringency of legal requirements related to safety, transforms food product and process innovation into a highly complex, time-consuming and risky endeavour, and hence one not to be lightly undertaken. However, recent important changes in the nature of both food demand and supply, coupled with a high level of competitiveness, have rendered innovation not only an unavoidable corporate activity, but also one that is increasingly vital for overall agri-business profitability.

Contemporary consumers demand unique flavours and singular foods, guilt-free convenience in cooking and eating, and an increasingly health-promoting diet closely tailored to their individual needs and preferences (Costa *et al.*, 2001, 2007). Such demand requires a type of product development that necessarily entails creating, or at the very least adopting, innovative technological solutions and new business models. On the other hand, recent general advances in areas such as biotechnology, nanotechnology and preservation technology offer an unprecedented number of opportunities for added-value applications in the food industry, many of which have the potential to adequately meet modern consumer demand (Juriaanse, 2006).

Unavoidable as it may be, innovation remains a highly challenging and complex process for the food processing industry to manage. The number of actors of different sectors involved in food production, together with their difficulty in single-handedly meeting all the heterogeneous (and often contradictory) requirements of intermediate customers, end-users and legislators, determines that innovation activities must be carefully co-ordinated. This, in turn, compels innovation processes to be managed both within and across organizational boundaries along the value chain (Costa and Jongen, 2006; Grunert *et al.*, 2005; Mikkelsen *et al.*, 2005). Moreover, many of the emerging technologies that can potentially sustain (or complement) a wave of successful new food applications (e.g. nanotechnology) are being developed outside the processing industry. In order to leverage these on-going innovation processes, food industry actors must therefore enter into more or less formal arrangements with other entities in the innovation system. Formal agreements are likewise required for the adoption of externally-developed novel technologies (Maula *et al.*, 2006). Las,t but not least, the establishment of close relationships with regulatory bodies, intermediate and end-users throughout the innovation process is essential to improve public acceptance of emerging food technologies and the commercial success of the products thereof (De Jong *et al.*, 2006; Vanhaverbeke and Cloodt, 2006).

All of the above implies that innovation in the food industry is likely to increasingly rely upon the decisions and activities of other entities in the innovation system. As such, the sector should exhibit a significant number of open-innovation strategies, the purpose of which could range from merely securing access to external sources of human capital to actively taking part in the creation of interorganizational knowledge and skills.

9.2.2 The concept of collaborative product innovation (CPI)

As globalization moves forward, markets and technologies converge, product lifecycles shorten and the rate of technological innovation increases. This creates mounting pressure upon companies to produce more innovative products in shorter periods and commercialize them simultaneously in a higher number of geographic markets. In view of this, open-innovation initiatives in the area of new product development have often been singled out as a means to overcome some of the shortcomings associated to operating in global consumer markets (Costa and Jongen, 2006; Emden et al, 2006; Littler *et al.*, 1995; Sarkar and Costa, 2008).

Collaborative product innovation (CPI) has been conceptualized in a number of ways (see Emden *et al.*, 2006, for a more recent definition), but as with other open-innovation initiatives, there is yet little agreement on how to best define and characterize it. For the purpose of this chapter, and based on a previously developed characterization of open-innovation strategies (Sarkar and Costa, 2008), CPI is understood 'as a collaborative relationship between an innovating company and an external partner, established with the purpose of sustaining the development and/or commercialization of an innovative product or product line'. Collaborative relationships are here defined as cross-boundary, information-exchange linkages that are characterized by high levels of relational and structural embededdness (i.e., high levels of interaction, integration, transparency, mindfulness and synergy, as well as highly similar actionable knowledge bases), and in which each party contributes actively and significantly to the common goal or end solution (Emden *et al.*, 2006; Rindfleisch and Moorman, 2001). In this setting, the innovating company may collaborate with independent external organizations, communities or individuals located at various stages of the value chain (customers, competitors, suppliers) or even in the surrounding innovation system (user communities, private and public research organizations). The collaboration projects may take several, non-equity-based forms and have a varied time span, but they are likely to be relatively structured and focused (at least initially), and to involve some type of contract or written agreement (Dittrich and Duysters, 2007; Littler *et al.*, 1995).

9.2.3 Empirical evidence of CPI initiatives in the food industry

Knudsen (2007) analyzed the results of a survey on the employment of interorganizational relationships in product innovation by EU manufacturing and service companies active in the food and beverages sector. She observed that all surveyed companies (n=132) had partnered, on average, with at least one other organization for the development of their last important product innovation. Additionally, survey results indicated that these companies would rather co-operate with customers, suppliers and competitors than with private/public research organizations or consultants, and preferably at the initial research stage rather than during technical development. Finally, she was also led to conclude that food companies preferably formed alliances with organizations in their own sector, probably because of the high degree of overlapping between their knowledge bases, and, it was believed, to facilitate interorganizational interactions and thereby increase the chances of innovation success.

Huston and Sakkab (2006) described the successful development and launch of a new type of Pringles' potato crisps (printed with words and images), driven by the application of the open-innovation concept. The authors reported on how Procter & Gamble (P&G) was able to lower product development costs and time-to-market for the new line through the in-sourcing of a technology for printing edible images on cakes and cookies. This technology had been primarily developed by a baker in Italy and was discovered through the global network of potential sources of ideas and know-how that Procter and Gamble maintained as a part of its open-innovation programme.

Alternatively, Thomke and von Hippel (2002) revealed how International Flavors and Fragrances (IFF), a company supplying flavours to the food industry, managed to outsource part of its new product design to customers. IFF developed a customer innovation toolkit, consisting of an interactive, internet-based application with a large database of flavour profiles, with which it equipped its clients in the food processing industry. This tool allowed customers to design and alter flavour samples at will, enabling IFF to bypass costly market research activities and accelerate the trial-and-error cycles that inevitably accompany product innovation. By putting customer expertise to use, IFF was also able to expand its knowledge base and increase the level of customization of its product offer, while lowering its share of the innovation risk.

Finally, Vanhaverbeke and Cloodt (2006) explained how Calgene, a plant biotechnology R&D company, established a network of interrelationships with seed companies, farmers, packers, consumers and legislators to support the launch of a new, genetically modified tomato for the fresh market. Calgene was forced to co-operate with other companies and organizations in the innovation system in view of the uncertainties inherent to the development and commercialization of foods derived from gene technology. Such uncertainties compromised its ability to reap value from the commercial applications of the novel technologies it pioneered. The resulting value network allowed Calgene to cope better with the high levels of product innovativeness introduced by its gene-modification technology and the consequent low initial levels of public acceptance and consumer adoption (Ram, 1989).

Table 9.1 summarizes the main characteristics of the open-innovation strategies reported to have been employed in the food industry so far, based on previously developed innovation categorization schemes (Garcia and Calantone, 2002; OECD and Eurostat, 2005; Sarkar and Costa, 2008). Although IFF and Procter & Gamble's open-innovation activities are clearly cases of technological process innovations, introduced to increase the efficiency of product innovation and sustain new marketing strategies for existing products, the case of Calgene is substantially different. In the latter, a new marketing strategy is implemented,

Table 9.1 Main characteristics of open innovation strategies in the food industry

	Open innovation in the food industry		
Case study	New Pringles' potato (Huston and Sakkab, 2006)	Design of new food flavours (Thomke and von Hippel, 2002)	Launch of GM tomato (Vanhaverbeke and Cloodt, 2006)
Innovating company	Procter & Gamble	International Flavors and Fragrances	Calgene
External partner	Technology supplier	Customers in the food industry (e.g. Nestlé)	Seed producers, farmers, packers, retailers, consumers, legislators
Type of relationship	Dyad at non-arm's length	Vertically integrated dyad	Network across innovation system
Stage	Process development	Product design	Commercialization
Strategy	Technology in-sourcing	NPD out-sourcing	Creation of value network to sustain market launch
Goal	Reduce NPD costs and time-to-market	Reduce NPD costs and time-to-market High customization	Ensure acceptance and market success of novel technology
Supporting technology	New printing technology for food	New tool kit for flavour design	Plant biotechnology
Newness to company	Really new	Really new	Incremental
End product	Printed potato crisps for consumer markets	Custom flavours for the food industry	New tomato for the fresh market
Newness to market	Incremental	Incremental	Radical

involving the development of new sales channels and promotion tactics, to sustain the successful creation of an entirely new market for a radical product innovation. Although the commercialization of a tomato with enhanced flavour might have been a novel initiative to Calgene and its partners, the basic knowledge of plant biotechnology employed in the process was, nonetheless, not new to them.

As already seen in Section 9.2, empirical evidence of food companies engaging in open-innovation practices is scarce, and, in particular, in respect of specific cases of CPI initiatives. Most importantly, a detailed analysis of such strategies, their rationale and market outcome is, with the exception of the few case studies reviewed here, virtually absent from both academic and practice-oriented studies. This gap in academic research concerning the study of open innovation in general, and CPI in particular, is by no means exclusive to the food area, a fact that has been often pointed out by many other innovation-management scholars (Chesbrough *et al.*, 2006; Emden et al, 2006; Littler *et al.*, 1995). In view of this, the following section presents the results of a case study research project (Yin, 2008), undertaken with the aim of better understanding why and how CPI takes place in mature sectors, namely in the food manufacturing and foodservice areas.

9.3 Collaborative product innovation (CPI) in the food-service industry: the path of diffusion of sous vide cooking in the US

Some background knowledge about the sous vide cooking technology and its applications, as well as the details and relevant findings of the case study research undertaken, are now presented.

9.3.1 Background

The foodservice industry encompasses all commercial and non-commercial/institutional organizations supplying consumers with meals prepared outside their homes (Ottenbacher and Harrington, 2008). In the USA, as in most developed economies, this is a very broad sector with substantial variation in price, quality and service level of the offers, which ranges from educational, health, military and corporate catering to travel and tourism businesses, foodservice areas in retail stores, takeaway and home-delivery outlets, fast food and other limited service formats, casual dining establishments and upscale restaurants. According to the National Restaurant Association (NRA, 2011), sales in the USA restaurant industry, i.e., non-commercial, limited (cafeterias, buffets and quick service operations) and full (midscale, casual dining, upscale casual and fine dining) service restaurants, reached US$604 billion in 2011. In the same year, the industry employed 12.8 million people in 960 000 venues, being one of the largest private sector employers in the country, and accounted for a 49% share of the food dollar. Overall, foodservice activities have a great economic and social relevance in this country: their overall impact in the economy totals US$1.7 trillion and they employ almost 10% of its workforce, a large proportion of which are female and minority entrepreneurs and managers who own and/or run micro and small enterprises.

Sous vide cooking (or cooking under vacuum) is based on advanced food-packaging and processing technologies (Haas, 2006). It consists of a series of relatively complex and sophisticated food preparation steps carried out sequentially in purposely developed equipment, as shown in Fig. 9.1. This technique involves packaging raw, minimally processed or precooked foods under vacuum in sealed, laminated plastic pouches or containers, and cooking them during a precisely

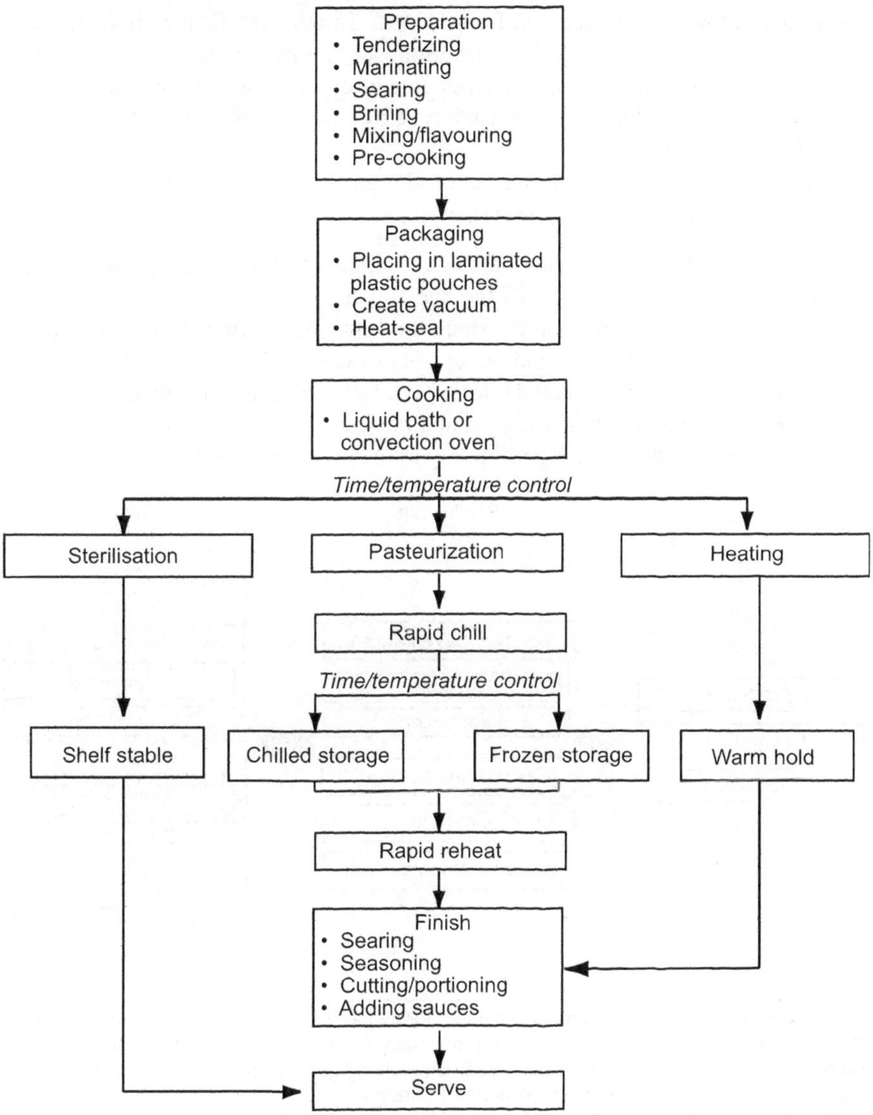

Fig. 9.1 The sous vide cooking process..

controlled heat treatment process in a water bath or a convection steam oven (Baldwin, 2012; Schellenkens, 1996). The essence of sous vide cooking resides in establishing, achieving and controlling the desirable core temperature of foods, in order to achieve their optimal palatability at acceptable safety levels (Baldwin, 2012).

Sous vide technology is used in the production of meals and meal components for a wide variety of customers by food manufacturers and foodservice operators alike (Tiampo, 2006), as depicted in Fig. 9.2. Different levels of heat treatment are used to cook foods under vacuum, depending on the safety standards required by regulatory authorities (Creed and Reeve, 1998; Lingle, 1991), the characteristics of the production, storage and distribution operations involved (Lingle, 1991; Creed, 2001a; Tiampo, 2006), and the degree of organoleptic quality and convenience in preparation demanded by each type of customer (Lingle, 1991; Costa *et al.*, 2001; Creed, 2001b).

Sous vide cooking methods are believed to present four main advantages over traditional interrupted catering systems:

1. reduce the degree of oxidation of food components, such as vitamins and other antioxidants, by packing, cooking and storing under vacuum;
2. prevent recontamination and reduce the loss of important food components (water, vitamins, flavour and odour volatiles normally lost in open container cooking), by using laminated plastic packaging resistant to high temperatures;
3. reduce the breakdown of vitamins and flavour and odour volatiles by cooking at low temperatures during relatively long periods of time;

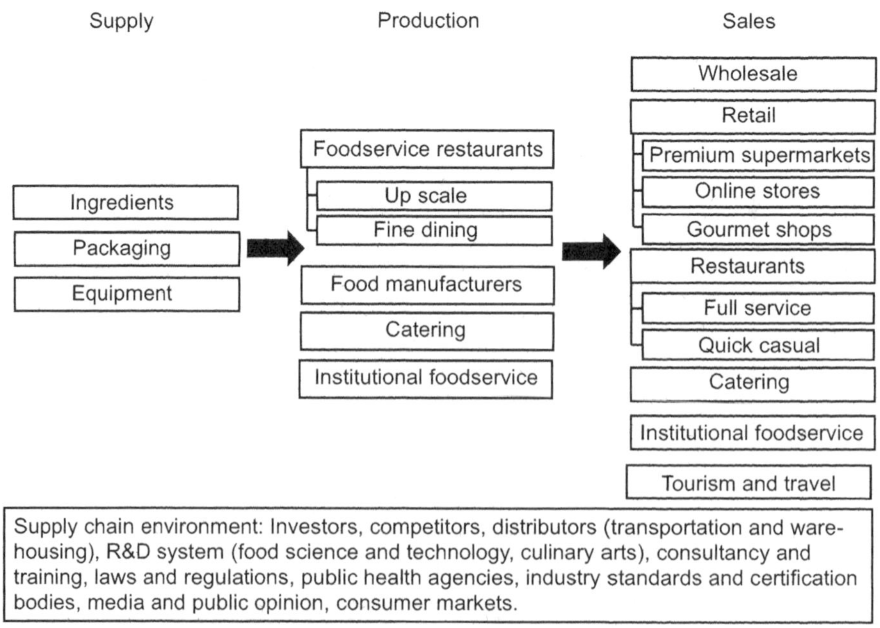

Fig. 9.2 The supply chain of sous vide foods

4. reduce the need for using salt, chemical and/or other artificial compounds to preserve foods for a longer time (Creed, 2001a).

The initial basis for the development of sous vide technology in the late 1960s was the 'industrialization' of foodservice operations through the adoption of food manufacturing processes such as centralized production, large-scale equipment, consistent safety and quality, and sophisticated packaging systems. The viability and success of these processes depended largely on the incorporation of a 'time buffer', a stage during which food could be safely and conveniently stabilized by storage at low temperatures, which interrupted the necessarily continuous flow of food through the traditional 'cook and serve' catering system. However, the efficacy of such a time buffer in terms of preserving the necessary levels of food safety was most often achieved at the expense of the sensory and nutritional quality of the reheated meal or meal component. That is, given the standing food safety requirements, the desired operational benefits came at the cost of poor perceived quality and low consumer acceptance. Consequently, the initial stages of development, commercialization and adoption of sous vide technology in the food area were largely driven by the promise it held of providing, for the first time, a highly positive balance between safety requirements, operational benefits and end-product quality.

A second reason for the development of sous vide technology was a growing consumer demand for convenience in meal preparation. Most of the active population increasingly felt that, at the end of the day, there was not much time left to eat, let alone to shop and cook. This translated itself into high growth rates of ready meals and other convenience and foodservice markets in the EU and USA all through the eighties and nineties, up until today (Costa *et al.*, 2001; Datamonitor, 2006). Given the operational benefits and the superior product quality provided by the technological sophistication of sous vide cooking, manufacturers and caterers worldwide turned to sous vide meals as the preferred means of satisfying the growing consumer demand for convenience (Otto, 1989). However, initial market acceptance of sous vide meals and the underlying technology was surprisingly low, with sales never really taking off and companies closing down their sous vide cooking operations only a few years, or even months, after start-up (Carlino, 1991). Both public and consumer organizations, especially in the USA, voiced concerns about the potential public health hazards involved in storing foods under anaerobic conditions, as well as doubts regarding the level of safety of sous vide meals (Martin, 1999). Moreover, foodservice customers were rightfully afraid of the markets' negative perception of the nutritional and sensory quality of sous vide meals. It was felt that, without a sufficiently intensive, informative and persuasive marketing strategy, consumers were simply not ready to come to terms with the notion of eating their dinner out of a vacuumized plastic pouch (Allen, 1991).

Last but not least, the development and widespread adoption of sous vide technology was significantly fuelled by the emergence of the highly popular, new millennium 'science of deliciousness', otherwise known as 'New Cookery' (Adriá *et al.*, 2006), 'Molecular Gastronomy' (This, 2005) or 'Hypermodern Cuisine'

(Pontin, 2005). Hypermodern Cuisine is aggressively technological, as it borrows the latest developments in the food science area (as well as its industrial applications), and turns them into 'haute-cuisine'. Sous vide cooking is considered to be its most remarkable, spectacular and well-known innovation, one that nowadays takes centre-stage in the kitchens of the best and most reputed restaurants in the world. World-renowned chefs such as Hestor Blumenthal, Ferran Adriá or Thomas Keller have picked a boring and fear-inspiring industrial cooking method and taken it to a new level by adding their creativity and art, a never-ending variety of applications and recipes, and much needed flair. Thanks to this, eating dinner out of a plastic pouch is no longer sad, lonely and suspicious, but a sign of sophistication and superior culinary craftsmanship. Most importantly, sous vide cooking currently epitomizes the 'slow food' trend in high-end cookery and is gradually spilling-over the technology and equipment to ordinary kitchens, as well as boosting sales of refrigerated, prepared meals in all distribution channels (Hesser, 2005; Newman, 2003).

9.3.2 Methodology

Open innovation in the service area remains a poorly understood topic (Tether and Tajar, 2008), particularly in mature sectors such as the hospitality (Den Hertog *et al.*, 2011) and the foodservice industries (Ottenbacher and Harrington, 2008). In view of this, a primarily exploratory research approach was undertaken. This type of approach is particularly suitable when the researcher intends to establish when, where, how and by whom certain decisions and actions are undertaken, as well as to provide a clearer understanding of the antecedents and consequences of such events (Creswell, 2008). A retrospective, longitudinal, multiple case study design (Yin, 2008) was hence employed to investigate the impact of open product innovation strategies on mature companies' innovation capabilities and market outcomes. Case study research analyses past or contemporary phenomena within its real-life context in a holistic but systematic manner. Critical or key cases, in particular, are often the subject of scrutiny because of their inherent interest for the understanding of the phenomenon under analysis and their level of embeddedness in the setting where the study takes place.

The study's subject of inquiry was the path of diffusion of sous vide technology in the USA between 1971 and 2008, while the object was the engagement of three key food manufacturing companies in projects of collaborative innovation involving renowned restaurant chefs, with the intent of creating and exploiting an emerging consumer market. Information was collected at four different levels of analysis (individual entrepreneurs, companies, industry and R&D system), to capture both micro and macro perspectives of the phenomenon under study and identify generative mechanisms at their precise source. Table 9.2 provides an overview of the main actors involved in the case and the relevant elements in their environment, particularly those related to the R&D system. It also depicts the corresponding periods of analysis considered, the type of data collected and the sources of information used.

Data analysis sought to:

1. identify the relevant industry actors and their motives to partake in collaborative innovation;
2. outline their inter-relationships;
3. establish the sequence of key, intertwined decisions, actions and events which triggered the diffusion of a new technology in such context;
4. analyse the corresponding outcomes.

The information collected was analysed with the aim of identifying the concepts, relationships and events relevant to the case, and subsequently structuring and organizing them into into datasets. These datasets described several aspects of the history of the companies investigated, such as market environment, birth, founders, changes in name, key personnel, product function (e.g. the shift from supplying production equipment to manufacturing foods), production technology, assortment, distribution channels and/or markets served, as well as evolution in performance, organizational structure, managerial processes and inter-firm relationships related to R&D and innovation (for instance, outsourcing, informal partnerships, contractual partnerships, equity-based joint ventures, and mergers and acquisitions) (Hagedoorn, 2002), and market exit if applicable. The quality of the analysis conducted was assured by:

1. triangulating findings between multiple sources of primary and secondary data, theoretical lenses and sense-making strategies;
2. having key informants checking the accuracy of the narrative and timeline produced;
3. having academic and industry experts review the case report to assess its face validity and the scientific and practical relevance of its contributions (Yin, 2008).

9.3.3 Case study findings

The first unit of analysis in the case study was W. R. Grace, a US company active in the area of food-grade materials and vaccum-packaging. In 1968, W. R. Grace started to allocate part of its R&D resources to the development of a method of sealing and pasteurizing ready-to-eat foods on an industrial scale for long storage times (Creed, 2001a). These R&D activities started to bear fruit in 1971, when the company filed a patent for sous vide cooking (as well as for the associated laminated plastic pouch), as a new and improved method of preparing and preserving ready-to-eat foods (Ready, 1971).

By the middle of the 1970s, W. R. Grace started to explore the possibility of diversifying its activities to include the large-scale production of sous vide meals and meal components for the catering industry. To this end, they hired two French consultants with very different backgrounds: Bruno Gossault, a scientist who worked in a food-safety laboratory and who had studied the vacuum-packaging of meat, and George Pralus, a restaurant chef who had developed a method for slow-

Table 9.2 Case information: actors, R&D system, broader environment, period of analysis, type of data collected and sources

	Period of analysis	Type of data	Sources
Entrepreneurs			
George Pralus: French chef, consultant, head of culinary schools, former consultant at W. R. Grace, developer of sous vide cooking	1974–1981	Secondary qualitative	General: • Electronic archives of companies, educational/training organizations, industry/trade magazines and press • Media appearances and personal accounts (audio/video interviews, websites, blogs, social networks) • Electronic archives of peer-reviewed, scientific publications and research reports • Authored books
Bruno Gossault: French food scientist and economist, consultant, head of a consultancy firm, Chief Scientist at Cuisine Solutions Inc., former consultant at W. R. Grace, developer of sous vide cooking	1974–2009		
Thomas Keller: world-renowned American chef and restaurateur, consultant, cookbook writer, theorizer of the New Cookery, lead-user of sous vide cooking, partner of Cuisine Solutions's Five Leaf Program	1994–2009		
Companies			
W. R. Grace (Columbia, MD): speciality chemicals and materials manufacturer, owner of Cryovac division (manufacturer of food packaging materials/equipment) and Grace Culinary Systems (subsidiary manufacturer/distributor of sous vide products)	1968–1996	Primary qualitative Secondary qualitative	• Telephone and e-mail interviews with key company and industry informants and former collaborators • General • Bloomberg Professional Service • Reports for the US Securities and Exchange Commission • Reports for the US Securities and Exchange Commission • Industry reports of market research and consultancy firms
Culinary Brands (Sausalito, CA): producer/distributor of sous vide foods	1986–1991		
Cuisine Solutions (Alexandria, VA): producer/distributor of sous vide foods	1987–2009		

Industry Foodservice in the US	1990–2011	Secondary quantitative	• General • Website of the National Restaurant Association • Industry reports of market research and consultancy firms
R&D system Food science and technology research Culinary arts	1990–2011	Primary qualitative	• Visits to R&D, consultants and manufacturers active in the European foodservice supply chain • Interviews with key company and industry informants
Broader Environment Investors, media and public opinion Customers, competitors and consumer markets Regulatory, standards and certification bodies, external evaluators Education, training and consultancy			• Participation in international symposia on sous vide and the home meal replacement (HMR) sector • Participation in applied research projects with European universities, research centres, R&D and manufacturing companies on product innovation for the HMR sector

cooking foie gras with a reduced weight loss. The latter was hired to improve the sous vide cooking method from the culinary viewpoint (operational costs, sensory quality, presentation and the development of new recipes), whereas the former was hired to optimize the time–temperature cooking combinations, in order to comply with the necessary food safety requirements. The two consultants worked closely together with W. R. Grace's own specialists, and with each other for about ten years, in the optimization of the sous vide cooking method. Their main goal was to achieve the right combination of plastic pouch composition, level of vacuum in the pouch, cooking and storage temperatures, and cooking and storage times, which allowed for maximum sensory and nutritional quality within the food safety limits imposed by legislation (Creed, 2001a; Hesser, 2005). By 1989, W. R. Grace had opened its own sous vide food production plant and a subsidiary, Grace Culinary Systems, that commercialized sous vide meals for the US market (Lingle, 1991). Both were sold in 1996 and changed activity.

The second unit of analysis in the case study was Culinary Brands, a US company launched in 1986 to produce and distribute chilled sous vide meals and meal components in California. Two former executives at a major foodservice company had come across the technique in France and managed to convince several venture capital companies in the USA to back a business based on the industrial production and distribution of sous vide foods. To that end, they recruited several French chefs familiar with the sous vide cooking method as consultants, to help develop the company's product line. At the time, Culinary Brands invested about US$10 million in research related to product innovation, resulting in the production of 65 different chilled food items and 100 000 dinners and entrees per day for the US foodservice market (Otto, 1989).

However, market diffusion of sous vide foods in the USA remained low throughout the nineties, mainly because of public concerns with food safety and poor consumer acceptance of the underlying technology and its benefits (Allen, 1991; Martin, 1999). This led Culinary Brands to switch from chilled to frozen storage and distribution of its products, to avoid food safety risks and increase customer confidence. But this switch came at a heavy cost of the end-product's overall sensory quality and convenience features, and the company would eventually cease its activities in 1992. Given the R&D effort made, such lack of success was attributed mainly to Culinary Brands' insufficient market-research efforts and a poor marketing and distribution strategy (Carlino, 1991).

The third unit of analysis was Cuisine Solutions, a US company founded in 1987 to produce and market frozen sous vide foods in Europe and US. This company currently operates plants in the USA, South America and Europe and markets its meals and meal components through channels such as airlines, passenger trains, cruise lines, hotels, retail, military, restaurant chains, and on-line consumer markets. Figure 9.3 shows the R&D effort and the net sales volume of Cuisine Solutions between 2004 and 2009 (Cuisine Solutions, 2009). The company has been listed on the American Stock Exchange since 2005.

Throughout its history, Cuisine Solutions has based its innovation strategy on a consciously crafted combination of in-house R&D and open-innovation initiatives.

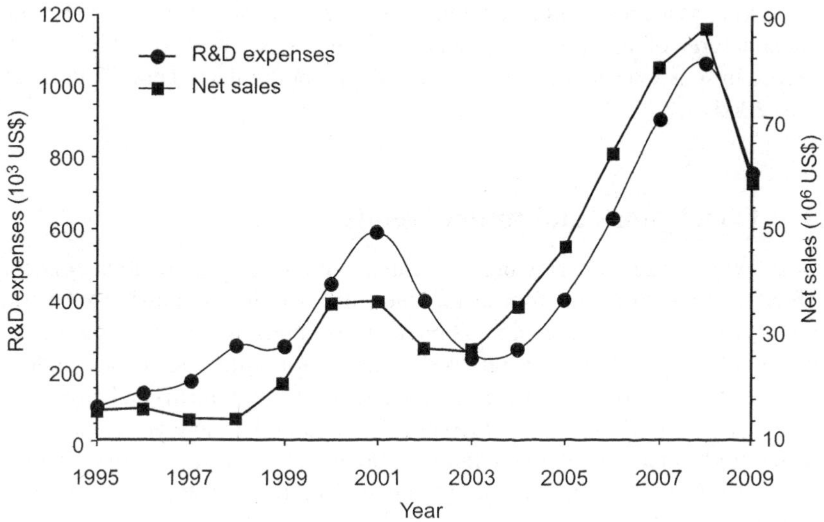

Fig. 9.3 R&D effort and net sales of Cuisine Solutions between 1995 and 2009 (Cuisine Solutions Inc., 1995–2009).

In 1983, while still operating in the bakery area, it started a research collaboration with a related French company to capitalize on their knowledge of frozen bread products and the quick-freezing process. In this way, the company came across the sous vide technology and realized the potential of its application in European and US foodservice markets, leading to sous vide becoming its sole area of business activity in 1987. Since 2002, and partly in reaction to the negative impact brought by the 9/11 events upon the world's travel and tourism industry, Cuisine Solutions regularly establishes partnerships with high-end retailers for the development of sous vide products for the French and US consumer markets. This gives the company the opportunity to test their products directly with the end-customer and helps increase consumer and public acceptance of the underlying sous vide technology. In 2004, Cuisine Solutions also established a strategic research alliance (Dittrich and Duysters, 2007; Yoshino and Rangan, 1995) with one of its suppliers in South America (an equity agreement where Cuisine Solutions is a minority stockholder), with the aim of developing and producing high quality, valued priced, fish-based sous vide products for the global retail and foodservice markets.

Concerning specific CPI projects, two of Cuisine Solutions' most recent initiatives stand out as prime examples of partnering between the foodservice industry and the end-user community of restaurant chefs. The first was the Five Leaf Program, which consisted in the development of an on-line sales channel to supply high-end sous vide meals at home to the gourmet customer segment (Thrush, 2007; Wine, 2003). The online order menu is composed by sous vide signature dishes created exclusively for Cuisine Solutions by world-renowned restaurant chefs such as Thomas Keller. The second was the development of their trade-marketed, salmon product developed by a French restaurant chef while he was working as a consultant at the company. The chef used custom tools made by a manufacturer of medical equipment in Germany, as well as a high level of expertise in sous vide cooking, to create a 'shank' out of the tail end of the salmon (Poris, 2005).

9.4 Conclusions and future trends

This chapter started by reviewing applications of the open-innovation concept in the food manufacturing and foodservice industry. The first conclusion to be drawn from this review is that open innovation does take place within the food sector, in spite of it being known as a relatively more traditional and mature industry. Moreover, open-innovation strategies come in a variety of forms and, as such, are also met with a wide variety of outcomes. Consequently, there is a clear need for a better understanding of open innovation in the food sector that should be addressed by the performance of further, and more focused, case studies and empirical research.

Next, the impact of the open-innovation strategies employed so far on the innovation capabilities and market outcomes of food manufacturing and foodservice industries was analysed. Namely, case study research was employed with the aim of understanding why and how CPI takes place in such areas. To this end, the innovation path of sous vide technology in the USA was traced back to the late 1960s and the R&D and commercialization endeavours of several corporations, restaurant chefs, and food scientists located on both sides of the Atlantic Ocean (France and the USA). This highlighted that several actors have collaborated closely on cross-boundary process- and product-innovation projects in the sous vide industry over the last 35 years, thus demonstrating the importance and usefulness of open innovation in mature business areas, such as the food manufacturing and foodservice sectors.

The findings obtained show that open-innovation initiatives frequently take place in food manufacturing and foodservice companies, encompassing both the exploration and exploitation stages of product innovation. In line with earlier studies (Sarkar and Costa, 2008; Knudsen 2007), they also indicate that despite different collaborative strategies being employed, expert user communities, in this case, accomplished, haute cuisine chefs in European and US restaurants (Adriá *et al.*, 2006; Fauchart and von Hippel, 2008), constitute preferred innovation partners. Collaboration with highly innovative chefs created value to manufacturers by aiding technical development, accelerating technology adoption and diffusion among customers (through the development of novel recipes and applications), and increasing technology acceptance and perceived product quality. In return, chefs benefited from efficiency gains, certified cooking methods, improved professional reputation and higher consumer patronage.

Overall, it can be concluded that companies stand a better chance of escaping the law of diminishing returns to innovation efforts if they can improve the effectiveness of both their technological and marketing capabilities in a concerted manner. This supports the long-standing call for a higher level of integration between R&D and marketing activities within agri-business companies (Costa and Jongen, 2006). Likewise, when effects on technological capabilities and market outcomes are analysed simultaneously within an integrated framework, research into (open) innovation stands a better chance to become more meaningful to academics and practitioners alike.

9.5 References

ADRIÁ, F., BLUMENTHAL, H., KELLER, T. AND MCGEE, H. (2006). Statement on the 'new cookery. *The Observer UK*, 10–12–2006.

ALLEN, R.L. (1991). Sous-vide's popularity still lukewarm in U.S. – vacuum packed cooked foods. *Nation's Restaurant News*, (December), 49–51.

BALDWIN, D.E. (2012). Sous-vide cooking: a review. *International Journal of Gastronomy and Food Science*, **1**, 15–30.

CARLINO, B. (1991). Lack of cash flow closes Culinary Brands. *Nation's Restaurant News*, (June), p. 25.

CHESBROUGH, H.W. (2003). The era of open innovation. *MIT Sloan Management Review*, **44**, 35–41.

CHESBROUGH, H.W. AND CROWTHER, A.K. (2006). Beyond high tech: early adopters of open innovation in other industries. *R&D Management*, **36**, 229–236.

CHESBROUGH, H.W., VANHAVERBEKE, W. AND WEST, J. (eds.) (2006). *Open innovation: researching a new paradigm*, Oxford: Oxford University Press.

COSTA, A.I.A. AND JONGEN, W.M.F. (2006). New insights into consumer-led food product development. *Trends in Food Science and Technology*, **17**, 457–465.

COSTA, A.I.A, DEKKER, M., BEUMER, R.R., ROMBOUTS, F.M. AND JONGEN, W.M.F. (2001). A consumer-oriented classification system for home meal replacements. *Food Quality and Preference*, **12**, 229–242.

COSTA, A.I.A., SCHOOLMEESTER, D., DEKKER, M. AND JONGEN, W.M.F. (2007). To cook or not to cook: a means-end study of the motivations behind meal choice. *Food Quality and Preference*, **18**, 77–88.

CREED, P.G. (2001a). Chilling and freezing of prepared consumer foods. In: Sun, D.-W. (ed.) *Advances in Food Refrigeration*, pp. 422–455. Leatherhead: Woodhead Publishing.

CREED, P.G. (2001b). The potential of foodservice systems for satisfying consumer needs. *Innovative Food Science & Emerging Technologies*, **2**, 219–227.

CREED, P.G. AND REEVE, W. (1998). Principles and applications of *sous vide* processed foods. In: Ghazala, S. *Sous vide and cook-chill processing for the food industry*, Great Britain, Aspen, 25–56.

CRESWELL, J.W. (2008). *Research design: qualitative, quantitative and mixed methods approaches – 3rd ed.* Thousand Oaks, Sage.

CUISINE SOLUTIONS INC. (1995–2009). USSEC Annual Reports and Financial Statements, available at http://quicktake.morningstar.com/stocknet/secdocuments.aspx?symbol=cusi > (Accessed 14 March 2012).

DATAMONITOR (2006). *Savoury snacks: global industry guide*. Research Report ref. 46918, 225 pp. UK: Datamonitor.

DE JONG, J., VAN KLEEF, E., FREWER, L.J. AND RENN, O. (2006). Perceptions of risk, benefit and trust associated with consumer food choice. In: Frewer, L.J. and van Trijp, H.C.M. (eds.) *Understanding consumers of food products*, pp. 125–150. Cambridge (England): Woodhead Publishing.

DEN HERTOG, P., GALLOUJB, F. AND SEGERS, J. (2011). Measuring innovation in a 'low-tech' service industry: the case of the Dutch hospitality industry. *The Service Industries Journal*, **31**, 1429–1449.

DITTRICH, K. AND DUYSTERS, G. (2007). Networking as a means to strategy change: the case of open innovation in mobile telephony. *Journal of Product Innovation Management*, **24**, 510–521.

EMDEN, Z., CALANTONE, R.J. AND DROGE, C. (2006). Collaborating for new product development: selecting the partners with maximum potential to create value. *Journal of Product Innovation Management*, **23**, 330–341.

FAUCHART, E. AND VON HIPPEL, E. (2008). Norms-based intellectual property systems: the case of French chefs. *Organization Science*, **19**, 187–201.

GARCIA, R. AND CALANTONE, R.J. (2002). A critical look at technological innovation typology and innovativeness terminology: a literature review. *Journal of Product Innovation Management*, **19**, 110–132.

GATIGNON, E.H., TUSHMAN, M.L., SITH, W. AND ANDERSON, P. (2002). A structural approach to assessing innovation. *Management Science*, **48**, 1103–1122.

GRUNERT, K.G., JEPPESEN, L.F., JESPERSEN, K.R., SONNE, A-M., HANSEN, K., TRONDSEN, T. AND YOUNG, J.A. (2005). Market orientation of value chains. *European Journal of Marketing*, **39**, 428–455.

HAAS, S. (2006). Better dining through chemistry. *Gastronomica: The Journal of Food and Culture*, **6**, 74–77.

HAGEDOORN, J. (2002). Inter-firm R&D partnerships: an overview of major trends and patterns since 1960. *Research Policy*, **31**, 477–492.

HESSER, A. (2005). Under Pressure. *New York Times*, 14–08–2005.

HUSTON, L. AND SAKKAB, N. (2006). Inside Procter & Gamble's new model for innovation. *Harvard Business Review*, **84**, 58–66.

JURIAANSE, A.C. (2006). Challenges ahead for food science. *International Journal of Dairy Technology*, **59**, 55–57.

KNUDSEN, M.P. (2007). The relative importance of interfirm relationships and knowledge transfer for new product development success. *Journal of Product Innovation Management*, **24**, 117–138.

LINGLE, R. (1991). New generation foods? Just say Grace – Company profile: Grace Culinary Systems – cover story. *Prepared Foods*, **160**, 58–64.

LITTLER, D., LEVERICK, F. AND BRUCE, M. (1995). Factors affecting the process of collaborative product development: a study of UK manufacturers of information and communications technology products. *Journal of Product Innovation Management*, **12**, 16–32.

MAULA, M., KEIL, T. AND SALMENKAITA, J.-P. (2006). Open innovation in systemic innovation contexts. In: Chesbrough, H.W., Vanhaverbeke, W. and West, J.(eds.) *Open innovation: researching a new paradigm*, pp. 241–257.

MARTIN, R. (1999). Experts warn of sous-vide dangers. *Nation's Restaurant News* (January), p. 24.

MIKKELSEN, B.E., KRISTENSEN, N.H. AND NIELSEN, T. (2005). Innovation processes in large scale public foodservice. *Journal of Foodservice Business Research*, **8**, 87–105.

NEWMAN, C.M. (2003). Sealed, signed and delivered. *Art Culinaire* (Winter), 71–74.

NRA (2011). *2011 Restaurant Industry Forecast*, United States of America, National Restaurant Association.

OECD AND EUROSTAT (2005). *Olso Manual: Gguidelines for collecting and interpreting innovation data – 3rd edition*. Paris: OECD.

OTTENBACHER, M.C. AND HARRINGTON R.J. (2008). Institutional, cultural and contextual factors: potential drivers of the culinary innovation process. *Tourism and Hospitality Research*, **9**, 235–249.

OTTO, A. (1989). Sous-vide or not sous-vide? French vacuum cooking technology – company profile. *Prepared Foods* (June), 58–61.

PONTIN, J. (2005). Technology and hypermodern cuisine. *MIT Technology Review* (October), pp.1–3.

PORIS, J. (2005). Techniques. *Food and Art Magazine* (March), 94–95.

RAM, S. (1989). Successful innovation using strategies to reduce consumer resistence: an empirical test. *Journal of Product Innovation*, **6**, 20–34.

READY, C.A. (1971). Method of preparing and preserving ready-to-eat foods. *U.S. Patent* 3,607,312, 4 pp.

RINDFLEISCH, A. AND MOORMAN, C. (2001). The acquisition and utilization of information in new product alliances: a strength-of-ties perspective. *Journal of Marketing*, **65**, 1–18.

ROMER, P.M. (1990). Endogenous technological change. *Journal of Political Economy*, **98**, S71–S102.

SARKAR, S. AND COSTA, A.I.A. (2008). Dynamics of open innovation in the food sector. *Trends in Food Science and Technology*, **19**(11), 574–580.

SCHELLEKENS, W. (1996). New research issues in sous-vide cooking. *Trends in Food Science & Technology*, **7**, 256–262.

SIMARD, C. AND WEST, J. (2006). Knowledge networks and the geographic locus of innovation. In: Chesbrough, H.W., Vanhaverbeke, W. and West, J. (eds.) *Open innovation: researching a new paradigm*, pp. 220–240.

TETHER, B. AND TAJAR, A. (2008). The organisational-cooperation mode of innovation and its prominence amongst European service firms. *Research Policy*, **37**, 720–739.

THIS, H. (2005). *Molecular gastronomy: exploring the science of flavor*. New York, NY: Columbia University Press.

THOMKE, S. AND VON HIPPEL, E. (2002). Customers as innovators: a new way to create value. *Harvard Business Review*, **80**, 74–81.

THRUSH, E. (2007), 'Délices & Saveurs: The Big Chill', *France Magazine*, **83**. Available at <http://www.francemagazine.org/articles/issue83/article193.asp?issue_id=83&article_id=193> (Accessed 24 March 2012).

TIAMPO, J. (2006) *Seal appeal: the nutrition, food safety, and operational benefits of sous vide technology for North American restaurants*. Available at <http://www.techne-calibration.com/adminimages/Sous_Vide_Information%281%29.pdf> (Accessed 14 March 2012).

VANHAVERBEKE, W. AND CLOODT, M. (2006). Open innovation in value networks. In: Chesbrough, H.W., Vanhaverbeke, W. and West, J. (eds.) *Open innovation: researching a new paradigm*, pp. 258–281. Oxford: Oxford University Press.

WINE, T. (2003). For gourmet at-home dining, the Internet delivers. *USA Today*, 03–11–2003.

YIN, R.K. (2008). *Case study research: design and methods (4ed.)*. CA: Sage.

YOSHINO, M.Y. AND RANGAN, U.S. (1995) *Strategic alliances – an entrepreneurial approach to globalization*, USA: Harvard Business School Press.

10

Effectiveness of cluster organizations in facilitating open innovation in regional innovation systems: the case of Food Valley in the Netherlands

S. W. F. Omta, Wageningen University and Research Centre, The Netherlands and F. T. J. M. Fortuin, Food Valley Organization, The Netherlands and Wageningen University and Research Centre, The Netherlands

DOI: 10.1533/9780857097248.2.174

Abstract: The perceived effectiveness of cluster organizations in fostering open innovation in regional innovation systems is examined by focusing on Food Valley Organization in the Netherlands. Forty member companies assessed Food Valley's support in demand articulation, network formation and innovation-process support. The Food Valley case clearly shows that, in accordance with previous results of open-innovation studies, the networking function is most important. Seen in the light of the important role of small to medium-sized enterprises (SMEs) in regional innovation systems, it is further concluded that the current cluster organization framework should be extended by adding internationalization support for SMEs.

Key words: regional innovation systems, cluster organizations, Food Valley Organization, demand articulation, network formation, internationalization, innovation process support.

10.1 Introduction

This chapter aims to shed light on the perceived effectiveness of a cluster organization (CO) in facilitating open innovation in a regional innovation system (RIS). The case study focuses on Food Valley Organization (in short Food Valley,

FV), an important CO in the RIS of the Dutch agri-food industry, located close to Wageningen University and Research Centre. It was evaluated by 40 of its member companies (23 small to medium-sized enterprises (SMEs) and 17 large companies), which were asked to assess the value of its open-innovation supporting functions, including services, activities and information sources, using Likert 7-point scales. The FV case clearly shows, that, in accordance with the results of previous open-innovation studies, the networking function of a CO, helping finding innovation partners, bringing them together, and playing a facilitating role, is clearly a central role for a CO in a RIS. Seen in the light of the important role of SMEs in RISs, it is further concluded that the current CO framework should be extended by adding the internationalization support for SMEs.

10.2 Theoretical background

According to Chesbrough (2003), innovative companies increasingly realize that the 'closed' model of innovation, in which the internal R&D department exclusively provides for new products and processes to foster the company's growth, no longer works in the current highly dynamic business environment. He, therefore, advocates an open form of innovation in which companies work together in innovation systems. Innovation can no longer be regarded as a stand-alone activity of a single company; it is, to a large extent, context (innovation system) dependent. RIS can be defined as all regional actors contributing to the emergence of innovations (e.g. Bergek *et al.,* 2008, Hekkert *et al.*, 2007). The innovation performance of a RIS merely depends on the quality of its actors and how they connect to each other. Gaps in connectivity and collaboration may reduce the (open) innovation performance of a RIS. Therefore, a connecting role is defined for specialized COs within RISs, also called intermediary organizations, innovation intermediaries or innovation brokers (Klerkx and Leeuwis, 2008a). Despite the generally assumed importance of COs, earlier studies have not (e.g. Howells, 2006; Winch and Courtney, 2007), or only to a limited extent (e.g. Batterink *et al*, 2010; Klerkx and Leeuwis, 2008a), analyzed the effectiveness of CO functions (services) to enhance open innovation in a RIS.

Innovation is often approached from an innovation system (IS) perspective, that argues that innovations should not be seen as stand-alone activities but as an evolutionary, complex, non-linear and interactive process, in which a large number of co-evolutions in the scientific, technological, and social system occur (Tödtling and Trippl 2005). The consequence of this approach is that firms are not considered to innovate in isolation, they are part of the IS, which consists of e.g. entrepreneurs, researchers, consultants, policy makers, suppliers, processers, retailers, and customers. These IS actors form networks, to engage in a process of joint learning and negotiation to shape an innovation (Malerba, 2004).

The IS approach has first been applied on the national level. The concept has been used since to develop, analyze and benchmark national innovation policies. The term 'national innovation system' is not only derived from technology and innovation policy but also includes a shared culture or language and the focus of government policy, legislation and funding, the infrastructure, and general market developments that condition the IS environment. In the last two decades increasingly attention has been paid by scientists and policy makers to RIS as sites of innovation and competiveness in the globalized economy (Carlsson, 2006). Most studies draw on the common rationale that territorial agglomeration provides the best context for an innovation-based globalized economy (Asheim and Coenen, 2005). Emphasizing the role of interaction, localization and embedding, the RIS provides a conceptual explanation of the resurgence of regional economies as structuring elements in global competition, as exemplified by alleged regional success stories such as Silicon Valley (Asheim and Coenen, 2005, De Bruijn and Lagendijk, 2005).

Studies employing the IS perspective increasingly pay attention to several types of COs, also referred to as innovation brokers, intermediating organizations, third parties, and bridge and superstructure organizations (Howells, 2006). The reasons why COs emerge are diverse, but generally they are created in response to a perceived suboptimal degree of connectivity between the network actors owing to market or innovation system failures. In addition, they contribute to reducing uncertainty in the early stages of innovation processes when there is a high risk of failure, which would preclude private parties from innovating (Klerkx and Leeuwis, 2009; Van Lente *et al.*, 2003). They aim to overcome the information, managerial, cultural and cognitive gaps that exist in the innovation processes of individual firms within RISs. A number of CO studies have been conducted so far related to the process of innovation (Howells, 2006), the sector (Klerkx and Leeuwis, 2008b), specific CO roles (Batterink, 2009), CO relationships (Johnson, 2008) and specific CO functions (Boon *et al.*, 2008). Howells (2006) defined the concept of the CO (that he calls intermediary organization) as: 'An organization or body that acts as agent or broker in any aspect of the innovation process between two or more parties'.

Other authors (e.g. Van Lente *et al.*, 2003; Winch and Courtney, 2007) define COs as: 'Facilitators of innovation acting as a member of a network of actors in an industrial sector that are focused on enabling the other actors in the network to innovate'.

Three main functions are used by various authors to identify the roles of COs within ISs: demand articulation, network formation and innovation process management (Batterink, 2009; Klerkx and Leeuwis, 2008b, 2009; Van Lente *et al.*, 2003). According to Howells (2006), the following specific type of services can be provided by COs: foresight and diagnostics, scanning and information processing, knowledge processing, generation and combination, gate keeping and brokering, testing, validation and training, accreditation and standards, regulation and arbitration, intellectual property (IP) protection, commercialization, exploiting the outcomes and assessment and evaluation. Furthermore, in this chapter the innovation process focus of Cooper (1990) and Mc Grath (1995) is used to relate the activities, services and information sources of COs to the different stages of the innovation process: idea/concept development, engineering and release to market.

10.2.1 The role of cluster organizations in supporting open innovation

In this subsection the special role of COs in fostering open innovation by providing network formation support is considered. As Quinn (2000) points out, in order to compete in current markets, co-operation within a network of partners is becoming more and more essential. As proven empirically by Caloghirou *et al.* (2004), interacting with external partners enables a firm to access a variety of new knowledge, a phenomenon they have termed 'enhanced absorptive capacity', which increases innovative performance. The ability to identify potential network partners and therewith the network formation function of COs is thus of crucial importance. Recently, many technology-based firms have formed alliances with start-up firms and have built up their own internal venturing groups of senior managers who scout for new ideas, products and processes to fill the R&D pipeline. Huston and Sakkab (2006) refer to this new paradigm of open innovation as 'connect and develop', instead of 'research and develop'. There is only limited research to support the idea of the beneficial impact of knowledge sharing in R&D but what does exist is compelling. Omta (1995) related R&D performance in the pharmaceutical industry to the openness of the organization's information and knowledge culture. He found that the best-performing pharmaceutical firms were characterized by less management concern about the leaking of company information and greater openness to outside information, including higher attendance at scientific conferences. Omta's research suggests that the more an organization wants to share its information with the external environment, the more it gets in return. A company that spends most of its energy hoarding and protecting its own knowledge is less open to new knowledge from the outside world. Innovative companies are generally those that do not rest on their intellectual laurels, but instead are constantly on the lookout for new innovative ideas they can use to develop new products, processes and services.

10.3 The Dutch agri-food sector and Food Valley Organization

Features of innovation in the Dutch agri-food sector and the Food Valley Organization are now described.

10.3.1 The Dutch agri-food sector

The Netherlands is one of the three largest exporters of agri-food products in the world, next to the United States and France, with Germany and the United Kingdom as its major buyers. In 2009, the export of agri-food products was 17.2% of the total Dutch export (EC Agriculture and Rural Development, 2010), whereof 12.3% consisted of trade in floricultural products, followed by meat with 10.6%. Owing to successful innovation, the Dutch agri-food sector went through major developments in the preceding few decades, e.g. through the vast emphasis on biotechnological research (e.g. mapping of plant genomes). Currently, the Dutch agri-food firm's R&D expenditure on biotechnology, although still lower, is

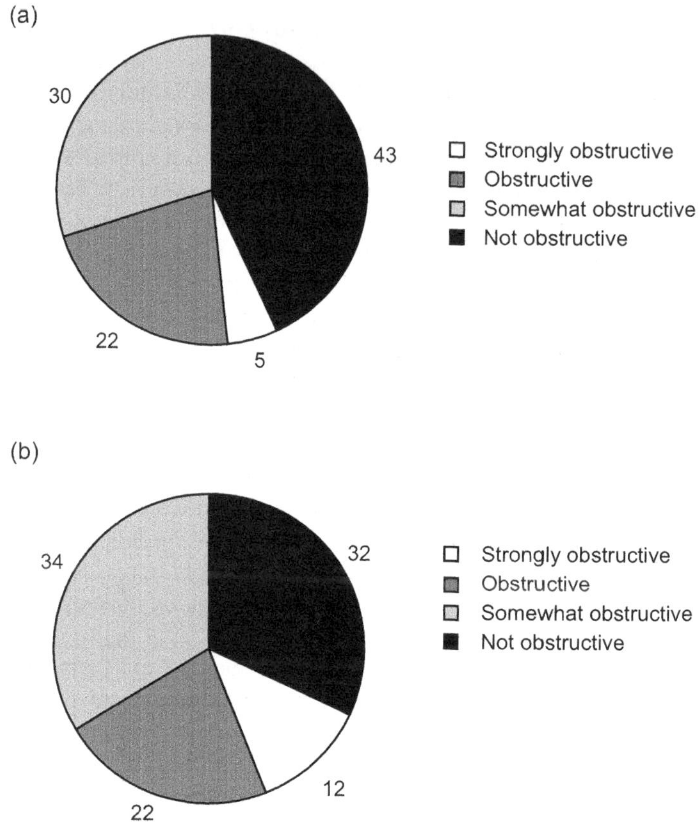

Fig. 10.1 Demand uncertainty obstructing innovation. (a) Total industry % of innovative companies, which experience uncertainty about demand as obstructing to innovation. (b) Food and beverages industry % of innovative companies, which experience uncertainty about demand as obstructing to innovation. *Source:* Federatie Nederlandse Levensmiddelen Industrie, © FNLI 2010.

approaching that of the pharmaceutical industry (10 versus 25%, CBS Statline, 2010).

Perhaps one of the greatest challenges for innovation in the Dutch agri-food industry is customer demand uncertainty. As Fig. 10.1 shows, the percentage of companies which experience demand uncertainty as obstructive to innovation is higher in the agri-food sector than the industry average. Innovative food companies indicate that uncertainty about demand is one of the most important impeding factors for innovation (CBS Statline, 2010). This uncertainty is created through the difficulty in assessing whether consumers are willing to adopt the new product. They are considered to be reluctant to adopt innovations e.g. because of fear for health hazards or quality loss (e.g. Bartels *et al.*, 2009). Although knowledge about customers/consumers is important for every industry; in the food industry this is particularly important because consumer health and safety and quality criteria need to be matched.

10.3.2 Food Valley Organization

Food Valley Organization is one of the major COs in Europe. Other important COs in the EU are the Skane Food Innovation Network in Sweden, Flander's Food and Wagralim (Wallonia) in Belgium and the Nutrition Health Longevity Cluster in France. With 15 000 scientists, 20 research institutes, 1440 food-related and 70 science-related companies, the Dutch Food Valley Region around Wageningen University and Research Centre can be regarded as one of the most important RISs in the world. Without much exaggeration one can say that the Dutch Food Valley region is the Silicon Valley of the agri-food sector. Food Valley Organization (FV) is the CO of the Food Valley region. Founded in 2004, it started organizing activities, offering services to and sharing information with its members. It is the mission of FV to improve the innovative strength of the Dutch agri-food sector in general and the member companies in particular by opening up and translating the knowledge base in and around Wageningen University and Research Centre. The main objective is to stimulate demand-driven innovation in the Dutch agri-food sector. The primary focus is on the agri-food cluster in the region around Wageningen in the Netherlands, although in recent years the scope of its activities and services widened to include the national level as well. FV is a public–private partnership, part of its funding stems from government, whereas companies, which can only become members by invitation, pay an annual membership fee. Member companies are provided with privileged activities and information sources which non-members do not have. The 105 member companies of FV include 62% SMEs and 38% large companies, such as Friesland Campina, DSM and Nutreco. As a consequence, the companies differ in size considerably, from 10 to over 10 000 employees. Four member types can be identified: food processors, technology suppliers, ingredient suppliers and service providers e.g. consultants advising about IP protection.

10.3.3 Food Valley functions

For this study, FV functions were categorized according to their nature. Next to the functions defined in previous studies, demand articulation, network formation and innovation process support, internationalization support was also defined as an important function of FV. Thus, the main functions, including fifteen FV services, activities and information sources, are the following:

- Demand articulation support: including market trend reports, dedicated one-to-one market insights and the FV Innovation Award.
- Network formation support: including the FV website, newsletter, society meetings, FV conference, open-innovation seminars and innovation partner search.
- Internationalization support: including exposure at international food exhibitions; international business missions and the FV Ambassador program (researchers and business managers from other countries advertising the skills of FV member companies in their home country).

- Innovation process support: including world-wide innovation alerts, the innovation link (to Wageningen University researchers and engineers) and support for obtaining innovation subsidies.

10.3.4 Data and methods

In 2009, a web-based questionnaire was sent to all members to evaluate FV's activities, services and means of information provision, as well as to indicate FV's contribution to their innovation processes. The respondents were asked to rate the importance of FV's functions, including fifteen services, activities and information sources (see Appendix 1) for their business and/or for their innovation processes using 7-point Likert Scales (1 = not at all important; 7 = very important).

After two weeks, non-responding companies received a reminder, and one week later they were telephoned in order to increase the response rate. It turned out that a number of companies joined the organization only in the course of 2009, stopped their membership in December 2009, or had never joined any activities or made use of the services. This group of companies was labeled non-eligible. Forty of the 70 eligible companies responded to the questionnaire, a response rate of 57%. Eleven food processors, 13 technology, eight ingredient and eight service providers answered the questionnaire. Table 10.1 shows the response rate per type of companies.

The response rate of large companies was higher than the response rate of SMEs. This could be explained by the fact that, for SMEs, the questionnaire was typically sent to the owner/director, whereas in the case of large companies, innovation or relationship managers dealt with the questionnaire. Owner/directors of SMEs are often under pressure responding to questionnaires, having many responsibilities, and innovation and relationship managers are expected to be more directly involved with FV. Furthermore, the response rate of the food processors was relatively high.

Table 10.1 Response rates for various types of companies

	Total	Eligible	Response	Total %	Eligible (%)
Large companies	40	24	17	43	71
SMEs	58	46	23	40	50
Total	98	70	40	41	57
Food processors	18	12	11	61	92
Technology suppliers	28	21	13	46	62
Ingredient suppliers	31	24	8	26	33
Service suppliers	21	13	8	38	62
Total	98	70	40	41	57

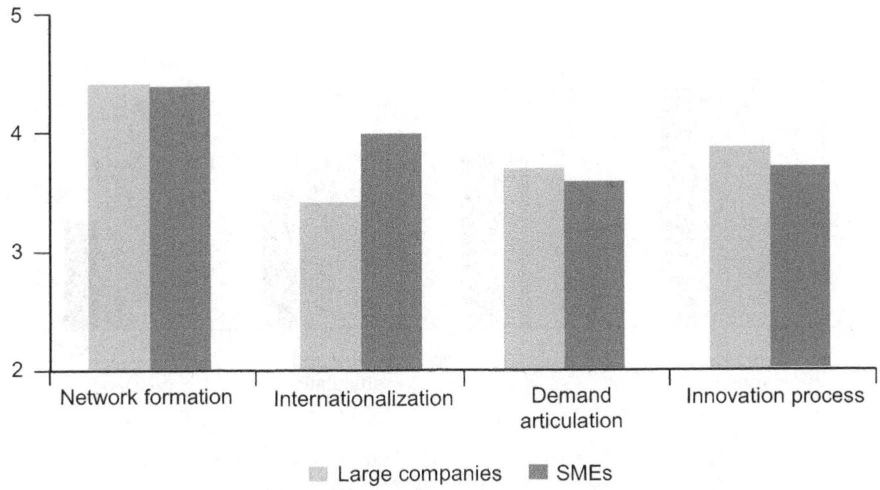

Fig. 10.2 FV functions as assessed by large companies and SMEs, for the clarity of presentation only the 2 to 5 part of the Likert 7-point scores (1 = not at all important; 7 = very important) are presented on the y axis.

10.3.5 Results

Figure 10.2 shows the assessments of the SME and large member companies of the importance of FV's different functions. In addition, Appendix 2 provides an overview of the answers on the individual services, activities and information sources for each function for the whole sample and for the SMEs and large companies separately. The results show that the highest importance is given by the companies to FV's function that is directly related to open innovation, namely FV's network formation services: the FV newsletter, FV website, the society meetings, the FV conference, the open-innovation seminars and the innovation partner search, which are all assessed above 4 on the 7-point Likert scale. Only support for finding innovation partners is regarded somewhat below 4 by the SMEs, apparently because they regard their innovation network as sufficient to find the appropriate partners. Membership-only network formation support is clearly regarded of more importance than the services that are also available to non-members. For example, the members-only society meetings, which are even regarded of high importance by SMEs (4.82), receive a higher appreciation than the FV conference. Also the members-only FV newsletter is regarded of higher importance than the FV website.

SMEs and large companies regard the importance of internationalization support quite differently. Both exposure at international food exhibitions (e.g. in the USA) and international business missions (e.g. to Japan) get a one point higher assessment by the SMEs compared with the large companies. This is understandable; the large member companies are mostly multinationals, not dependent on a CO for building international relationships. In contrast to this, the large companies give a higher priority to the ambassador program than the SMEs. Being multi-

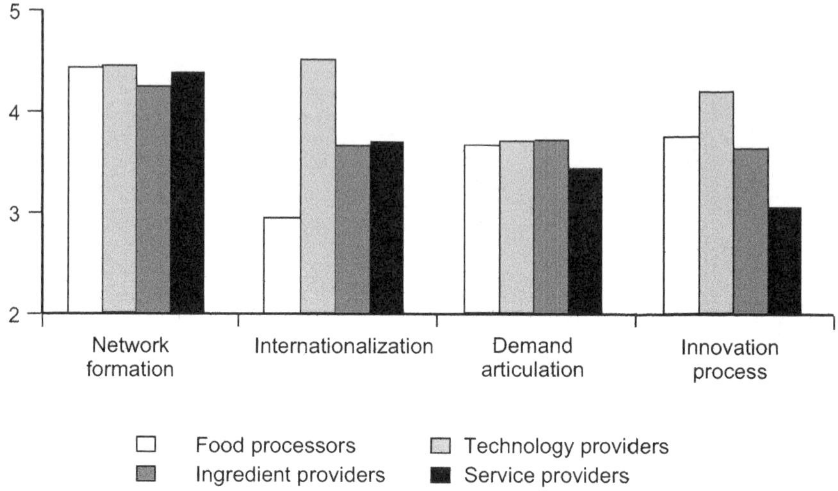

Fig. 10.3 FV functions as assessed by the different types of companies, for clarity of presentation only the 2 to 5 part of the Likert 7-point scores (1 = not at all important; 7 = very important) are presented on the y axis.

nationals, the ambassadors can provide more visibility to their activities in the home countries of the ambassadors, and also some of the ambassadors might at some point in their career become employed in one of their subsidiaries.

Demand articulation and innovation process support are regarded as comparatively less important by large companies and SMEs alike. The clear exceptions are again the one-to-one support services in obtaining innovation subsidies and receiving dedicated market insights, which are both regarded above 4 on a 7-point Likert scale by the big companies. Apparently, they have the money and want to spend it for dedicated FV services.

Figure 10.3 shows the assessment of the importance of FV functions by the different company types, food processors, technology and ingredient suppliers and service providers. In addition, Appendix 3 provides an overview of the answers on the individual services, activities and information sources for each company type. The findings show relatively high scores for the open innovation related network-formation support services, the assessments of all types of companies being above 4 on a 7-point Likert scale. Only dedicated partner search is regarded as much more important by the technology and ingredient suppliers than by the food processing companies and the service providers. Apparently, supplier companies are actively searching for innovation partners and consider FV a valid partner to do so. With 4.88 on a 7-point scale, service providers clearly perceive interactive activities such as the FV conference of high importance. They typically are part of the FV network to enhance co-operation and interaction with the processing companies.

Technology suppliers in particular are enthusiastic about FV's internationalization support, with a 4.65 on a 7-point scale for exposure at international exhibitions and even a 5.23 for FV's international business missions. The importance of helping to internationalize their business is thought to stem from their high level of specialization, for which reason they have a great need for a larger market than the national one. Being SMEs, it is problematic for them to reach international markets on their own. The large food processors clearly have no need for this kind of services. As was already indicated this group contains a number of multinational companies that clearly do not need a CO to internationalize.

Demand articulation support is generally regarded as less important by the FV member companies. Only the importance of receiving dedicated market insights is regarded above 4 by both food processors and ingredient suppliers. The FV services, activities and information sources regarding innovation process support are considered quite differently. The innovation link is only graded above 4 by the food processors, whereas support for obtaining innovation subsidies is regarded as very important (4.95) by the technology suppliers and important (4.13) by the ingredient suppliers that also regard the innovation link as important (4.0). In contrast, the service providers, conducting no technical innovation processes, are not interested in either of these services.

10.4 Conclusions

It turned out that a CO such as FV can be of great importance to its members. Although their expectations might be different, both SMEs and large companies show similar and relatively high evaluations of the importance of FV functions, including its services, activities and information sources for their business and/or innovation processes, except for international business missions, the importance of which are evaluated significantly higher by SMEs. Although the differences are larger (but still not significant owing to the relatively lower number of cases compared with the comparison large companies versus SMEs), the same holds for the evaluation of food processors and technology, ingredient and service providers.

If we look at the three main functions of COs as described in literature (e.g. Klerkx and Leeuwis, 2008a), network formation, demand articulation and innovation process management support, it is clear that, according to the theory, linking actors in a RIS is one of the core functions of FV. In accordance with open-innovation studies (e.g. Chesbrough, 2003) the networking function is mentioned as being of the highest importance by all the member companies. The food processors and service providers in particular are very interested in the networking possibilities of FV. For food processors, FV provides possibilities to contact innovation partners for the idea-generation and concept phases of the innovation process, whereas for service providers it is of great importance to contact the agri-food processing companies to conduct coinnovation trajectories. The demand-articulation and innovation-process management needs are clearly different for the three member types of FV. Where the technology suppliers, being dependent on knowledge-based innovation for their future competiveness, are clearly searching for innovation-process (management) support, the food processors are more interested in demand articulation as the starting point of their innovation processes, FV turns out to play an important role for them providing dedicated and independent market information outside the supply chain. The high level of competition in the agri-food sector, in particular between retailers and food processors, might play a role here. There was one type of function provided by FV that was not included in the theoretical function framework, namely help in internationalization, which turned out to be especially important for SMEs in general and technology providers in particular.

It is therefore specifically concluded, that:

- The FV case clearly shows, that, in accordance with open-innovation studies (e.g. Chesbrough, 2003), the networking function of a CO, helping finding innovation partners, bringing them together, and playing a facilitating role, is clearly a central role for a CO in a RIS.
- Seen in the light of the important role of SMEs in RISs, it is further concluded that the current CO framework should be extended by adding the internationalization support for SMEs.

It is expected that the results presented in this chapter will help COs to increase their effectiveness by focusing on their functions to effectively support the open-innovation trajectories of their member companies.

10.5 Future trends

It is expected that the number of companies working together in innovative clusters will continue to grow in the coming years. Clusters are also increasingly working together in national as well as international networks. Network co-operation of national clusters are mainly focused on stimulating learning among industries, to grasp the advantages of companies in different industries working closely together in open-innovation projects, without the threat of direct competition. Internationally there is a clear trend towards supernational clusters in which individual companies and clusters from different countries co-operate, grasping the advantages of cross border learning. Seen in the light of these trends, the functions of CO to build effective networks and create the right conditions for open innovation will become even more important in the future.

10.6 Sources of further information and advice

An introductory book to innovation management: Tidd *et al.*, 2001.
An important book on open innovation: Chesbrough, 2003.
Important articles on RISs: Asheim and Coenen, 2005; Batterink *et al.*, 2010.
An important article on COs: Winch, and Courtney, 2007,

An important article on innovation networks: Zeng *et al.*, 2010.
An important EU agrifood cluster: European Food Alliance Initiative.
Important EU agrifood cluster organizations: Food Valley Organization (Netherlands), Skane Food Innovation Network (Sweden), Flander's Food and Wagralim (Wallonia, Belgium), Nutrition Health Longevity Cluster (France).

10.7 References

ASHEIM, B.T. AND COENEN, L. 2005. Knowledge bases and regional innovation systems: comparing Nordic clusters. *Research Policy*, **34**(8): 1173–1190.

BARTELS, J., ONWEZEN, M.C., RONTELTAP, A., FISCHER, A.R.H., KOLE, A.P.W., VAN VEGGEL, R.J.F.M. AND MEEUSEN, M.J.G. 2009. Eten van Waarde; Peiling Consument en Voedsel, LEI, Den Haag.

BATTERINK, M. 2009. *Profiting from external knowledge. How firms use different knowledge acquisition strategies to improve their innovation performance*. Wageningen Academic Publishers, Innovation and Sustainability Series, Volume 3, ISBN: 978-90-8686-101-9, p. 190.

BATTERINK, M.H., WUBBEN, E.F.M., KLERKX, L. AND OMTA S.W.F. 2010. Orchestrating innovation networks: the case of innovation brokers in the agri-food sector. *Entrepreneurship and Regional Development*, **22**(1): 47–76.

BERGEK, A., JACOBSSON, S., CARLSSON, B., LINDMARK, S. AND RICKNE, A. 2008. Analyzing the func-tional dynamics of technological innovation systems: a scheme of analysis. *Research Policy*, **37**(3): 407–429.

BOON, W.P.C., MOORS, E.H.M., KUHLMANN, S. AND SMITS, R. 2008. Demand articulation in intermediary organisations: the case of orphan drugs in the Netherlands. *Technological Forecasting and Social Change*, **75**(5): 644–671.

CALOGHIROU, Y., VONORTAS, N.S. AND IOANNIDES, S. 2004. *European collaboration in research and development: business strategies and public policies*, Edward Elgar Publ., Northampton, MA.

CARLSSON, B. 2006. Internationalization of innovation systems: a survey of the literature, *Research Policy*, **35**(1): 56–67.

CBS STATLINE 2010. Research en Development (R&D) door bedrijven, SBI 2008 en bedrijfsomvang. Copyright © Centraal Bureau voor de Statistiek.

CHESBROUGH, H.W. 2003. *Open innovation: the new imperative for creating and profiting from technology*, HBS Publ., Mass.

COOPER, R.G. 1990. Stage-gate systems: a new tool for managing new products. *Business Horizons*, **33**(3): 44–54.

DE BRUIJN, P. AND LAGENDIJK, A. 2005. Regional innovation systems in the Lisbon strategy. *European Planning Studies*, **13**(8): 1153–1172.

EC AGRICULTURE AND RURAL DEVELOPMENT 2010. Agriculture in the European Union – Statistical and economic information: http://ec.europa.eu/agriculture/agrista/index_en.htm

HEKKERT, M.P., SUURS, R.A.A., NEGRO, S.O., KUHLMANN, S. AND SMITS, R. 2007. Functions of innovation systems: a new approach for analysing technological change. *Technological Forecasting and Social Change*, **74**(4): 413–432.

HOWELLS, J. 2006. Intermediation and the role of intermediaries in innovation. *Research Policy*, **35**(5): 715–728.

HUSTON, L. AND SAKKAB, N. 2006. Connect and develop: inside Procter & Gamble's new model for innovation. *Harvard Business Review*, March, 60–72.

JOHNSON, W.H.A. 2008. Roles, resources and benefits of intermediate organizations supporting triple helix collaborative R&D: the case of Precarn. *Technovation*, **28**(8): 495–505.

KLERKX, L. AND LEEUWIS, C. 2008a. Balancing multiple interests: embedding innovation intermediation in the agricultural knowledge infrastructure. *Technovation*, **28**(6): 364–378.

KLERKX, L. AND LEEUWIS, C. 2008b. Matching demand and supply in the agricultural knowledge infrastructure: experiences with innovation intermediaries. *Food Policy*, **33**(3): 260–276.

KLERKX, L. AND LEEUWIS, C. 2009. Establishment and embedding of innovation brokers at different innovation system levels: insights from the Dutch agricultural sector. *Technological Forecasting and Social Change*, **76**(6): 849–860.

MALERBA, F. 2004. *Sectoral systems of innovation*, Cambridge University Press, Cambridge.

MCGRATH, M.E. 1995. Product strategy for high-technology companies: how to achieve growth, competitive advantage, and increased profits. McGraw-Hill, New York.

OMTA, S.W.F. 1995. *Critical success factors in biomedical research and pharmaceutical innovation*. Kluwer Academic Publications, Dordrecht, Boston, London, p. 285.

QUINN, J.B. 2000. Outsourcing innovation: the new engine of growth. *Sloan Management Review*, Summer, 13–28.

TIDD, J., BESSANT, J. AND PAVITT, K. 2001. Managing innovation: integrating technological, market, and organizational change, John Wiley & Sons, New York.

TÖDTLING, F. AND TRIPPL, M. 2005. One size fits all? Towards a differentiated regional innovation policy approach. *Research Policy*, **34**(8): 1203–1219.

VAN LENTE, H., HEKKERT, M., SMITS, R. AND VAN WAVEREN, B. 2003. Roles of systemic intermediaries in transition processes. *International Journal of Innovation Management*, **7**(3): 247–280.

WINCH, G. M. AND COURTNEY, R. 2007. The organization of innovation brokers: an international review. *Technology Analysis & Strategic Management*, **19**(6): 747–763.

ZENG, S.X., LIE, X.M. AND TAM, C.M. 2010. Relationship between cooperation networks and innovation performance of SMEs. *Technovation*, **30**(3): 181–194.

10.8 Appendix 1: the four main functions of Food Valley, including fifteen services, activities and information sources

1. Network formation support

- Newsletter
- Website
- Society meetings
- FV conference
- Open-innovation seminars
- Innovation partner search

2. Demand articulation support

- Market trend reports
- Dedicated market insights
- FV Innovation Award

3. Internationalization support

- Exposure at international food exhibitions

- International business missions
- Ambassador program

4. Innovation process support

- Worldwide innovation alerts
- Innovation link
- Support in obtaining innovation subsidies

10.9 Appendix 2: assessment of small to medium-sized enterprises and large companies of the importance of Food Valley functions.

	Large companies	SMEs
Network formation support		
Newsletter	4.79 (0.98)[1]	4.85 (1.44)
Website	4.63 (1.09)	4.07 (1.58)
Society meetings	4.29 (1.05)	4.82 (1.56)
FV conference	4.12 (1.22)	4.50 (1.51)
Open innovation seminars	4.47 (0.94)	4.15 (1.41)
Innovation partner search	4.18 (1.81)	3.93 (1.84)
Demand articulation support		
Market trend reports	3.56 (1.46)	3.85 (1.41)
Dedicated market insights	4.06 (1.71)	3.57 (1.75)
FV Innovation Award	3.57 (1.75)	3.39 (2.06)
Internationalization support		
Exposure at international food exhibitions	3.41 (1.62)	4.41 (1.72)
International business missions	3.29 (1.65)	4.27 (2.12)[2]
Ambassador program	3.53 (1.60)	3.34 (1.70)
Innovation process support		
Worldwide innovation alerts	3.80 (1.42)	3.43 (1.47)
Innovation link	3.71 (1.11)	3.65 (1.34)
Support in obtaining innovation subsidies	4.06 (1.73)	3.98 (1.81)

[1]Mean and standard deviation (in parentheses).
[2]$p < 0.10$.

10.10 Appendix 3: member company assessment of the importance of Food Valley functions by company type

	Food processors	Technology suppliers	Ingredient suppliers	Service providers
Network formation support				
Newsletter	4.89 (1.36)*	4.81 (1.60)	4.50 (0.93)	5.14 (0.90)
Website	4.55 (1.37)	4.27 (1.67)	4.25 (1.28)	4.00 (1.27)
Society meetings	4.45 (1.51)	4.85 (1.28)	4.14 (0.90)	4.75 (1.75)
FV conference	4.36 (1.29)	4.04 (1.66)	4.38 (1.19)	4.88 (1.36)
Open innovation seminars	4.73 (1.27)	4.12 (1.29)	4.00 (0.76)	4.25 (1.49)
Innovation partner search	3.64 (2.25)	4.65 (1.55)	4.25 (1.49)	3.38 (1.77)
Demand articulation support				
Market trend reports	3.82 (1.66)	3.88 (1.42)	3.50 (1.69)	3.57 (0.79)
Dedicated market insights	4.27 (2.01)	3.46 (1.66)	4.00 (1.77)	3.38 (1.51)
FV Innovation Award	2.91 (1.58)	3.83 (2.13)	3.63 (1.77)	3.38 (2.07)
Internationalization support				
Exposure at international food exhibitions	3.27 (1.62)	4.65 (1.89)	4.00 (1.69)	3.88 (1.55)
International business missions	2.55 (1.70)	5.23 (1.92)	3.86 (1.57)	3.38 (1.41)
Ambassador program	3.09 (2.07)	3.65 (1.43)	3.14 (1.07)	3.83 (1.94)
Innovation process support				
Worldwide innovation alerts	4.09 (1.70)	3.85 (1.41)	2.83 (0.98)	2.83 (0.98)
Innovation link	3.55 (1.29)	3.77 (0.93)	4.00 (1.41)	3.38 (1.51)
Support in obtaining innovation subsidies	3.64 (1.69)	4.95 (1.27)	4.13 (1.81)	3.00 (2.00)

*Mean and standard deviation (in parentheses).

11

The importance of networks for knowledge exchange and innovation in the food industry

B. Kühne, V. Lefebvre, C. Cochez and X. Gellynck, Ghent University, Belgium

DOI: 10.1533/9780857097248.2.189

Abstract: The various types of knowledge flows that are important in the various types of innovations are investigated using three case studies on networks in the Flemish food sector. Knowledge flows in networks are shown to depend on various types of ties, context-dependent variables, and the intermediation functions of the network, i.e. demand articulation, network formation and innovation process management.

Key words: networks, knowledge exchange, information exchange, learning, innovation, food industry.

11.1 Introduction

Open innovation is the exchange of knowledge through in- and out-flows of knowledge in a company (Chesbrough, 2003). In particular, small and medium-sized enterprises (SMEs) profit from applying an open-innovation strategy and participating in networks to achieve access to the vast range of new knowledge that is created outside their firm's boundaries, but which is necessary to successfully innovate (Gellynck *et al.*, 2007; Omta, 2002). Open innovation has been applied in a variety of forms in the food sector (Sarkar and Costa, 2008). However, studies focusing on SME networks are limited and, in particular, with regards to the role of networks in knowledge exchange and innovation. Therefore, e.g. Sarkar and Costa (2008) conclude that more focused case studies and empirical research is needed.

Consequently, this chapter focuses on how networks facilitate knowledge flow through the creation of different types of ties and the influence of the networks' activities on the innovativity of the network members. Using case studies, this chapter identifies different strategies of networks to increase knowledge exchange and innovation and to determine the types of knowledge and innovations that are stimulated through networking.

The chapter is arranged as follows, the subsequent section a literature review of studies in knowledge exchange and innovation, and the role of networks in these two aspects. In Section 11.3 methodological aspects are provided, followed by the results and discussion of three case studies in Flanders (Belgium). The chapter closes with conclusions and recommendations for science and businesses.

11.2 Knowledge exchange and innovation and the importance of networks

11.2.1 Knowledge exchange and innovation

Knowledge exchange is a prerequisite for learning and consequently for innovation (Lundvall *et al.*, 2002; Powell *et al.*, 1996). Knowledge exchange is considered as the transfer of knowledge by interaction between different parties through linkage and exchange; this can result in mutual learning (based on Argote and Ingram, 2000; Inkpen and Tsang, 2005; Nonaka and Takeuchi, 1995; and on the terms and definitions for knowledge mobilization given in http://www.oise.utoronto.ca/rspe/KM_Products/Terminology/index.html). Information and knowledge are two different concepts: information is a flow of messages, whereas knowledge is created and organized by the very flows of information, anchored on the commitment and beliefs of its holder (Nonaka, 1994, p.15).

Knowledge is usually distinguished into two main categories: tacit knowledge and explicit knowledge (Hall and Andriani, 2002; Nonaka *et al.*, 2000). Tacit knowledge is not articulated or codified and is generally harder to imitate by other organizations. Because it requires sharing of experiences through observation and imitation (Hall and Andriani, 2002), it has a greater competitive value. Explicit knowledge is more tangible and easy to transfer between individuals. It can be processed, transmitted and stored relatively easy in form of data, manuals, formulae etc. (Hall and Andriani, 2002; Nonaka *et al.*, 2000). Firms need to use both tacit and explicit knowledge to create innovations (Cowan *et al.*, 2000).

The process of learning is characterized by identifying, gathering, exchanging and interpreting relevant information and subsequently using this information to develop and apply new competencies for business improvement and innovation (Keursten *et al.*, 2006). As learning can be considered a social construction process, the conditions under which it is learned strongly affect what is learned (Cohen and Levinthal, 1990; Powell *et al.*, 1996). Networks can shape the conditions for optimal learning and fostering learning mechanisms (Powell *et al.*, 1996). Learning in networks refers to the organization's ability to combine tacit and explicit knowledge resources through interaction with other organizations in the network in a dynamic process. This knowledge exchange can be divided into four main categories of processes: socialization, combination, articulation, and internalization (Nonaka, 1994; Nonaka and Takeuchi, 1995):

- **Socialization** is the transfer of 'tacit to tacit' knowledge through observation, imitation and practice.Socialization skills are difficult to articulate and to transfer, but are learned by doing and observing.
- **Combination** is the transfer of 'explicit to explicit' knowledge by combining discrete pieces of explicit knowledge into a new whole. For innovation, new knowledge can be created by combining existing knowledge about markets and customers/consumers with existing knowledge on product characteristics into, e.g., a new form of market innovation, that comprises new ways to reach the market or to approach consumers.
- **Articulation** is the transfer of 'tacit to explicit' knowledge and thus making tacit knowledge explicit. Nonaka refers to the example of a software developer who trained with one of the Japan's best bakers to explicate the secret of the baker's outstanding but difficult to codify baking technique. After a year of learning and trial-and-error, the software developer succeeded in devising a new kitchen appliance machine that was able to replicate the baker's special kneading technique.
- **Internalization** is the transfer of 'explicit to tacit' knowledge and appears when new explicit knowledge is disseminated/shared within an organization to broaden the employees own tacit knowledge.

The different learning processes are of different importance with regards to the novelty of the innovation under concern. Depending on the degree of novelty of the innovation (e.g. new to the firm or to the world), different steps of the learning process as well as of the level of learning are required (Cohen and Levinthal, 1990; Tödtling *et al.*, 2009). For instance, incremental innovation has lower learning requirements because it involves 'learning by doing'. In contrast, radical innovation requires substantial learning processes, e.g. for developing completely new technologies or ways of doing business.

In this chapter, innovation is defined as the successful exploitation of new combinations of existing resources into new or improved products, methods of production, sources of supply, and ways to organize business, and the exploitation of new markets or new ways to reach existing markets (Lundvall, 1995; Pittaway *et al.*, 2004).

In order to achieve innovation, different and functionally distinct but interdependent processes of knowledge exchange are necessary, as described in the innovation production process (Chen and Guan, 2011) or innovation value chain (Roper *et al.*, 2008). The innovation production process usually starts with the process of knowledge accumulation, after which knowledge is transformed and finally exploited (Chen and Guan, 2011; Hansen and Birkinshaw, 2007; Lefebvre and Gellynck, 2012; Roper *et al.*, 2008):

- **Knowledge accumulation** is the acquisition and accumulation of knowledge and innovation experiences.

- **Knowledge transformation** is the step where the accumulated knowledge is translated into innovation outputs such as new or improved technology or product prototypes, based on the allocation of resources related to both R&D and non-R&D to the innovation development.
- **Knowledge exploitation** is the final phase when the firm derives economic profits from the innovation outcomes.

However, the firm's openness for conducting the innovation production process and, in particular, for tapping into external knowledge is greatly dependent on the extent of its internal innovation activities. Firms that are more actively involved in innovation are more open to external knowledge and, in a self-reinforcing cycle, they become more innovative because they are more aware of external opportunities (Cohen and Levinthal, 1990).

11.2.2 Knowledge exchange and innovation in the food sector and small and medium-sized enterprises (SMEs)

For small and medium-sized enterprises (SMEs) in the food sector, the increasing complexity of the innovation process has led to a greater dependency on interaction for innovation within networks (Edwards *et al.*, 2005; Zeng *et al.*, 2010). In the food industry, innovation is not purely based on R&D, but rather involves a learning process and interaction between different actors, as described in the theory of the 'new economy' (Avermaete and Viaene, 2002; Weaver, 2008).

In relation to the introduction of innovations, SMEs also often face internal problems, such as lack of human and financial resources, which result in limited organizational capabilities, poor motivation toward innovation, and lack of strategic vision (Avermaete *et al.*, 2003; Scozzi *et al.*, 2005). Furthermore, it has been proven that SMEs are often unable to efficiently allocate and co-ordinate their resources, do not have access to relevant information and knowledge, and do not focus on learning, but rather exhibit a day-to-day work mentality (Avermaete *et al.*, 2003; Scozzi *et al.*, 2005). Most of these problems can be overcome if SMEs are open towards collaboration and involved in networks (Avermaete *et al.*, 2003; Scozzi *et al.*, 2005; Tödtling *et al.*, 2009). In addition, recent trends and developments, such as increasing costs and complexity of R&D, shorter technology life cycles, increasingly knowledgeable suppliers and clients, the growth of venture capital and growing diffusion of leading-edge knowledge in universities and research labs around the world, have led to a greater dependency of SMEs on knowledge and information available externally to the firm (Avermaete *et al.*, 2004a; Enzing *et al.*, 2008; O'Reilly *et al.*, 2003; Sarkar and Costa, 2008; Stewart-Knox and Mitchell, 2003).

Because most of the new knowledge is created outside of the firm's boundaries, it is of utmost importance that firms actively tap into these external sources to strengthen their business. Open innovation implies the extensive use of interorganizational ties to in-source this external knowledge into the firm, but also includes marketing internal ideas through channels outside a firm's current business area (Chesbrough, 2004). Through open innovation, the innovating firms establish ties with other organizations, as firms are increasingly forced to collaborate with other organizations in order to develop or absorb new technologies, commercialize new products or simply stay up-to-date with the latest developments and trends (Avermaete *et al.*, 2004a; Enzing *et al.*, 2008; Sarkar and Costa, 2008; Stewart-Knox and Mitchell, 2003). Networks can play a very important role in the formation and maintenance of such ties and collaborations (Pittaway *et al.*, 2004; Sawhney *et al.*, 2006).

11.2.3 Role of networks in knowledge exchange and innovation

The role of networks in the stimulation of innovation has been described widely (e.g. Brennan and Dooley, 2005; Daskalakis and Kauffeld-Monz, 2005; MacKinnon *et al.*, 2002; Omta, 2004). Within the 'new economy' theory, many scholars acknowledge the network, within which a firm is embedded, to be more important for the development and implementation of innovation than the firm itself (e.g. Avermaete and Viaene, 2002; Edwards *et al.*, 2005; Matthyssens *et al.*, 2006; Omta, 2002; Pittaway *et al.*, 2004; Weaver, 2008). A network serves as the locus of innovation because it provides access to knowledge and resources, that are otherwise unavailable (Grunert *et al.*, 2008; Omta, 2002, 2004; Pittaway *et al.*, 2004; Powell *et al.*, 1996).

A network can be defined as a set of actors connected by a set of repeated interactions of formal and/or informal ties (Borgatti and Foster, 2003; Granovetter, 1973; Hamdouch, 2010; Owen-Smith and Powell, 2004). The actors are firms (competitors, suppliers, customers, auxiliary businesses etc.), individuals (boundary spanners etc.), knowledge centres (universities, research centres etc.) and other actors (network organizations, governments, special interest groups, industrial organizations etc.). The ties are the relationships between the actors. They may be formal (contractual, institutionalized) or informal (social, trust-based). The network is thus the place where actors within one, or between several related industrial sectors interact and collaborate to add value for the customer (Omta, 2004).

Networks increase the flow of information and have thus an important role in the diffusion and adoption of innovations (Ng *et al.*, 2003; Pittaway *et al.*, 2004). Facilitating collaboration or networking for innovation is important, because it offers opportunities for new or alternative relationships, links or markets, and allows access to new or complementary competencies and technologies (Lazzarini *et al.*, 2001; Pittaway *et al.*, 2004). Thus, the value of collaboration or networking for innovation is the rapid establishment of a complex knowledge base and diffusion system through streamlining information flows (Pittaway *et al.*, 2004; Sawhney *et al.*, 2006).

For SMEs, in particular, network building is an important strategy. SMEs have been found to be more innovative when they are able to join and manage networking activities (Avermaete and Viaene, 2002; Gellynck *et al.*, 2007; Omta, 2002; Sarkar and Costa, 2008). Previous studies have shown that SMEs collaborate in particular with customers, suppliers, competitors, government organizations and research institutions to achieve more successful innovation (Edwards *et al.*, 2005; Gellynck and Kühne, 2010; Pittaway *et al.*, 2004; Ritter and Gemünden, 2003; Saad *et al.*, 2002).

11.2.4 Role of networks in the creation of ties and relationships for knowledge exchange and innovation

Previous studies have found that learning and knowledge exchange is significantly influenced by the structure of the network and the quality of the network member's relationships (Caniels and Romijn, 2008). Collaboration in networks can be based on formal or informal interactions among the network members creating deep or wide ties.

Formal interactions or relationships

Formal interactions or relationships are systematically established and organizationally structured (Eisenhardt and Schoonhoven, 1996; Owen-Smith and Powell, 2004; Powell *et al.*, 1996; Simard and West, 2006). Usually it is contractually agreed which knowledge flows are intended to be controlled, e.g. through licensing agreements and alliances. This type of relationship can easily be included into an open-innovation strategy: subsequent to the identification of in-house knowledge gaps, potential suppliers of that knowledge can be sought for collaboration. Within formal relationships, unexpected knowledge spillover may also appear, because they form channels for informal knowledge flow as well (Owen-Smith and Powell, 2004). The challenge is to develop capabilities to recognize those unanticipated spillovers and capture their potential benefits.

Informal interactions or relationships

Informal interactions or relationships are based on unplanned interactions which can lead to knowledge spillovers in particular when individuals are members of a community that spans multiple organizations (Agrawal and Henderson, 2002; Granovetter, 1985; Murray, 2002; Simard and West, 2006; Uzzi, 1996). Similary, just as formal interactions may lead to informal relationships, informal interactions may lead to the formation of formal relationships (Gulati and Westphal, 1999).

Deep ties

Deep ties, also called exploitative ties, are related to the exchange of explicit or existing knowledge and resources (March, 1991; Simard and West, 2006). The organization's needs to achieve access to key sources for innovation can be addressed by seeking deep embeddedness in a network, thus obtaining innovation advantages through a unique and difficult-to-imitate network position. Firms can deepen their position in a network through joining densely populated networks, building their own value networks or by strengthening their ties in the network through trust building (Gulati, 1995). Furthermore, geographic proximity has been proved to be a key facilitator of knowledge transfer and the creation of deep ties, because it facilitates frequent interactions (Chesbrough, 2003). Usually deep ties allow the sharing of tacit and informal knowledge and provide valuable access to knowledge and resources for incremental innovation (Freeman, 1991).

Wide ties

Wide ties, also called explorative ties, are ties that are based on occasional rather than frequent interactions and that span structural holes (Granovetter, 1973; March, 1991; Simard and West, 2006). Such ties offer an organization access to different networks and, thus, different sources for new technologies and markets, provide opportunities to explore new technologies, and link the innovating firm with a diverse technological environment providing access to non-redundant information. Wide ties bear a greater potential for innovation as they provide access to new knowledge, different views, and non-redundant information.

In order to gain optimal innovation benefits from the interactions in a network, an organization needs to have a balance between deep and wide ties (Rese and Baier, 2011; Uzzi and Gillespie, 1999).

Networks can support knowledge exchange among members through creating an environment of co-operative behaviour, clear communication of goals and expectations and facilitating the sharing of risks and benefits amongst members (Gellynck and Kühne, 2010; Pittaway *et al.*, 2004). The creation of such an environment includes also the facilitation of the right types of relationships and ties related to the kind of knowledge to be exchanged and the type of innovation targetted. Networks have to fulfil an act of intermediation by connecting actors, managing a network of organizations and its activities in order to enable organizations to learn and to innovate (Lefebvre and Gellynck, 2012). Three main functions of intermediation by networks have been identified: demand articulation, network formation and innovation process management (Klerkx and Leeuwis, 2009):

- **Demand articulation**, which is also called innovation initiation (Batterink *et al.*, 2010), relates to the diagnosis and analysis of possible problems faced by the target group and to the articulation of the needs of firms in demand for innovation inputs (Howells, 2006, Klerkx and Leeuwis, 2009).
- **Network formation** includes network brokerage, network design, and network construction (Dhanaraj and Parkhe, 2006; Klerkx and Leeuwis, 2009; Lee *et al.*, 2010). Through network formation, the establishment of interactions between the demand and supply side for innovation inputs is facilitated through scanning, scoping, filtering and matching possible partners for collaboration.
- **Innovation process** management or network management and network orchestrating aims at developing and improving collaboration through alignment, communication and fostering learning among partners (Klerkx and Leeuwis, 2009; Lee *et al.*, 2010).

11.3 Network methodology: a case study approach

Three case studies were analysed in order to understand more about the role of

Table 11.1 Sections of semistructured interview guide and main respondent categories

(a) Sections	(b) Main categories of respondents
1. General profile of the network	• Network co-ordinator or network founder
2. Network inception	• Companies
3. Network evolution	• Research centres and organizations
4. Network membership	• Public bodies supporting or having supported the network
5. Network configuration and network ties	
6. Network activity	
7. Network governance and management	
8. Performance	

networks in knowledge exchange and innovation in food SMEs. The focus lies on triple-helix networks, consisting of companies, research institutions and policy makers, the latter functioning as the funding organizations in our cases. All networks focus on the stimulation of learning and innovation of food SMEs.

A case-study approach is followed, whereby different actors in food networks are interviewed regarding their learning and innovation in networks. The selection of case-networks was based on an initial specification of structural indicators and an identification of an array of networks that could assure a wide variety of network alternatives. The initial indicators include: age (<2 years versus >2 years), source of finance (public, private, mixed), spatial orientation (local/domestic, global/international), targets in value chain (horizontal, vertical), focus of innovation (product, process, organization, market), network driver (industry, research, focal company, public body or group, SMEs), and scope (food, non-food). For each network, about 10 face-to-face interviews with different network members were conducted; a number large enough to ensure stable results and network diversity. Depending on the type of network, the emphasis was on those triple-helix players that are most important for the type of network.

The data collection took place within the FP7 project NetGrow.[1] The interviews were carried out simultaneously from December 2010 until May 2011 to avoid the effects of changing economic environments. A semistructured interview guide was developed and tested in advance. The interview guide was structured in sections and was adjusted to the main categories of respondents, as indicated in Table 11.1a and b, because of the relevance and formulation of questions for each respondent type. Each interview was audio-recorded and transcribed subsequently. The method of analysis used was 'open coding' based on 'grounded theory' (Glaser, 1978). Accordingly, the data were first broken down into corresponding paragraphs or quotes with a similar context. Secondly, they were grouped and categorized again by codes, following the open-coding methodology.

[1]The research leading to these results has received funding from the European Union Seventh Framework Program (FP7/2007-2013) under grant agreement no 245301 NetGrow – 'Enhancing the innovativeness of food SMEs through the management of strategic network behaviour and network learning performance'. The information in this document reflects only the authors' views and the Community is not liable for any use that may be made of the information contained therein. More information about the project: www.netgrow.eu.

Table 11.2 Summary of Belgian networks studied

	Flanders' FOOD	PLATO	VLAZ
Focus	Product and process innovation	Organizational innovation	Product and process innovation
Strategy	Bringing people (firms and research institutions) together, mostly within collaborative projects, but also individual advice is provided and there is strong emphasis on knowledge dissemination	Group sessions of SME managers guided by experienced employees from large firms	Bilateral projects between a firm and a university
Sources of topics	Steering committee with large firms questionnaires	The topics are discussed during the sessions	Technological problems of single firms
Funding	80% government funded, 20% member fees	75% government funded, 25% participant fees	80% government funded, 20% industry federation
Founded by	Government in co-operation with the food industry	Chamber of Commerce	University
Sector	Food and beverages sector and related companies	All sectors, including food and beverage sector	Specific branch of the food industry
Age (years)	6	21	8

In Belgium, three networks were selected based on the criteria mentioned above. An overview of the explored networks is summarized in Table 11.2.

11.4 Results of the three Flemish case studies

The three networks investigated in Belgium are stimulating learning and innovation within firms and especially within SMEs, but they have a different strategy to accomplish this. In Flanders' FOOD and PLATO networks, firms share knowledge and experiences and work together in small groups. However, their focus is on different aspects. In Flanders' FOOD, the focus lies on basic research, whereas in PLATO, the focus is on sharing experiences about management practices. In VLAZ, there is only bilateral technological collaboration between a firm and a research institution.

In the following sections, each network is briefly described and, subsequently, we specify how knowledge exchange is facilitated in the different networks and to which innovations the networks contribute.

11.4.1 Process of knowledge exchange in networks

When considering knowledge exchange as a social construction process (Powell *et al.*, 1996) it is important to pay attention to the conditions under which knowledge is exchanged as they strongly influence what is learned. Thus, in this section, first a description of the conditions under which knowledge is exchanged in the different networks is provided followed by the type of knowledge exchanged and the openness of the relationships.

Flanders' FOOD aims at strengthening the competitive power of Flemish food firms by stimulating innovation. The focus lies on the collaboration of food firms, especially SMEs, and research institutions by means of collaborative projects of 1–2 years. In most of these projects, basic research is conducted within research institutions in close collaboration with the participating firms. The topics for the projects are identified by the steering committee or based on up-front assessed demand of the firms. At a later stage, the results are implemented in the participating firms. All kinds of firms related to the food sector are working together, competitors as well as non-competitors, e.g. producers, technology providers, raw material suppliers and packaging firms. Thus, the network facilitates the development of knowledge of its member firms. Furthermore, Flanders' FOOD aims also at knowledge valorization, i.e. to spread the learned knowledge and ascertain that the food firms actually use the knowledge. Last, but not least, Flanders' FOOD actively supports access to, and dissemination of, knowledge generated in the collaborative projects and elsewhere through seminars, books/booklets, newsletter, workshops/training and consultancy.

When the networks started their first collaborative projects, the communication and relationship between the participants, in particular among the firms, was not very open. In the beginning, communication was directed only towards the network co-ordinator. One reason for this was the presence of competitors in the same collaborative project. Another reason was the barrier between firms and scientists, because they did not have the same language and cognitive background. However, as the firms had more and repeated contact and more collaborative projects together, trust and hence knowledge exchange about rather sensitive information has improved, though it is still limited owing to the presence of competitors. On the other hand, the presence of firms that represent different aspects throughout the entire food-processing chain (e.g. suppliers, processing companies, users) enrich the project, adding different aspects of the topics. In addition, communication among firms and scientists has improved over time towards more direct and open contacts, because the firms realized that science can offer them information and knowledge which is useful for their company.

In this network, the focus is mainly on knowledge combination and internalization. During knowledge combination, explicit knowledge from both business and science is combined into a new whole of explicit knowledge within the collaborative projects. Subsequently, the new explicit knowledge is internalized by the firms through translation of generic knowledge into specific knowledge both at a collective and individual basis through the dissemination and valorization activities of the network.

The network co-ordinator plays an important role towards the 'translation' of the knowledge into the firm's language, so that even the smallest companies are able to grasp the theory of the basic research in order to be able to internalize the relevant knowledge.

The objective of **PLATO** is to facilitate knowledge exchange by CEO's or key managers of SMEs by sharing their experiences and raising the problems they struggle with in relation to managerial issues, within a small group of 10–15 people over a period of two years. The group is guided and moderated by two 'godfather(s)' and/or 'godmother(s)' who work in larger firms and have already a lot of field experience. First, godfathers/godmothers are trained in advance, to be sure that they have a clear understanding about what is expected from them and to teach them how to address certain issues. Then, the first group activity is a start-up weekend, with all participating firms of the group together with the godfathers/godmothers, devised as a relationship-, trust- and team-building activity. Subsequently, the meetings take place monthly and the themes discussed in these meetings are chosen by the participants. Every participant is free to voice his/her opinion, give suggestions and share experiences within the group. At the monthly meetings, guest speakers provide background information about the topic in question. An important issue is the composition of the groups. The network co-ordinator carefully composes the group out of non-competing firms mostly from different sectors. After the start-up weekend and some monthly meetings, the group members get to know each other and establish a more open relationship. During the 21 years of existence, the project concept has been proven to be successful and thousands of Flemish entrepreneurs took part in the project.

The openness of the participants to exchange knowledge and also sensitive information is stimulated through the specific relationship building activity at the start-up weekend and a disclosure agreement within the group, but also through the very clear rule on confidentiality 'what is said in the group, stays in the group'. Furthermore, the careful selection of the members of each group, assuring the absence of competitors is an important point for facilitating open knowledge exchange.

Among the group members, deep ties are developed allowing the exchange of very sensitive or confidential information. This also allows the transfer of tacit knowledge among the participants when exchanging their know-how and experiences on firm management and improvement of management skills. This process is also called socialization when tacit knowledge from one firm is transferred to another through observation, imitation and practice. However, knowledge combination, articulation and internalization also readily take place in this kind of setting.

The success of this type of network activity largely depends on the quality and skills of the godfathers/godmothers, i.e. their ability to motivate the SME managers to listen and to share information with the other participants. Furthermore, the parti-cipating SME managers appreciated very much that they get open and honest feedback about their (daily) problems. Usually the participating SME managers are working alone or have only a few employees, and the group is a valuable 'board of advice' and place of regular self-evaluation and reflection to them. A disadvantage might be that there is no exchange between the different groups, neither among god-fathers/godmothers nor among SME managers. Nevertheless, after a group has finished the SME managers can join another starting group on the same or another topic. On some occasions, existing groups start a new group together but on another topic. This group is then also open to new participants, but the difference in openness among the group members that already knew each other and the newcomers is obvious.

VLAZ established a low-threshold platform for SMEs providing access to technical advice and to conduct research in collaboration with a university with the aim of reinforcing innovation in SMEs. Thus, the theoretical knowledge of the university is translated into more practical knowledge for the firm. If a firm has a certain technological problem, it contacts the network co-ordinator (university). In most cases, the network co-ordinator first carries out a study visit in the firm and then tries to provide a solution to the technological problem. If the problem is too complex, a small research project is set up where the network co-ordinator brokerages the contact between the firm and a specific research institution, both members of the network. Thus, the network activities are mainly bilateral between only one firm and the university (network co-ordinator), group projects are very rare. The co-operation between the firm and the university/research institution is kept highly confidential.

The networks is set up such in a way that there is no exchange between the firms that are members of this network because they are all members of the same branch of the food industry and therefore they are mainly competitors in a mostly highly competitive environment. The bilateral approach in this network between an individual firm and a research institution allows the open exchange of sensitive and even confidential information.

In these bilateral co-operations, deep ties are developed between the individual firm and the network co-ordinator (university). This allows the exchange of explicit knowledge from research made applicable to the company's situation creating new explicit knowledge (knowledge combination). For the more complex problems, knowledge socialization and articulation are also taking place, i.e. when the research institute is learning more about the specific problem of the company and creating a specific solution to this problem.

The focus of this network is foremost on providing SMEs with easy access to scientific knowledge tailor-made to their problems. The bilateral approach seems necessary owing to the (highly) competitive environment that the network members (competitors) are operating in. Even though there is no exchange in this network among the members, the firms are able to openly share their knowledge with the research institution they contacted. The network co-ordinator has a central role in this network as all contacts are mediated via him.

11.4.2 Role of the network in innovations in the food industry

Knowledge exchange is an important antecedent for innovation. As stated above knowledge exchange is part of the first step in the innovation production process, which is composed of three phases: knowledge accumulation, transformation and exploitation (Chen and Guan, 2011; Hansen and Birkinshaw, 2007; Roper *et al.*, 2008). In previous studies, the focus has mainly been on the knowledge-accumulation phase, and in particular, on knowledge exchange as this process seems to be the most difficult in its implementation (Lin *et al.*, 2012). Within this chapter, we extend this focus to the other two phases, namely transformation and exploitation into innovation outcomes, and the role of the network in these three phases. Furthermore, in this section, we investigate how the networks fulfil the three basic functions of intermediation with the focus on innovation, i.e. demand articulation, network formation and innovation process management (Klerkx and Leeuwis, 2009).

Flanders' FOOD is clearly focusing on all three aspects of the innovation-production process, with the main focus on overcoming the barriers of SMEs with regards to knowledge transformation and exploitation. For knowledge accumulation, the network co-ordinator monitors demand and starts up market-driven, application-oriented research based on the innovative needs of the companies. Thus, the network uses its contacts with domestic as well as foreign knowledge institutions such as universities, colleges and private research centres. Furthermore, existing scientific and technological knowledge is disseminated with the focus on the widest possible access for SMEs. The knowledge is disseminated to the SMEs via newsletters, seminars, workshops and individual consultancies. Within the market-driven, application-oriented research (collaborative projects), individual company visits (consulting) and the other network activities, attention is paid to knowledge translation with the aim of translating generic knowledge and expertise, both on a collective as well as an individual level, into specific knowledge. This knowledge translation mainly results in product and process innovation in the SMEs. Knowledge exploitation, i.e. when the firm derives economic profits from the innovation outcomes, is also strongly stimulated at both the individual and the collective level. For instance, collaborative projects must include an activity for knowledge exploitation, e.g. in the form of pilot-scale tests in some of the participating companies. However, the results need to be presented and shared with the whole of the collaborative project group. It is important that all participating companies have access and can exploit the knowledge accumulated in the project. Owing to the nature of these collaborative projects, only common problems can be tackled, while allowing each company to develop a specific innovation based on the knowledge derived from the project. At the individual level, e.g. on company visits, the firms are encouraged to think about what they have learned and how they can use that knowledge to fix or prevent (future) problems.

This network is thus clearly targeting its activities towards the demand and innovation needs of their members, fulfilling the demand-articulation task. Interactions between demand (firms) and supply (research institutions and knowledge centres) of innovation inputs are facilitated rather with regard to common problems in the food industry, and partners (participants of the collaborative projects) are not actively matched as every interested firm can subscribe to the projects. Network formation thus primarily involves the provision of the occasion for interaction. The innovation process is managed to the extent that learning is actively fostered on the collective as well as the individual level. Communication and alignment are areas which have been improved over time among the participating firms towards their peers and the researchers, and which might need further support from the network co-ordinator to function as a translator between businesses and researchers.

In **PLATO** the focus is on facilitating, in particular, knowledge transformation through the exchange of experiences of SME managers and the transfer of management know-how from guest speakers and mentors to SME managers and among SME managers. The aim is to establish a network between (very) small and larger companies, as well as among SMEs in particular, leading to organizational innovation. Owing to the nature of this type of innovation, knowledge exploitation only occurs over time and may only have a minor impact in the short term, while proving to be of high value in the long term. However, the monthly meetings and an open atmosphere allow the participants to accumulate knowledge on specific topics from peers and guest speakers and to obtain feedback on possible ways for knowledge transformation.

In this network, the participants of the mentored groups can choose the topics they want to tackle during the two-year period by themselves. However, the overall themes are set by the network co-ordinator allowing firms or groups of firms to articulate their demand for new themes as well. Within PLATO, special attention is paid to the composition of each group (network formation) in not allowing competitors in the same group, as well as carefully selecting participants with similar backgrounds and motivations for joining such a group. With regards to the innovation-process management, the network is careful to develop and improve collaboration among the participants through providing a mentor (two or three godfathers and/or godmothers per group) who safeguard further alignment and communication among the participants and foster learning from each other and the guest speakers invited to some of the monthly meetings.

In **VLAZ**, knowledge is accumulated and translated to solve (current) problems of the individual member companies. Thus, knowledge is accumulated between an individual firm and a research institution and translated for the former, as described in the previous section. In this network, knowledge is provided that may lead to product and process innovations, but there is no follow up to see whether the provided knowledge is exploited for economic profit from the innovation outcomes. However, the knowledge is tailor-made to the specific problem of the individual company and, owing to the high level of confidentiality assured, firms can achieve competitive advantage when exploiting the knowledge into economic profits.

Owing to the structure of this network, demand is articulated directly from the member firms, whenever they contact the network co-ordinator with regard to a technological problem they face. However, this type of demand articulation is facilitated through the low barrier of getting in contact with the network co-ordinator and the guarantee of confidentiality. Furthermore, although within this network only bilateral contacts are taking place, the network co-ordinator matches the firm in need of a solution to the right research institute that is able to provide an appropriate solution. Thus, for the network formation, the possible partners are carefully matched through scanning, scoping, and filtering. Finally, in this network, collaboration is not developed among the member firms, but between individual firms and research institutes. The management of the innovation process is thus focusing on the access of the member firms to scientific knowledge and expertise and the translation into tailor-made solutions for the firm in need. Over time an innovation culture could be (re)created in particular in SMEs.

11.4.3 Discussion of the results

Access to knowledge and information seems to be the most critical factor in particular for SMEs in the food sector in order to innovate (Avermaete *et al.*, 2004b; Cassiman and Veugelers, 2006; Gellynck *et al.*, 2007). Networks are more and more seen as the locus of innovation as they facilitate, stimulate and provide access to the necessary knowledge and the exchange of knowledge (Pittaway *et al.*, 2004; Powell *et al.*, 1996).

Our findings confirm the importance of networks in the process of knowledge exchange and innovation for SMEs in the food sector. The main findings of our case studies are summarized in Table 11.3, highlighting the similarities and differences between the three case studies on the different aspects investigated.

The most important role of the networks in knowledge exchange and innovation for SMEs in the food sector is to create the appropriate environment according to the type of knowledge and the step(s) in the innovation production process focused on. For example, in VLAZ, where the context is highly competitive, the focus is on bilateral contacts between a firm and a research institute to allow open exchange of rather tacit knowledge and sensitive information. In PLATO instead, the issue of competitors is tackled by the careful selection of the participants, paying most attention to the absence of competitors in order to facilitate open and honest

Table 11.3 Summary of main findings, highlighting similarities and differences between the three case studies

	Flanders' FOOD	PLATO	VLAZ
Context			
Level of competitiveness in environment/sector	Competitive	Not competitive	Highly competitive
Sectors	Food and food-related companies	Food and non-food	Specific branch of food sector
Type and members of network activity	• Businesses (SMEs and multi-national corporations) and researchers • Presence of competitors in groups • 1–2 year research projects	• Groups of SMEs guided by larger firms • No competitors • Business network	• SMEs and research • Bilateral co-operation • Short-term/ direct support
Objectives of network	Access to and valorization of scientific knowledge	Exchange platform for experiences and (daily) problems on managerial issues	Improve innovation culture in SMEs; low-threshold platform to access scientific, technological knowledge
Dyadic or collective	Collective	Collective	Dyadic/bilateral
Openness (explanation)	Open (presence of competitors and all other actors of the food chain)	Very open (clear rule of confidentiality and high trust within the group)	Very open (but only in dyad, between one firm and the network co-ordinator)
Ties formed	Mostly wide ties among members	Mostly deep ties among members	Deep ties with network co-ordinator

Knowledge exchange			
Main type of knowledge[1]	Explicit	Tacit	Tacit
Processes of knowledge exchange	Combination and internalization	Socialization, combination, articulation and internalization	Combination, socialization, and articulation
Focus in innovation production process	Active stimulation of all three steps: knowledge accumulation, transformation, and exploitation	Focus on transformation; inherent accumulation; no explicit stimulus of exploitation, appearance of effects only in the long term	Focus on accumulation and transformation; no follow up on exploitation, but can be expected because of tailored accumulation and transformation
Intermediation functions of the network	a) Up-front demand articulation b) Network formation in form of providing the occasion, no specific matching of participants c) IPM:[2] learning actively fostered	a) Active demand articulation of specific topics at occasion, up-front demand articulation of general topics b) Active network formation: careful matching of participants c) IPM:[2] focus on development and facilitation of collaboration	a) Direct demand articulation b) Careful matching of partners c) IPM:[2] focus on access to scientific knowledge and translation into tailor-made solutions

[1]Main type of knowledge exchanged in this network. It does not exclude the exchange of the other type of knowledge; [2]IPM: innovation process management.

exchange of even confidential information and rather tacit knowledge. Finally, in Flanders' FOOD, rather explicit knowledge is exchanged and on a less open level than in the other networks investigated. However, the involvement of all actors throughout the food chain is enriching the knowledge base of each participant.

With regard to the innovation production process, a very important task of the network is to actively stimulate knowledge transformation into innovation outputs such as new or improved technology or product prototypes. In particular, micro and small enterprises were found to face the greatest problems in grasping all relevant knowledge and in possessing sufficient human and financial resources for transforming the knowledge into tangible innovation outputs. In Flanders' FOOD, SMEs are accompanied also on an individual level if necessary in order to transform and exploit the scientific knowledge e.g. accumulated in the collective research projects. However, transformation and exploitation might not result in short-term effects, but effects show only after a certain time depending on the kind of innovation, as illustrated in the case of PLATO where the focus is on organizational innovation in contrast to the product and process innovation created in Flanders' FOOD and VLAZ.

The role of the network in knowledge exchange and innovation thus lies in the areas/aspects where food firms and in particular SMEs would have difficulties in accessing the relevant knowledge by themselves. Two of the networks under investigation focus on the transfer of scientific knowledge towards businesses (Flanders' FOOD and VLAZ), whereas, in the other network (PLATO) the focus is on transferring management skills among businesses. The first two networks thus tackle the problems related to the often occurring incongruity between the outcome of scientific research and the applicability to a business setting. The other network is thus focusing on the improvement of the (lack of) managerial skills of SME managers in many different areas, which leads to an overall improvement in their business success and enhances their sustainable competitive advantage in the long term.

As stated by Cowan *et al.* (2000), a balance between explicit and tacit knowledge is necessary in order to successfully innovate. In our three case-networks, both types are exchanged. It appears that the contextual aspects (i.e. level of competitiveness, sectors, types and members, objective) seem to determine which type of knowledge is more focused on, and is, thus, possibly easier to exchange. Moreover, the contextual aspects seem also to define the type of ties that can be developed between network members. In Flanders' FOOD, the focus is rather on explicit knowledge exchanged in rather wide ties in joint projects owing to the limitations of openness and the differences in cultural backgrounds of businesses and researchers, which is countered by the involvement of all actors throughout the food chain enriching the knowledge base of each participant in the joint projects. In contrast, in the other two networks, the focus is rather on tacit knowledge exchanged in deep ties, owing to openness in the relationships as confidentiality is assured. However, in VLAZ, ties are only formed with the network co-ordinator and in contrast, in Flanders' FOOD, deep ties are also developed among the participants in some joint projects where implicit knowledge might be exchanged as well.

In order to successfully facilitate and stimulate knowledge exchange, it is also very important that the network co-ordinators pay careful attention to the three basic functions of intermediation, i.e. demand articulation, network formation and innovation process management. In all three networks, it was stated that the network members, in particular the companies, participate in the network activities if it suits their needs for knowledge and innovation. The network members need to see the benefits of spending their time in the network's activities and sharing openly sometimes sensitive information and knowledge. For instance, in Flanders' FOOD, the benefits of the activities are further highlighted during the subsequent knowledge transformation, both at a collective as well as an individual level. On the other hand, in PLATO, where demand is articulated at the beginning of a network activity and the network is carefully formed, the information exchange is very open, leading to a greater potential for innovation. Even though it is mainly incremental, innovation has an important impact on the network member's business success. Last, but not least, in VLAZ, no collective activities take place owing to the highly competitive environment that the network members operate in. Nevertheless, competitive advantage is created for the individual network member through careful matching of partners, assuring confidentiality and translation of scientific research results into tailor-made solutions for the firm.

11.5 Conclusions and future trends

Three networks for knowledge exchange and innovation have been investigated in this study. All three networks are formal networks focusing on enhancing the competitive advantage of SMEs in the food sector through knowledge exchange and innovation. Nevertheless, all three networks follow very different approaches in order to facilitate, stimulate and support knowledge exchange and innovation among their members.

Our main conclusion is that networks have to adapt their approach and activities depending on the type of innovation sought, and on the impact of their activities on the competitive advantage of their members. The main role of networks for knowledge exchange and innovation is to provide access to knowledge in areas where SMEs, in particular, are facing difficulties in accessing the relevant knowledge by themselves, and to facilitate knowledge transformation and exploitation into tangible innovation outcomes. The networks under investigation each have a different approach in fulfilling this role. However, most important is that the network serves the needs of its members and, therefore, proper demand articulation, careful network formation and appropriate support in the innovation process management are found to be the more important tasks of the network co-ordinator. These intermediation activities are very important in overcoming the barriers faced by the SMEs related to knowledge exchange and innovation.

Therefore, networks have to adapt to the environment their members are operating in as well as to the type of knowledge that needs to be exchanged. Tacit knowledge seems to be more easily exchanged in a setting that allows open knowledge exchange based on deep ties. In a less open setting where mainly explicit knowledge is exchanged, the firms with the capability of grasping all relevant knowledge, might be more successful in the eventual transformation and exploitation of the knowledge into innovation, although this aspect can be actively tackled by the network co-ordinator as in the case of Flanders' FOOD. The creation of an environment that allows open exchange of both tacit and explicit knowledge is positively influencing the innovation outcome of the network members.

In conclusion, scientists need to adopt a more entrepreneurial culture in order to be able to provide results applicable to a business setting. On the other hand, businesses need to be more open in sharing the relevant knowledge necessary for the development of successful innovation. However, businesses also need to be more open towards tapping into knowledge external to their firm and transforming and exploiting this knowledge as well. Networks play an important role as facilitators and translators in the knowledge exchange process of converting external, scientific or business knowledge into internal knowledge, useable for the creation and implementation of successful innovations.

11.6 References

AGRAWAL, A. AND HENDERSON, R. (2002) Putting patents in context: exploring knowledge transfers from MIT. *Management Science*, **48**, 44–60.

ARGOTE, L. AND INGRAM, P. (2000) Knowledge transfer: a basis for competitive advantage in firms. *Ornanizational Behavior and Human Decision Processes*, **82**, 150–169.

AVERMAETE, T. AND VIAENE, J. (2002) On innovation and meeting regulation – the case of the Belgian food industry. *DRUID Summer Conference on 'Industrial dynamics of the new and old economy – who is embracing whom?'*. Copenhagen/Elsinore, 6–8 June 2002.

AVERMAETE, T., VIAENE, J., MORGAN, E. J. AND CRAWFORD, N. (2003) Determinants of innovation in small food firms. *European Journal of Innovation Management*, **6**, 8–17.

AVERMAETE, T., VIAENE, J., MORGAN, E. J. AND CRAWFORD, N. (2004a) The impact of firm characteristics and macroeconomic performance on innovation in small food firms: Case study from Belgium, Ireland and UK. In de Noronha Vaz, T., Viaene, J. and Wigier, M. (Eds.) *Innovation in small firms and dynamics of local development.* Warsaw, Scholar Publishing House, 79–95.

AVERMAETE, T., VIAENE, J., MORGAN, E. J., PITTS, E., CRAWFORD, N. AND MAHON, D. (2004b) Determinants of product and process innovation in small food manufacturing firms. *Trends in Food Science & Technology*, **15**, 474–483.

BATTERINK, M. H., WUBBEN, E. F. M., KLERKX, L. AND OMTA, S. W. F. (2010) Orchestrating innovation networks: the case of innovation brokers in the agri-food sector. *Entrepreneurship and Regional Development*, **22**, 47–76.

BORGATTI, S. P. AND FOSTER, C. F. (2003) The network paradigm in organizational research: a review and typology. *Journal of Management*, **29**, 991–1013.

BRENNAN, A. AND DOOLEY, L. (2005) Networked creativity: a structured management framework for stimulating innovation. *Technovation*, **25**, 1388–1399.

CANIELS, M. C. J. AND ROMIJN, H. A. (2008) Actor networks in strategic niche management: insights from social network theory. *Futures*, **40**, 613–329.

CASSIMAN, B. AND VEUGELERS, R. (2006) In search of complementarity in the innovation strategy: internal R&D and external knowledge acquisition. *Management Science*, **52**, 68–82.

CHEN, K. AND GUAN, J. C. (2011) Mapping the innovation production process from accumulative advantage to economic outcomes: a path modelling approach. *Technovation*, **31**, 336–346.

CHESBROUGH, H. (2003) *Open innovation: the new imperative for creating profiting from technology*, Boston, Harvard Business School Press.

CHESBROUGH, H. (2004) Managing open innovation. *Research & Technology Management*, January–February, 23–26.

COHEN, W. M. AND LEVINTHAL, D. A. (1990) Absorptive capacity: a new perspective on learning and innovation. *Administrative Science Quarterly*, **35**, 128–152.

COWAN, R., DAVID, P. A. AND FORAY, D. (2000) The explicit economics of knowledge codification and tacitness. *Industrial and Corporate Change*, **9**, 211–253.

DASKALAKIS, M. AND KAUFFELD-MONZ, M. (2005) Trust and knowledge in the behavioural dynamics of innovation networks. *EMAEE 2005 4th European Meeting on Applied Evolutionary Economics.* Utrecht.

DHANARAJ, C. AND PARKHE, A. (2006) Orchestrating innovation networks. *Academy of Management Review*, **31**, 659–669.

EDWARDS, T., DELBRIDGE, R. AND MUNDAY, M. (2005) Understanding innovation in small and medium-sized enterprises: a process manifest. *Technovation*, **25**, 1119–1127.

EISENHARDT, K. M. AND SCHOONHOVEN, C. B. (1996) Resource-based view of strategic alliance formation: strategic and social effects in entrepreneurial firms. *Organization Science*, **7**, 136–150.

ENZING, C. M., JANSZEN, F. H. A. AND OMTA, O. S. W. F. (2008) The impact of the openness of the innovation process on the short term and the long term market performance of new products: evidence from new product announcements of the Dutch food and drinks industry. *8th International conference on management in agrifood chains and networks.* Ede, The Netherlands, 28–30 May 2008, Wageningen Academic Publishers.

FREEMAN, C. (1991) Networks of innovators: a synthesis of research issues. *Research Policy*, **20**, 499–514.

GELLYNCK, X. AND KÜHNE, B. (2010) Horizontal and vertical networks for innovation in the traditional food sector. *International Journal on Food System Dynamics*, **1**, 123–132.

GELLYNCK, X., VERMEIRE, B. AND VIAENE, J. (2007) Innovation in food firms: contribution of regional networks within the international business context. *Entrepreneurship & Regional Development*, **19**, 209–226.

GLASER, B. G. (1978) *Theoretical sensitivity: advances in the methodology of grounded theory*, Mill Valley, CA: Sociology Press.

GRANOVETTER, M. (1973) The strength of weak ties. *American Journal of Sociology*, **78**, 1360–1380.

GRANOVETTER, M. (1985) Economic action and social structure – the problem of embeddedness. *American Journal of Sociology*, **91**, 481–510.

GRUNERT, K. G., JENSEN, B. B., SONNE, A.-M., BRUNSØ, K., BYRNE, D. V., CLAUSEN, C., FRIIS, A., HOLM, L., HYLDIG, G., KRISTENSEN, N. H., LETTL, C. AND SCHOLDERER, J. (2008) User-oriented innovation in the food sector: relevant streams of research and an agenda for future work. *Trends in Food Science & Technology*, **19**, 590–602.

GULATI, R. (1995) Does familiarity breed trust? The implications of repeated ties for contractual choice in alliances. *Academy of Management Journal*, **38**, 85–112.

GULATI, R. AND WESTPHAL, J. D. (1999) Cooperative or controlling? The effects of CEO–board relations and the content of interlocks on the formation of joint ventures. *Administrative Science Quarterly*, **44**, 473–506.

HALL, R. AND ANDRIANI, P. (2002) Managing knowledge for innovation. *Long Range Planning*, **35**, 29–48.

HAMDOUCH, A. (2010) Conceptualising innovation networks and clusters. In Laperche, B., Sommers, P. and Uzundis, D. (Eds.) *Innovation networks and clusters: the knowledge backbone.* Brussels, P.I.E. Peter Lang S.A.

HANSEN, M. T. AND BIRKINSHAW, J. (2007) The innovation value chain. *Harvard Business Review*, **85**, 121–130.

HOWELLS, J. (2006) Intermediation and the role of intermediaries in innovation. *Research Policy*, **35**, 715–728.

INKPEN, A. C. AND TSANG, E. W. K. (2005) Social capital, networks, and knowledge transfer. *Academy of Management Review*, **30**, 146–165.

KEURSTEN, P., VERDONSCHOT, S., KESSELS, J. AND KWAKMAN, K. (2006) Relating learning, knowledge creation and innovation: case studies into knowledge productivity. *International Journal of Learning and Intellectual Capital*, **3**, 405–420.

KLERKX, L. AND LEEUWIS, C. (2009) Establishment and embedding of innovation brokers at different innovation system levels: insights from the Dutch agricultural sector. *Technological Forecasting and Social Change*, **76**, 849–860.

LAZZARINI, S. G., CHADDAD, F. R. AND COOK, M. L. (2001) Integrating supply chain and network analyses: the study of netchains. *Journal on Chain and Network Science*, **1**, 7–22.

LEE, S., PARK, G., YOON, B. AND PARK, J. (2010) Open innovation in SMEs – an intermediated network model. *Research Policy*, **39**, 290–300.

LEFEBVRE, V. AND GELLYNCK, X. (2012) Intermediary organizations and development of social capital in interorganizational networks: a Belgian case study in the food sector. *10th Wageningen international conference on chain and network management 'multi-stakeholder dynamics in chains and networks'.* Wageningen, The Netherlands, 23–25 May 2012.

LIN, T.-C., WU, S. AND LU, C.-T. (2012) Exploring the affect factors of knowledge sharing behavior: the relations model theory perspective. *Expert Systems with Applications: An International Journal*, **39**, 751–764.

LUNDVALL, B.-A. (1995) *National systems of innovation: towards a theory of innovation and interactive learning*, London, UK, Pinter Publishers.

LUNDVALL, B.-A., JOHNSON, B., ANDERSEN, E. S. AND DALUM, B. (2002) National systems of production, innovation and competence building. *Research Policy*, **31**, 213–231.

MACKINNON, D., CUMBERS, A. AND CHAPMAN, K. (2002) Learning, innovation and regional development: a critical appraisal of recent debates. *Progress in Human Geography*, **26**, 293–311.

MARCH, J. G. (1991) Exploration and exploitation in organizational learning. *Organization Science*, **2**, 71–87.

MATTHYSSENS, P., VANDENBEMPT, K. AND BERGHMAN, L. (2006) Value innovation in business markets: breaking the industry recipe. *Industrial Marketing Management*, **35**, 751–761.

MURRAY, F. (2002) Innovation as co-evolution of scientific and technological networks: exploring tissue engineering. *Research Policy*, **31**, 1389–1403.

NG, D., SONKA, S. AND WESTGREN, R. (2003) Co-evolutionary processes in supply chain networks. *Journal of Chain and Network Science*, **3**, 45–58.

NONAKA, I. (1994) A dynamic theory of organizational knowledge creation. *Organization Science*, **5**, 14–37.

NONAKA, I. AND TAKEUCHI, H. (Eds.) (1995) *The knowledge-creating company: how japanese companies create the dynamics of innovation*, Oxford, University Press.

NONAKA, I., TOYAMA, R. AND KONNO, N. (2000) SECI, Ba and leadership: a unified model of dynamic knowledge creation. *Long Range Planning*, **33**, 4–34.

O'REILLY, S., HAINES, M. AND ARFINI, F. (2003) Food SME networks: process and governance – the case of Parma ham. *Journal of Chain and Network Science*, **3**, 21–32.

OMTA, O. (2004) Management of Innovation in chains and networks. In Camps, T., Diederen, P., Hofstede, G. J. and Vos, B. (Eds.) *The emerging world of chains and networks. Bridging theory and practice.* 's-Gravenhage, Reed Business Information.

OMTA, O. S. W. F. (2002) Innovation in chains and networks. *Journal of Chain and Network Science*, **2**, 73–80.

OWEN-SMITH, J. AND POWELL, W. W. (2004) Knowledge networks as channels and conduits: the effects of spillovers in the Boston biotechnology community. *Organisation Science*, **15**, 5–21.

PITTAWAY, L., ROBERTSON, M., MUNIR, K., DENYER, D. AND NEELY, A. (2004) Networking and innovation: a systematic review of the evidence. *International Journal of Management Reviews*, **5–6**, 137–168.

POWELL, W. W., KOPUT, K. W. AND SMITH-DOERR, L. (1996) Interorganizational collaboration and the locus of innovation: networks of learning in biotechnology. *Administrative Science Quarterly*, **41**, 116–145.

RESE, A. AND BAIER, D. (2011) Success factors for innovation management in networks of small and medium enterprises. *R&D Management*, **41**, 138–155.

RITTER, T. AND GEMÜNDEN, H. G. (2003) Network competence: its impact on innovation success and its antecedents. *Journal of Business Research*, **56**, 745–755.

ROPER, S., LOVE ,J. H. AND DU, J. (2008) Modelling the innovation value chain. *Research Policy*, **37**, 961–977.

SAAD, M., JONES, M. AND JAMES, P. (2002) A review of the progress towards the adoption of supply chain management (SCM) relationships in construction. *European Journal of Purchasing & Supply Management*, **8**, 173–183.

SARKAR, S. AND COSTA, A. I. A. (2008) Dynamics of open innovation in the food industry. *Trends in Food Science & Technology*, **19**, 574–580.

SAWHNEY, M., WOLCOTT, R. AND ARRONIZ, I. (2006) The 12 different ways for companies to innovate. *MIT Sloan Management Review*, **47**, 75–81.

SCOZZI, B., GARAVELLI, C. AND CROWSTON, K. (2005) Methods for modeling and supporting innovation processes in SMEs. *European Journal of Innovation Management*, **8**, 120–137.

SIMARD, C. AND WEST, J. (2006) Knowledge networks and the geographic locus of innovation. In Chesbrough, H., Vanhaverbeke, W. and West, J. (Eds.) *Open Innovation: researching a new paradigm.* Oxford, Oxford University Press, 220–240.

STEWART-KNOX, B. AND MITCHELL, P. (2003) What separates the winners from the losers in new food product development? *Trends in Food Science & Technology*, **14**, 58–64.

TÖDTLING, F., LEHNER, P. AND KAUFMANN, A. (2009) Do different types of innovation rely on specific kinds of knowledge interactions? *Technovation*, **29**, 59–71.

UZZI, B. (1996) The sources and consequences of embeddedness for the economic performance of organizations: The network effect. *American Sociological Review*, **61**, 674–698.

UZZI, B. AND GILLESPIE, J. (1999) Corporate social capital and the cost of financial capital: an embeddedness approach. In Leenders, R. and Gabbay, S. (Eds.) *Corporate social capital and liability*, Boston, Kluwer Academic Publishers.

WEAVER, R. D. (2008) Collaborative pull innovation: origins and adoption in the new economy. *Agribusiness*, **24**, 388–402.

ZENG, S. X., XIE, X. M. AND TAM, C. M. (2010) Relationship between cooperation networks and innovation performance of SMEs. *Technovation*, **30**, 181–194.

Part III

Establishing and managing open-innovation partnerships and networks

12

Managing open-innovation communities: the development of an open-innovation community scorecard

I. Blohm, University of St. Gallen, Switzerland, J. M. Leimeister, Kassel University, Germany and H. Krcmar, Technische Universität München, Germany

DOI: 10.1533/9780857097248.3.215

Abstract: This chapter presents a holistic approach for systematically managing open-innovation communities (OICs) on the Internet. Based on existing innovation and virtual community management approaches, an open-innovation community scorecard is developed that consists of an innovation process, a community member, an organizational learning and a financial perspective. The scorecard helps operators of OIC to systematically manage their OIC in order to harness the full potential of this mode of customer integration. An innovation performance measurement perspective is used and the most important goals of such OICs are transformed into operations that can be measured and managed. Based on its implementation at the software producer SAP, suggestions for its implementation are offered.

Key words: innovation community, virtual community, open innovation, balanced scorecard, management, performance measurement.

12.1 Introduction

In the twentieth century, many leading companies generated and commercialized ideas for innovations mainly through in-house research and development (R&D) laboratories. Today, companies are increasingly rethinking the ways of managing their innovation activities. Companies, regardless of whether they sell products or services, increasingly open up not only their innovation process but, also their production and sales processes to customers, who are seen as one of the biggest resources for innovations (Chesbrough, 2006; von Hippel, 2005). Open innovation and crowdsourcing are thus gaining recognition in research and practice. The positive impact of customer integration on company success has been demonstrated in various studies of open innovation (Gassmann, 2006; von Hippel, 2005). The increasing popularity of integrating customers into innovation development has led to the rise of innovation platforms on the Internet, as an effective means for integrating customers into the innovation process (Ebner *et al.*, 2009). As a consequence, various open-innovation communities have been started by food and life science companies. Prominent examples include MyStarbucksIdea in which more than 80 000 innovation ideas have been submitted or the Austrian supermarket Spar who has gathered more than 5000 user-generated designs for its shopping bags. Additionally, new approaches show how users of Facebook-based innovation groups can also create new prototypes of mustard and chutney for restaurants and food companies. These open-innovation communities (OICs) demonstrate this approach's enormous potential for the development of new products and various other business functions of a company such as marketing and human resources. Such communities now play a pivotal role for customer relationship management, public relations and the recruiting of new employees.

OICs develop in a highly dynamic way and, for most companies, they usually represent a process of innovation itself. Managing these communities needs a flexible management instrument that can cope with these dynamics as traditional approaches for managing R&D are rarely applicable and have reached their limits, e.g., most applied success measures can determined in the long run only and are outpaced by the community's speed of development. Moreover, many important performance indicators for OICs such as the quality of the submitted ideas cannot be measured with these approaches. Moreover, various approaches from the field of building and managing virtual communities exist but are not tailored to new product development. In order to utilize the entire potential of OICs, a holistic management instrument that is focused on a company's strategic objectives from various business units such as R&D, marketing and human resources is required. However, methods that are appropriate for managing open innovation are rarely available (Hilgers and Piller, 2009).

This chapter presents a balanced scorecard, specifically designed for the management of OICs, that incorporates established research on managing virtual communities and innovation development. Using Leimeister and Krcmar's (2006) community building and community management model, we define aims for managing OICs from the viewpoint of the innovation process, community users, organizational learning and finance. For each of these dimensions, we derive indicators for success measurement. Finally, we discuss how this approach has been applied by other companies and make suggestions for its implementation. The remainder of the chapter is structured as follows: In Section 12.2, we introduce and define OIC and highlight central foundations of managing virtual communities and innovation development. In Section 12.3, we present our approach and pinpoint the single dimensions as well as a strategy map unravelling central interrelations among the previously defined success measures. In Section 12.4, we discuss the implementation of the scorecard and in Section 12.5 our findings are summarized and we discuss the practical implications.

12.2 Introduction to open-innovation communities and their management

12.2.1 Open-innovation communities

Open innovation is the renunciation of the traditional innovation process that can largely be located within the boundaries of companies and that exclusively commercializes ideas developed by the internal R&D department (Chesbrough, 2006). By applying open innovation, companies overcome these boundaries, integrating various external stakeholders such as customers or business partners into the innovation development. One of the underlying principles of this approach is the utilization of the 'wisdom of crowds'. According to this principle of collective intelligence, the quality of a decision that is jointly made within a community, involving the contribution of every single member, can be superior to decisions made by single experts or individuals (Leimeister, 2010). This point is exactly where OICs are rooted; they are initiated and operated by companies to integrate customers and end-users in the early stages of the innovation process. Customers can submit innovation ideas and can collaborate on them, as the OIC enables its users to comment on ideas proposed by other people, rate them and/or jointly develop innovations. The community character promotes the creativity and quality of contributions because customers from different backgrounds, with different areas of expertise, and different skills and experiences can work together. Since OICs are based on the assumption that customer ideas represent their wishes and needs, companies can use them as an efficient strategy to gain needs information from their customers. Depending on the configuration of an OIC, customer ideas can also contain precise information concerning the practical implementation of an innovation. Thus, apart from needs information, solution information can also be acquired (von Hippel, 1994). The usage of this information can be described as a sequential process. During idea generation community members submit ideas using pre-structured input forms. Using commenting, 'wiki' and rating functionalities, the ideas can be discussed, further developed and rated by other users (Blohm *et al.*, 2011b, Riedl *et al.*, 2010).

12.2.2 Managing virtual communities

Generally, virtual communities can be defined as groups of people with common interests and needs who come together online (Hagel and Armstrong, 1997). From that perspective, OICs are virtual communities whose purpose for the community operator and the community members is to collaboratively generate and elaborate on innovation ideas. The community building and community management model (Leimeister and Krcmar, 2006) illustrates how virtual communities can systematically be built up and operated (Fig. 12.1). Community controlling that directly refers to the management of virtual communities based on performance indicators

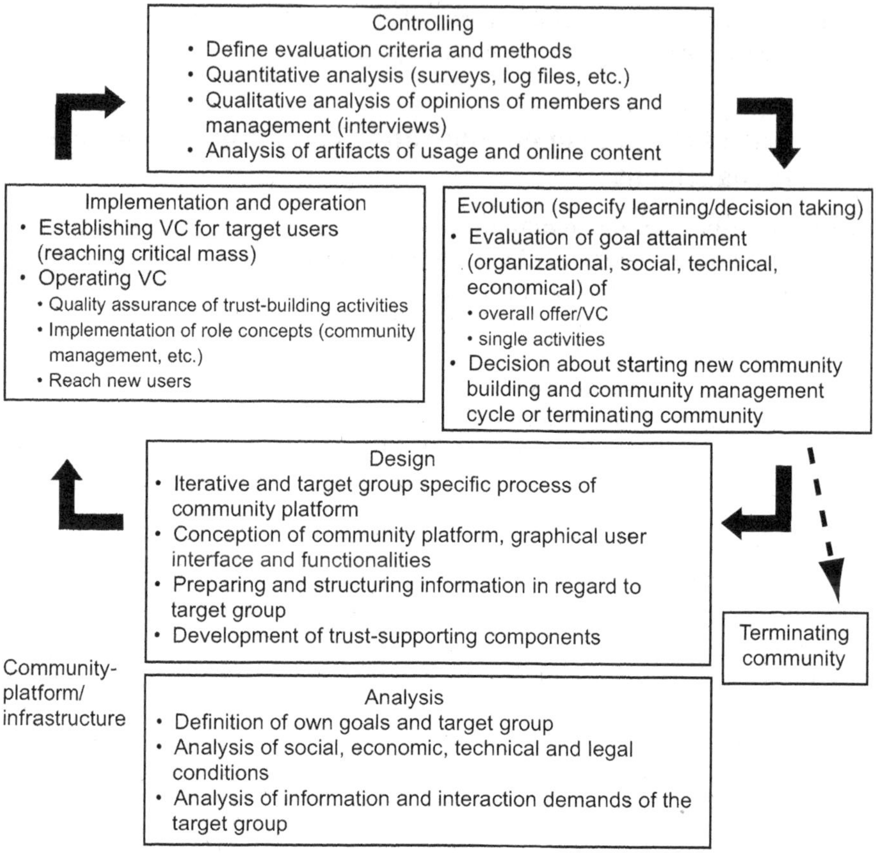

Fig. 12.1 Community building und community management model (Leimeister and Krcmar (2006)).

is a central element of this model. According to this model, the aim of this task is to estimate and review the success of activities for stimulating the virtual communities and to start counteractions in case *a priori* specified goals are not met. In this context, community controlling comprises the definition of appropriate evaluation criteria and methods. Usually the required data is directly collected within the virtual commnuities. By means of quantitative log file analysis, the behaviour of community members, e.g., the time users spent online can be investigated. Additionally, content analysis of user artefacts such as comments and ongoing member surveys are important instrument for identifying current problems and opportunities for improvement (Leimeister and Krcmar, 2006).

In general, the success of virtual communities can be measured from an operator or a user perspective. In this regard, the operator is the focal company that runs the OIC in order to integrate its customers in innovation development whereas the users can be considered as the participating customers. However, for operators, virtual communities can only be successful (e.g., in terms of customer retention), if the users perceive the virtual community offers as sufficiently attractive that they continue participating. Thus, success factors such as the number of members, the number of comments, the mode and number of user interactions, and the quality of contributions are pivotal from the user perspective (Preece, 2001). In a similar vein, the number of page views, the visit length, and the frequency of visits are important from an operator's perspective. Other important indicators for providers of successful virtual communities such as improvement of brand image, customer retention, and profits from additional sales or advertising revenues are often originated in the operator's marketing department (Cothrel, 2000).

12.2.3 Innovation-performance measurement

Innovation-performance measurement is the profit-oriented focus of all processes (planning, control, evaluation) that aim at the development of new technical knowledge regarding products, processes and applications (Gerybadze *et al.*, 2010). In this regard, research and development need to be distinguished. Research activities aim at the generation of new insights that will be integrated in the development of new products. Whereas research dominates in the early phase of the innovation process, the domain of its late phases is development where traditional management approaches are well applicable. Development processes are regularly well-structured and give easily measurable results. By contrast, success control and success management is far more difficult in the course of research. The highly variable and unstructured processes involved in research at the fuzzy front end of innovation development are hard to assess and often show no defined outcome. However, in the context of performance measurement, variables such as labour costs, number and quality of ideas, and learning new knowledge are suggested in order to measure this performance at the fuzzy front end (Janssen and Möller, 2009). These measures focus either on the inputs or the outputs, because the activities performed in the stage of research are frequently not well defined and ever changing. Thus, such measures can in principle be well applied to OICs.

12.3 Development of an open-innovation community scorecard

For many companies, the establishment of an OIC can be considered as an organizational innovation itself. For this reason, this process is often accompanied by uncertainty and a high level of variability. Moreover, crucial processes and routines for community management and idea implementation have still to be developed and have thus a rather unstructured character (Blohm *et al.*, 2011a). Moreover, many performance indicators for OICs such as the quality of the submitted ideas are difficult to assess and the objectives of various stakeholders such as marketing and human resources have also to be taken into account. Approaches for managing virtual communities are not applicable for systematically developing innovations as they serve the purpose of creating and building a

Table 12.1 Shortcomings of existing management approaches

Domain	Virtual community	Research and development
Main goals	• Customer retention • Increase sales	• Development of innovations • Gaining new knowledge
Central management instruments	• Log file analysis • User survey	• Performance measurement • Activity-based costing • Capital budgeting • Project planning and evaluation
Shortcomings regarding open-innovation community	• Not related to innovation development • Inconsistent performance indicators and variables • Little knowledge on interdependencies • Lack of systematic methods	• Performance indicators for OICs can not, or can only, be collected in the long run • Dynamic of OICs is not assessable • Limited possibility to assess pivotal performance indicators of OIC (e.g., idea quality) • Lacking consideration of marketing and of recruitment goals

vital community. In contrast, approaches from innovation management would be capable of doing so, but they completely lack the social focus of traditional virtual community management approaches (Table 12.1). In this context of OICs, it is often impossible to collect data on performance measures of product innovations or open-innovation projects. If at all, this information can only be aggregated in the long run. Moreover, the narrow focus of these approaches on innovation development often neglects crucial performance indicators of OICs.

The OIC scorecard developed in this chapter is grounded in a balanced scorecard (BSC) that is an approach for the holistic management of companies. A BSC converts a business strategy into operations and is usually derived from a company's vision with the four management perspectives of finance, customers, internal business processes as well as learning and growth (Kaplan and Norton, 1996). We adapted these dimensions to the context of OICs and refer to them as finance, members, innovation process and learning. In the following sections, we present our OIC scorecard (Fig. 12.2) and, for each of the dimensions, the most important generic objectives and performance indicators are defined that are important for virtually all OICs.

12.3.1 The innovation-process perspective

For new product development, the quantity and quality of the submitted ideas is one of the prevalent performance measures of an OIC because the creative potential of an OIC is reflected by its submissions. Thus, this potential grows with an increasing quantity and quality of ideas, as more substantial needs and solutions information can be collected by community operators (von Hippel, 1994, 2005). Customer ideas are creative products that are usually unspecific and show a low

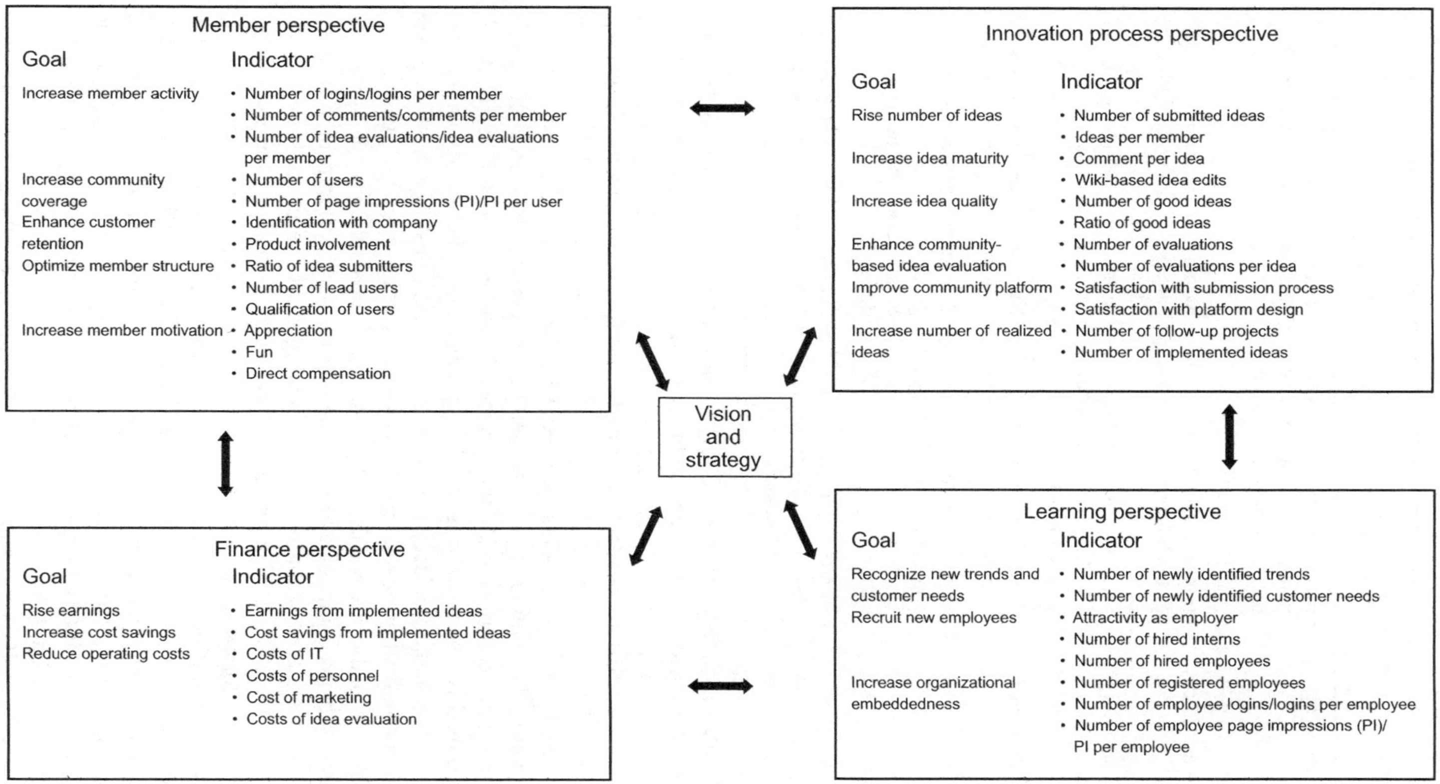

Fig. 12.2 Open innovation community scorecard.

Table 12.2 Goals and performance measures from the innovation process perspective

Goal	Definition	Indicator
Increase number of ideas	• Increase the OIC's creative potential by continuously receiving new ideas from community members	• Number of submitted ideas • Number of ideas per member
Increase idea maturity	• Improve idea quality through user collaboration	• Number of comments per idea • Number of wiki-based idea edits
Increase idea quality	• Increase effectiveness of OIC through rising ratio of good ideas	• Number of good ideas • Ratio of good ideas
Enhance community-based idea evaluation	• Assessing customer acceptance of singe ideas	• Number of idea evaluations • Number of evaluations per idea
Improve community platform	• Support user creativity with appropriate toolkit design	• Satisfaction with submission process • Satisfaction with platform design
Increase number of realized ideas	• Improving internal idea absorption and implementation processes	• Number of follow-up projects • Number of implemented ideas

degree of elaboration (Blohm, *et al.*, 2011b). Thus, methods of creativity research such as the consensual assessment technique (Amabile, 1996) that have been successfully used for the evaluation of customer-generated new product ideas (Blohm, *et al.*, 2011b) are needed for a timely quantification of an idea's quality. As ideas are often poorly elaborated, the enhancement of idea quality is a major goal from an innovation process perspective. The number of performed idea revisions within the community and its members is a crucial indicator for this, as these efforts significantly increase the quality of ideas (Blohm *et al.*, 2011b).

For idea quality, the ratio of 'good' ideas, those ideas that satisfy an *a priori* defined quality standard, is decisive (Reinig *et al.*, 2007) as only these ideas are considered for innovation development. Poor ideas, already known suggestions, or ideas of average quality are equally valueless, as their realization is not pursued by the operator. When a company adopts an idea for further development, follow-up projects that assess the feasibility of implementing ideas are set up (Blohm *et al.*, 2011a). Thus, the number and share of good ideas as well as the number of consecutive development projects are pivotal success measures. As well as the possibility of submitting or revising ideas, users are able to evaluate ideas of other community members (Berg-Jensen *et al.*, 2010; Riedl *et al.*, 2010). In this context, the community rating is an important indicator of the perception and acceptance of ideas among potential customers and the more customer ratings are aggregated, the higher are the possibilities for generalization to overall customer segments (Di Gangi and Wasko, 2009; Riedl *et al.*, 2010).

OICs provide toolkits that support idea generation and submission in order to promote the creativity of its users (Piller and Walcher, 2006, von Hippel and Katz, 2002). An effective design of the toolkit and the entire submission process can activate and motivate community users (Leimeister *et al.*, 2009) and defines shape and content of the submitted ideas (Blohm *et al.*, 2011a, Di Gangi *et al.*, 2010). Thus, the continuous improvement of the toolkit is a further goal that can be measured in terms of user satisfaction with the idea submission process and the platform (Goodhue and Thompson, 1995). Table 12.2 pinpoints all strategic goals and its performance indicators from the innovation process perspective.

12.3.2 The member perspective

In addition to the innovation process perspective, the customer dimension is of particular importance for the success of an OIC. Activity and coverage are decisive target variables for the management of virtual communities (Cothrel, 2000; Preece, 2001). Activity measures the number of user actions such as frequency of communication (Hilgers and Piller, 2009) and coverage describes, e.g., the number of page views or page impressions. In this regard, a critical mass of users has to be gained until a self-energizing circuit develops that leads to a continuous growth of the virtual community (Hagel and Armstrong, 1997). Indicators for measuring activity can be directly generated from the log files and comprise, e.g., the number

Table 12.3 Goals and performance measures from the member perspective

Goal	Definition	Indicator
Increase member activity	• Stimulate innovation relevant activities within the community	• Number of logins/logins per member • Number of comments/number of comments per member • Number of idea evaluations/ number of idea evaluations per member
Increase community coverage	• Rise number of members and page impressions in order to create self-energizing cycle of growth	• Number of community members • Number of page impressions (PI)/PI per member
Enhance member retention	• Enhance member retention and minimize churn rate	• Identification with community operator • Product involvement of members
Optimize member structure	• Increase innovative potential through attracting more qualified members	• Ratio of idea submitters • Number of lead users • Qualification of members
Increase member motivation	• Satisfy motives of community members	• Direct compensation • Appreciation • Fun

of logins per member, the share of idea contributors (idea contributors relative to all members), or the share of ideas (ideas in relation to members). A further goal is increasing coverage, e.g., by increasing the number of community members and page impressions in order to achieve self-energizing effects (Hagel and Armstrong, 1997; Hilgers and Piller, 2009).

Furthermore, qualitative factors also need to be considered (Preece, 2001). From a customer perspective, a member's identification with the community operator and their involvement are of particular importance. As community members are potential clients of the community provider, the virtual community should increase member and customer retention (Cothrel, 2000). Moreover, the member structure should be optimized so that the share of active members is growing and as many qualified users as possible are registered (Cothrel, 2000; Preece *et al.*, 2004; von Hippel, 2005). In this regard, OICs are highly attractive to so-called lead users – highly innovative users that have specific needs that are relevant for the future mass market (von Hippel, 2005). Moreover, it is particularly important that community members remain motivated in the long run, as otherwise the interest in the OIC and, as a consequence, also its creative potential, declines. For OICs, fun, social recognition and direct compensation of effort are the central motives of participants (Leimeister *et al.*, 2009). Table 12.3 highlights all the strategic aims of the user perspective alongside its respective measures.

12.3.3 The learning perspective

Because learning is a crucial source of a company's knowledge base, it is a major task of R&D (Cohen and Levinthal, 1990). Taking a content analytical approach such as a netnography (Kozinets, 2002) ideas and online discussions, conclusions on customer needs can be drawn and, thus, new trends identified (Füller *et al.*, 2006).

For effective organizational learning and to enable spillover effects from the OIC on innovation development, it is pivotal to embed the OIC into the organizational context of the community provider (Blohm *et al.*, 2011a). As more employees of the community operator are registered within the OIC, more employees increase their knowledge of customer needs and knowledge transfer between the community and the company is improved (Smith *et al.*, 2005). In this context, the number of members of R&D staff that are registered with a community as well as number of their logins and page impression are indicators that are easy to collect.

Committed and qualified community members have a comprehensive understanding of the community operator's products. They carry needs and solutions information that is pivotal for the development of innovations (von Hippel, 1994, 2005). Moreover, they are highly identified with the community operator. Therefore, the recruitment of new staff from the community, that can be measured by the number of interns and permanent employees, as well as the general user perception of employer attractiveness, are additional important performance measures of virtual communities. Table 12.4 shows all strategic aims derived from the learning perspective alongside its respective measures.

Table 12.4 Goals and performance measures from the customer perspective

Goal	Definition	Indicator
Recognize new trends and customer needs	• Recognition of new trends and customer needs through meta-analysis of online content	• Number of newly identified trends • Number of newly identified customer needs
Recruit new employees	• Using the OIC as platform for recruiting new staff	• Number of recruited interns • Number of recruited employees • Attractivity as employer
Increase organizational embeddedness	• Improving exchange between community operator and community members through greater engagement of employees	• Number of registered employees • Number of employee logins/logins per employee • Number of employee page impressions (PI)/PIs per employee

Table 12.5 Goals and performance measures from the financial perspective

Goal	Definition	Indicator
Rise earnings	• Increase earnings from implemented ideas	• Earnings from implemented ideas
Increase cost savings	• Increase cost savings from implemented ideas	• Cost savings from implemented ideas
Reduce operating costs	• Save necessary costs for operating OIC	• Costs of IT • Costs of personnel • Costs of marketing • Costs of incentives • Costs of idea evaluation

12.3.4 The financial perspective

From a financial perspective, the costs and profits resulting from an OIC have to be considered. A long term goal of OICs is to increase earnings and cost savings associated with the realization of ideas (Chesbrough, 2003, 2007; Hilgers and Piller, 2009). At the same time, operating costs of the OIC such as developing and hosting the IT platform and labour costs for the community management should be reduced to a minimum. Furthermore, there are costs for accompanying advertising efforts, rewards and incentives for community members as well as the expenditure of time needed for the assessment of ideas. Table 12.5 demonstrates all strategic aims derived from the financial perspective alongside its respective measures.

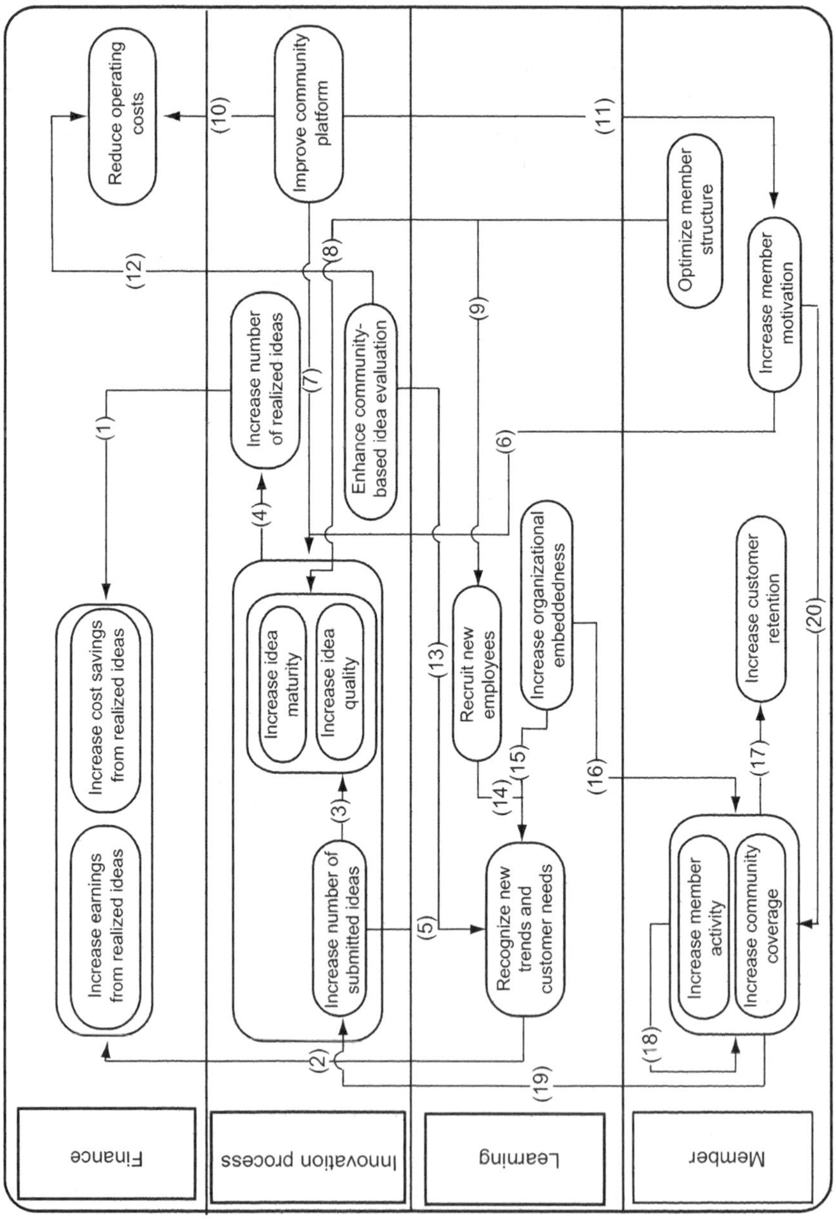

Fig. 12.3 Exemplary strategy map.

12.3.5 Cause and effect linkages between performance parameters

Strategy maps are a useful means to demonstrate the relationships of cause and effect between the single perspectives of a BSC and its performance measures. Thus, mutual dependencies and potential conflicts of varying sub-goals are highlighted. From these insights effective measures of goal achievement and strategy realization can be derived (Kaplan and Norton, 2004). The strategy map of the developed OIC scorecard is demonstrated in Fig. 12.3 and explained in Table 12.6.

Table 12.6 Relationships of the strategy map

Relationship	Description
1	With an increasing number of realized ideas, additional profits and cost savings can be realized (Chesbrough, 2003; Chesbrough, 2007; Hilgers and Piller, 2009)
2	Based on the identification of new trends and customer needs, the existing customer base can be served better and those departments with frequent customer contact will be able to realize additional profits (Kozinets, 2002; Füller *et al.*, 2006)
3	A growing number of submitted ideas increases the probability of gaining ideas that are worth implementing (Reinig and Briggs, 2008)
4	The higher the degree of an idea's elaboration, the higher its likelihood of realization (Blohm *et al.*, 2011b)
5	An increasing number of ideas gives rise to a larger pool of ideas, from which trends and needs can be derived by means of netnography (Kozinets, 2002; Füller *et al.*, 2006)
6	Increased member motivation increases quality and elaboration of ideas (Bretschneider, 2011)
7	Enhanced toolkits support the process of generating ideas so that the quantity and quality of ideas can be improved (von Hippel and Katz, 2002; Piller and Walcher, 2006)
8	Optimizing the member structure entails an enlarged share of lead users and higher qualified members and thus results in the production of a higher share of good ideas and a growing degree of elaboration (von Hippel, 2005; Yang *et al.*, 2009)
9	By optimizing the member structure, the community transforms into an attractive pool of potentially prospective employees (Hagel and Armstrong, 1997; Kim, 2000)
10	Advancement of the community platform reduces unnecessary costs of operation (e.g., easing the job of community management) (Preece, 2000; Di Gangi *et al.*, 2010)
11	An effectively designed toolkit activates the motives of the users so that a higher degree of motive attainment can be realized (Leimeister *et al.*, 2009)
12	Advancing the quality of idea evaluations of members eases the pre-selection of ideas and thus reduces costs of idea evaluation and selection (Leimeister *et al.*, 2009)

(Continued)

Table 12.6 Continued

Relationship	Description
13	Idea evaluations of members help to reveal new trends and customer needs (Di Gangi and Wasko, 2009; Riedl *et al.*, 2010)
14	When particularly appropriate members staff are recruited (e.g., lead users), a company employs competent personnel who can reveal trends and customer needs at an early stage (Cothrel, 2000; Kim, 2000)
15	The more members of staff of the community operator who actively take part in the community, the better is the understanding of ideas, trends and user needs as increased exposure enhances the community operator's absorptive capacity (Cohen and Levinthal, 1990; Blohm *et al.*, 2011a)
16	Direct feedback of representatives of the company is an important incentive for community members and directly affects their activity (Di Gangi *et al.*, 2010; Adamcyk *et al.*, 2011)
17	The more strongly a member feels connected to the community operator, the more he or she contributes to it (Kim, 2000; Preece and Shneiderman, 2009)
18	An increase in user activity up to a certain level causes a self-energizing cycle that, in turn, causes the continuous growth of a community (Hagel and Armstrong, 1997)
19	A more active and bigger community entails a larger number of submitted ideas (Yang *et al.*, 2009)
20	Every member has certain motives to contribute to a community. The better those motives are satisfied, the higher is his or her activity (Preece, 2000; Preece, 2001)

12.4 Implementation of the open-innovation scorecard

The OIC scorecard was empirically tested in the scope of the 'SAPiens' research project that aimed at using OICs for the development of new software at SAP, a global producer of enterprise resource planning (ERP) software solutions. The data needed for using the OIC can easily be collected from community log files, qualitative interviews with the community management and a survey of community members. Appendix 1 gives an overview over which indicators can be derived from which data sources and Appendix 2 lists the items of the member survey.

The application of the OIC at SAP could clearly demonstrate an OIC's various contributions to innovation development as well other business functions such as marketing reflecting SAP's high satisfaction with the community. Moreover, based on the application of the OIC scorecard, the community could be developed further and precise figures for the future management be obtained. The application of the scorecard pinpointed that collaborative activities among community members happened rarely and in an unsystematic fashion. However, the scorecard showed that the best ideas and concepts were usually developed by a group of participants. Based on this analysis, additional functionalities, which support collaboration among users, e.g. the possibility of group submissions, were developed in order to increase idea quality. Additionally, collaboration of community members was stimulated by additional organizational measures such as incentives.

The OIC scorecard implies that the OIC is well embedded into the processes and the organizational structure of the community operator. It is assumed that the submitted ideas are infused into the innovation system of the operator. However, at SAP the OIC was started as a 'pilot project' and was thus integrated only rudimentarily into the company. The implementation of the OIC pinpointed that a structural and process-based integration of OICs is a pivotal prerequisite for successfully operating OICs. In this regard, the community management had a pivotal role in integrating the OIC into SAP. It had to act as a boundary spanner between the company and the OIC. In several instances, ideas from the community were only followed up internally because community management convinced other employees of the potential of the customer ideas. As the community management plays such an important role it should generally be interdisciplinary in nature so that is has good access to all relevant business units. Moreover, it was recognized that the community management and the other participating employees had to learn the 'language of the community' in the first instance in order to unravel the value of the various contributions of the participants. Furthermore, from that perspective, interdisciplinary community management teams should be advantageous.

However, in implementing the OIC scorecard, it has to be acknowledged that a balanced scorecard is always an instrument for the execution and not for the formulation of business strategies. Thus, a common understanding among all participating business units as well as the top management of SAP had to be developed before its usage. However, in this regard, marketing and R&D had different goals with the OIC. Whereas R&D was mainly interested in the ideas and concepts originated by the customers, marketing's main involvement was about creating a positive attitude towards SAP. Thus, the OICs must generally be integrated into the superordinate corporate strategy in order to have a reference framework for the activities performed online together with the customers. In this regard, the generic OIC goals that are suggested in this chapter need to be adapted to the specific context of other companies before implementation of the OIC scorecard. Moreover, some data such as log data or member surveys can be collected quite easily, whereas other information such as information about the quality of the ideas can only be elicited with considerable effort. Implementing the OIC scorecard, community operators should be aware that holistic OIC management is not an end in itself. Thus, a clear focus on the most important strategic goals is necessary, and, having achieved this, a first step towards a continuously ongoing OIC management is complete.

12.5 Conclusion and future trends

This chapter presented a newly developed balanced scorecard approach for the management of OICs. This approach enables operators of OICs to manage their OICs effectively and can easily be accomplished, without limiting the potential of OIC as an 'innovation playground' that is essential for successfully using OICs.

The application of this approach in other contexts, revealed that the results that are obtained are highly dependent on the nature of the ideas developed in the OIC. In this regard, submitted ideas are highly influenced by the characteristics of the product that an idea strives to improve. In existing OICs, user ideas are represented in different formats. Frequently, users may submit their ideas as textual descriptions. However, there are also OICs that engage users to submit graphical illustrations or designs. Depending on whether users describe ideas in written form or submit graphical illustrations of their concepts, the characteristics of performance measures varies. Thus, our approach serves for the comparison of different OICs only if the submitted ideas of these virtual communities are of similar type.

This chapter has two important theoretical contributions. Firstly, the concept of balanced scorecards is applied to a new context. Our developed OIC scorecard helps to evaluate the effectiveness of OICs and most of our success measures are applicable for other open-innovation approaches such as idea competitions as well. Secondly, our approach views OICs from a holistic perspective, unravelling central cause and effects for various success criteria. This enhances our understanding about innovation communities and helps to develop a comprehensive theory of open innovation – a research stream that mainly derived from the observation of practice.

In practice, the chapter offers a theoretically grounded and empirically tested instrument for managing OICs. It helps to visualize the value innovation communities endow and provides practical advice for capturing this value. The continuous application of this instrument unravels the dynamic development of OICs, helps to assess the impact of activities performed for measuring the community and provides guidance for systematically evaluating the engagement in OICs. Additionally, the presented OIC scorecard is useful to justify an OIC's various contributions to the different stakeholders involved such as R&D, marketing, and human resources.

12.6 References

ADAMCYK, S., J. B. A. HALLER, A. C. BULLINGER AND K. M. MÖSLEIN. 2011. Knowing is silver, listening is gold: on the importance and impact of feedback in IT-based innovation contests. *Wirtschaftswinformatik Proceedings 2011*, Zurich.

AMABILE, T. M. 1996. *Creativity in context. Update to social psychology of creativity.* Westview Press, Oxford.

BERG-JENSEN, M., C. HIENERTH AND C. LETTL. 2010. Forecasting the attractiveness of user-generated design. An empirical study within the Lego user commnity. *2010 Academy of Management Annual Meeting Proceedings*, Montreal, Canada, p. 1.

BLOHM, I., O. KÖROGLU, J. M. LEIMEISTER AND H. KRCMAR. 2011a. Absorptive capacity for open innovation communities: learnings from theory and practice. *2011 Academy of Management Annual Meeting*, San Antonio, Texas/USA.

BLOHM, I., U. BRETSCHNEIDER, J. M. LEIMEISTER AND H. KRCMAR. 2011b. Does collaboration among participants lead to better ideas in IT-based idea competitions? An empirical investigation. *Int. J. Networking Virtual Organizations*, **9**(2), 106–122.

BRETSCHNEIDER, U. 2011. Die Ideen Community zur Integration von Kunden in die frühen Phasen des Innovationsprozesses. Empirische Analysen und Implikationen für Forschung und Praxis. PhD thesis, Technical University of Munich.

CHESBROUGH, H. W. 2003. The era of open innovation. *MIT Sloan Manage. Rev.*, **44**(3), 35–41.

CHESBROUGH, H. 2006. *Open innovation: the new imperative for creating and profiting from technology.* Harvard Business School Press, Boston, MA.

CHESBROUGH, H. W. 2007. Why companies should have open innovation business models. *MIT Sloan Manage. Rev.*, **48**(2), 22–28.

COHEN, W. AND D. A. LEVINTHAL. 1990. Absorptive capacity: a new perspective on learning and innovation. *Admin. Sci. Quart.*, **35**(1), 128–152.

COTHREL, J. P. 2000. Measuring the success of an online community. *Strategy & Leadership*, **28**(2), 17–21.

DI GANGI, P. M. AND M. M. WASKO. 2009. Steal my idea! Organizational adoption of user innovations from a user innovation community: a case study of Dell IdeaStorm. *Decision Support Systems*, **48**(1), 303–312.

DI GANGI, P. M., M. M. WASKO AND R. E. HOOKER. 2010. Getting customers' ideas to work for you: learning from DELL how to succeed with online user innovation communities. *MIS Quart. Executive*, **9**(4), 213–228.

EBNER, W., J. M. LEIMEISTER AND H. KRCMAR. 2009. Community engineering for innovations: the ideas competition as a method to nurture a virtual community for innovations. *R&D Manage.*, **39**(4), 342–356.

FÜLLER, J., M. BARTL, H. ERNST AND H. MÜHLBACHER. 2006. Community based innovation: how to integrate members of virtual communities into new product development. *Electr. Commerce Res.*, **6**(1), 57–73.

GASSMANN, O. 2006. Opening up the innovation process: towards an agenda. *R&D Manage.*, **36**(3), 223–228.

GERYBADZE, A., U. HOMMEL, H. W. REINERS, D. THOMASCHEWSKI, P. SCHENTLER, F. LINDNER AND R. GLEICH. 2010. Innovation performance measurement. *Innovation and international corporate growth*, Springer, Berlin Heidelberg, 299–317.

GOODHUE, D., AND R. THOMPSON. 1995. Task-technology fit and individual performance. *MIS Quart.*, **19**(2), 213–236.

HAGEL, J. AND A. G. ARMSTRONG. 1997. *Net gain: expanding markets through virtual communities.* McGraw Hill Professional, Boston, MA (USA).

HILGERS, D. AND F. T. PILLER. 2009. Controlling for open innovation – Theoretische Grundlagen und praktische Konsequenzen. *Zeitschrift für Controlling*, **21**(2), 5–11.

JANSSEN, S. AND K. MÖLLER. 2009. Performance measurement von Produktinnovationen. Konzepte, Instrumente und Kennzahlen des Innovationscontrollings. *Zeitschrift für Betriebswirtschaft*, **21**(2), 89–96.

KAPLAN, R. S. AND D. P. NORTON. 1996. *Balanced scorecard: translating strategy into action.* Harvard Business School Press, Boston, MA (USA).

KAPLAN, R. S. AND D. P. NORTON. 2004. *Strategy maps: converting intangible assets into tangible outcomes.* McGraw-Hill, Boston, MA (USA).

KIM, A. J. 2000. *Community building on the web: secret strategies for successful online communities*, Boston (MA), Addison–Wesley Longman.

KOZINETS, R. 2002. The field behind the screen: using netnography for marketing research in online communications. *J. Marketing Res.*, **39**(1), 61–72.

LEIMEISTER, J. M. 2010. Collective intelligence. *Business Inform. Syst. Eng.*, **52**(4), 239–242.

LEIMEISTER, J. M. AND H. KRCMAR. 2006. Community-engineering: Systematischer Aufbau und Betrieb Virtueller Communitys im Gesundheitswesen. *Wirtschaftsinformatik*, **48**(6). 418–429.

LEIMEISTER, J. M., M. HUBER, U. BRETSCHNEIDER AND H. KRCMAR. 2009. Leveraging crowdsourcing – activation-supporting components for IT-based idea competitions. *J. Manag. Inform. Systems*, **26**(1), 197–224.

PILLER, F. T. AND D. WALCHER. 2006. Toolkits for idea competitions: a novel method to integrate users in new product development. *R&D Manage.*, **36**(3), 307–318.

PREECE, J. 2000. *Online communities: designing usability, supporting socailibility*. Chichester, UK, Wiley.

PREECE, J. 2001. Sociability and usability in online communities: determining and measuring success. *Behavior Inform. Technol. J.*, **20**(5), 347–356

PREECE, J. AND B. SCHNEIDERMAN. 2009. The reader-to-leader framework: motivating technology-mediated social participation. *AIS Trans. Human-Comput. Interact.*, **1**, 12–32.

PREECE, J., B. NONNECKE AND D. ANDREWS. 2004. The top five reasons for lurking: improving community experiences for everyone. *Computers in Human Behavior*, **20**(2), 201–223.

REINIG, B. A. AND R. O. BRIGGS. 2008. On the relationship between idea-quantity and idea-quality during ideation. *Group Decision Negotiat.*, **17**, 403–420.

REINIG, B. A., R. O. BRIGGS AND J. F. NUNAMAKER JR. 2007. On the measurement of ideation quality. *J. Manage. Inform. Syst.*, **23**(4), 143–161.

RIEDL, C., I. BLOHM, J. M. LEIMEISTER AND H. KRCMAR. 2010. Rating scales for collective intelligence in innovation communities: why quick and easy decision making does not get it right. *Proceedings of the 2010 International Conference on Information Systems*, St. Louis, MI, USA.

SMITH, K. G., C. J. COLLINS AND K. D. CLARK. 2005. Existing knowledge, knowledge creation capability, and the rate of new product introduction in high-technology firms. *Acad. Manage. J.*, **48**(2), 346–357.

VON HIPPEL, E. 1994. 'Sticky information' and the locus of problem solving: implications for innovation. *Manage. Sci.*, **40**(4), 429–439.

VON HIPPEL, E. 2005. *Democratizing innovation.* MIT Press, Cambridge, MA.

VON HIPPEL, E. AND R. KATZ. 2002. Shifting innovation to users via toolkits. *Manage. Sci.*, **48**(7), 821–833.

YANG, Y., P.-Y. CHEN AND P. PAVLOU. 2009. Open innovation: an empirical study of online contests. In *International conference on information systems (ICIS'09)*, Phoenix, AR, AIS, pp. 1–15.

12.7 Appendix 1: data sources of success measures

Perspective	Goal	Indicator	Data source
Innovation process	Increase number of ideas	• Number of submitted ideas	• Log data
		• Number of ideas per member	• Log data
	Increase idea maturity	• Number of comments per idea	• Log data
		• Number of wiki-based idea edits	• Log data
	Increase idea quality	• Number of good ideas	• Expert evaluation
		• Ratio of good ideas	• Log data/expert evaluation
	Enhance community-based idea evaluation	• Number of idea evaluations	• Log data
		• Number of evaluations per idea	• Log data
	Improve community platform	• Satisfaction with submission process	• Member survey
		• Satisfaction with platform design	• Member survey

Perspective	Goal	Indicator	Data source
	Increase number of realized ideas	• Number of follow-up projects • Number of implemented ideas	• Interview community management • Interview community management
Member	Increase member activity	• Number of logins/logins per member • Number of comments/ number of comments per member • Number of idea evaluations/number of idea evaluations per member	• Log data • Log data • Log data
	Increase community coverage	• Number of community members • Number of page impressions (PI)/PI per member	• Log data • Log data
	Enhance member retention	• Identification with community operator • Product involvement of members	• Member survey • Member survey
	Optimize member structure	• Ratio of idea submitters • Number of lead users • Qualification of members	• Log data • Interview community management/ member survey • Member survey
	Increase member motivation	• Direct compensation • Appreciation • Fun	• Member survey • Member survey • Member survey
Learning	Recognize new trends and customer needs	• Number of newly identified trends • Number of newly identified customer needs	• Interview community management • Interview community management
	Recruit new employees	• Number of recruited interns • Number of recruited employees • Attractivity as employer	• Interview community management • Interview community management • Member survey
	Increase organizational embeddedness	• Number of registered employees • Number of employee logins/logins per employee • Number of employee page impressions (PI)/ PIs per employee	• Log data • Log data • Log data
Finance	Rise earnings	• Earnings from implemented ideas	• Management accounting
	Increase cost savings	• Cost savings from implemented ideas	• Management accounting

Perspective	Goal	Indicator	Data source
	Reduce operating costs	• Costs of IT	• Management accounting
		• Costs of personnel	• Management accounting
		• Costs of marketing	• Management accounting
		• Costs of incentives	• Management accounting
		• Costs of idea evaluation	• Management accounting

12.8 Appendix 2: member survey

Employer attractivity
I attended at the SAPiens innovation community because…
…I hoped that I could work for SAP after completing my university program.
…I hoped to call attention to SAP.

Identification
I attended at the SAPiens innovation community because…
… I identify with the SAP brand.
… I'm into SAP and because of that I wanted to support SAP.

Lead userness
I attended at the SAPiens innovation community because…
…I recognized a need that has not been fulfilled by SAP yet.
…I'm not satisfied with SAP software or the existing business strategies of SAP.

Involvement
SAP-Software is …
unimportant – important;
dull – exciting;
boring – interesting.

Knowledge
I have technical knowledge about SAP Software.
Compared to my friends I know pretty much about SAP software and the company in general.
The features of SAP software are well-known to me.

13

The evolution of partnering in open innovation: from transactions to communities

Kevin G. McFarthing Ph. D., Innovation Fixer Ltd, UK

DOI: 10.1533/9780857097248.3.235

Abstract: The selection, engagement and management of partnerships are crucial to the success of open innovation, and factors related to this are described. Increasingly partnerships are becoming more complex and sophisticated, and evolving to incorporate elements of community. Approaches to the building and structuring of relationships are addressed, together with the development of ecosystems. Embracing the intangible human factors described in this chapter as well as the tangible facts will be key to maintaining success in the future.

Key words: open innovation, partnership, community, management, human factors.

13.1 Introduction

Kraft Foods estimate that 98% of all food-related intellectual property (IP) exists outside their company. Procter & Gamble believe that there are two hundred times as many experts in relevant fields outside the company than there are inside. Leaders at General Mills state that employees must see 'the world as their lab, not the lab as their world'.

These statements and many more underline the reasons why the food and drink industry is adopting open innovation. The strategic reasons behind this adoption are not just to gain access to more IP and technical expertise in a transactional manner. They include a need to get to market fast; to tap into competencies that companies may not possess or in which they are not sufficiently competitive; to gain greater power in the creative process; and last, but not least, to build relationships with partners who can add significant value to the future flow of innovation.

In essence, companies need to adapt or change their business model. The route to market for 'outside-in' open innovation, the use of ideas, resources and technology from outside the company, may still be the same. However, the inbound route to the start of the innovation 'funnel' is significantly different, and has important implications for R&D, procurement and corporate culture.

Open innovation is defined elsewhere in this book. For the purposes of this chapter, the focus is exclusively on outside-in open innovation. Even though this book focuses on the food and drink industry, there are lessons to be learned from companies in other sectors, and these examples are discussed as appropriate.

All open innovation collaborations involve relationships between partners. The stronger that relationship, the more likely it is that successful business outcomes result. In this chapter, I describe how open innovation collaborations are evolving from single-project, transactional approaches to extended enterprise models with community spirit.

13.2 Identifying and securing partners

13.2.1 The strategic context

All innovation should flow from strategy and be placed in a context where it makes sense from the perspective of achieving strategic goals. This is true whether the innovation is classified as breakthrough, for example a major innovation that opens up a new category, or incremental, for example a short-term response to competitive activity.

All companies should have a clear corporate strategy, complemented by divisional/category/brand strategies as appropriate. Within this, an innovation strategy should be developed. Cascading from that, a technology strategy should inform the relevant people of technology needs within the timeframe of the strategic plan. Finally, the open innovation strategy should map out what is to be developed internally, what sourced outside, what areas are to be the subject of external search, the resources to be applied, etc.

Only when a strategic framework is applied to innovation is it clear whether the work done in open innovation is relevant to corporate strategic goals. This is important not only for the effectiveness of innovation, i.e. making sure the right choices are made but also for the efficiency of the innovation, i.e. ensuring valuable resources are applied to the areas where there is likely to be the most difference to growth.

Open innovation can also figure in strategic targets. Perhaps the most famous example is that of CEO AG Lafley's objective of 50% of Procter & Gamble's ideas coming from outside the company. Lafley not only wanted the additional innovation, he wanted to change P&G's culture to be more innovative and less inward facing.

Setting the target and formulating strategy is not enough. The path to achievement needs to be facilitated by the right management structures, resources, training and top-level support. The words must be matched by the actions.

13.2.2 Readiness to build partnerships

It may be a generalization that to be good at open innovation, a company must already have a successful track record of internal innovation. This is because many of the competencies required to succeed at open innovation are the same, for example having a collaborative mindset rather than a focus on a 'silo'; an ability to understand the consumer; rapid commercialization; enabling processes rather than a limiting bureaucracy; and a creative approach to risk taking rather than too much caution.

Therefore, if a company is not good at innovation per se, attention should be paid to putting its own house in order before devoting too many resources to open innovation. It is not a panacea, external ideas are not necessarily better than internal ones; they are simply a much greater source of quantity and diversity of opportunity. Equally, if a company is very formal and process-driven, it is less able to successfully implement open innovation than one that is driven by communication, networking and partnership.

The corollary is that many of the tools, processes and mindset used to accelerate open innovation can also be used to complement internal innovation. For example, General Mills has an internal social networking tool, with similar qualities to Facebook, to encourage internal collaboration, which brings diversity of source and thinking to the challenges of innovation. Other tools exist to connect companies internally, including Innocentive@Work, Spigit, Ideascale, Yammer and Sharepoint.

A key element to building innovation is to have both a structure and culture that is internally collaborative. Companies such as Reckitt Benckiser that work well internally, with strong formal and informal connections, are more likely to also be good at external collaboration and innovation.

13.2.3 Definition of a 'partner'

I apply a different definition for the word 'partner' than that to be found in dictionaries for the purpose of this chapter. The dictionary terms describe people or organizations associated with others in a shared endeavour, at the same time associating the same word with a marriage, so there is clearly a wide spectrum of definition. I believe that, in the context of open innovation, we should bring in the sense of commitment without the intention to have an unbreakable bond. I would therefore offer the following definition of an open-innovation partner: 'a company or individual with whom one or more projects are pursued together, characterized by a joint commitment to pursuing further opportunities in the future.' This definition incorporates the key elements of collaboration, projects, commitment and the longer term.

13.2.4 Partner identification and selection criteria

Choosing the right partner is probably the most important choice in open innovation. If you are fortunate enough to have options on the technical solution under consideration, deciding who is the best partner is crucial. Before the partner search starts, it is important to understand the criteria on which the final partner is chosen.

There is no right or wrong way to do this but, as with many factors to be considered in open innovation, the choice must make sense for the company and its way of working. One option is to have many different factors with complex sophisticated scoring systems, applying weighting to each factor depending on its importance. A second option is to list the factors under consideration and rank each potential partner in order (first, second, etc.). The latter option is often used as a checklist to prompt assessment and thinking so that a better decision is reached.

Whichever approach is taken, both the hard and soft factors involved should be considered. The 'hard' facts relate to company size, IP strength, and ability to deliver etc. The 'soft' factors are those such as culture fit, operational style and reputation (McFarthing, 2011). There is an old adage that a bad contract can still deliver a good outcome, but a bad relationship rarely does. If one subscribes to this, then it is crucial that the partner selection criteria address the soft factors, and time is taken to gather the information and experience that can help the decision.

13.2.5 Assessing cultural fit

It is often said that 'culture eats strategy for breakfast'. The perfect strategic rationale and implementation plans for an open innovation project and relationship may be decided, but if the partners do not get on, all this effort is in vain. The cultures do not need to match completely, but successful partnership depends on compatibility.

If only one criterion is used to assess the cultural fit between potential partners, simply ask yourself 'can we work with these people?' Note that the question relates to the people, not the company. There is always a risk that the people you meet initially in the open innovation context are not truly representative of the ones you will be working with more closely as projects progress.

If possible, it is good to ask the opinion of key contacts in your network, confidential disclosure agreement (CDA) permitting. Ask for references, to talk directly to some of their current partners, again respecting obligations of confidentiality.

There is a tendency among large companies to form partnerships with bodies from the same country (Mortara *et al.*, 2009). This can go beyond just nationality and result in partnerships with 'the usual suspects'. If this is genuinely and objectively the right thing for all involved, then it should not be criticized, but companies should continually challenge themselves to look more widely for potential partners.

Choosing a partner also needs consideration of 'path dependence' (Slowinski and Sagal, 2003). In essence, this concept means that once a company is part-way down the path of partnership with another company, its freedom of action becomes limited and it must either continue down the path at full speed, or seek a way out to a potentially disadvantaged position for both partners.

Path dependence also has a strong upside in the essence of partnership as the closer partners become, the more they learn about each other and the more mutually beneficial possibilities emerge.

13.3 Building and structuring relationships

Building and structuring open innovation relationships requires careful consideration of many factors. There is much to learn from the experience of large strategic alliances, as well as having a clear view of the boundaries of the collaboration. Each partner needs to understand how they will make a return, and how intellectual property will be handled. It is then appropriate to design the relationship structure. Finally, large companies must consider the perspective of small companies with whom they may be working.

13.3.1 The experience of strategic alliances

Much work has been done analyzing the structure and operation of strategic alliances between corporations (Darby, 2006; Doz and Hamel, 1998; Slowinski and Sagal, 2003; Spekman *et al.*, 2000). There are great similarities with open innovation in their approach; in effect a strong open innovation partnership is a strategic alliance. It is therefore logical that we may learn from this existing body of knowledge. A full assessment of these learnings should be the subject of another text; for the purposes of this chapter, I simply summarize some key aspects.

The most basic point is to ensure strategic alignment in any kind of partnership. The companies involved must first understand their strategic plans and the ties between them and the potential alliance. Next, both companies should reveal the strategic plan clearly to the other party. Finally, they must agree on intentions, commitments, rights and limitations that satisfy the strategic plans of both firms (Slowinski and Sagal, 2003).

The alliance framework as described by Slowinski and Sagal (2003) has been found by many companies to be a very useful approach, consisting of a disciplined process and an iterated document. The process recognizes that internal elements of both firms need to be bonded together as well as the interfaces between the two firms, the so-called 'three alliances in one'.

Each element in the alliance framework process involves multidisciplinary teams, effectively creating the internal alliance necessary to establish support for the external alliance, and gaining momentum for the alliance in operation. The process also ensures that the correct strategic thinking is done before partners are identified or approached. Many alliances fail because the strategic rationale was incorrect at the start and consequently the alliance was built on foundations of sand.

Most companies are proud of the way they work. They believe their operating systems are both effective and efficient. In the context of an alliance this is not always good news, if one or both parties are determined to stick to their ways of working, leading to disagreement over how the alliance will work. Incompatibility at such an early stage is not a good sign for future success. It is therefore important to remain flexible, and ensure the focus is on the substantive issues rather than the detail of how they are to be addressed. For example, if both companies have different ways of approaching the alliance framework, but there are ways to align the basic elements such as strategic goals, division of value, roles and responsibilities, the alliance can move forward.

Table 13.1 Some factors to consider for partnership governance

Factor	Consideration
Leadership	Who will represent each company? Will the levels be compatible?
Management	Who is the main 'go to' person within each company?
Project management	Who manages which elements of which projects?
Direct contacts	Which areas have direct contact e.g. legal to legal, technical to technical, marketing to marketing, sales to sales?
Meeting structure	How often will teams meet?
Dispute resolution	How will disagreements be settled?

It is a common belief in business that success comes from under-promising and over-delivering. It is no different with partnerships. Partners should not overplay their capabilities and resources. This works both ways, in that you should not take a rosy view of your partner. It is better to conduct a good due-diligence process and take a realistic view on both sides of the potential for success. This process of discovery also helps to bond the parties together.

The governance of the partnership is also important. Some of the many elements to be considered are shown in Table 13.1.

Multiple strategic alliances are also well documented and can help point the way for open innovation relationships with more than two partners involved. Doz and Hamel (1998) outline the key strategic issues in building a multilateral alliance. These include choosing the alliance composition, perhaps the most important step, the size, growth path and elements of governance.

The leadership and management of the multiple alliance are best done by the company at the hub, also referred to by Doz and Hamel (1998) as the nodal company. The hub takes a view across all the competing and complementary interests to ensure that the strategic and project aims of the multilateral alliance members are being met. The hub is likely to be the company that gains most, not necessarily in proportion to its size, but also the one most likely to take products and services to market in an open innovation environment.

There are many factors to consider beyond those in strategic goals, project management and output. Conflict, competing interests, internal alignment and external communication all need to be managed closely. This is where mature alliances and communities such as Nestlé's Sharing is Winning (SiW) appear to succeed (Traitler and Saguy, 2009).

13.3.2 Defining the boundaries

When mapping out an open innovation partnership all the partners involved should define the full scope. It is just as important to define the areas in which there will be no co-operation. These boundaries include project-specific parameters such as work to be done, by whom and by when. They may also include business exploitation elements such as geography, business sector, trade channels and financial return to each partner. Over time this scope will evolve, but getting off to the right start is important.

13.3.3 Sharing the spoils

The traditional approach to negotiating deals simplistically assumes that 'we want a bigger slice of the pie'. The open innovation approach should be to work closely together to ensure that the pie is bigger to start with. The ultimate objective is to create a product or service that consumers want to buy in preference to what is already available. The better the product, the higher the price and volume.

In the spirit of win–win, once the route to maximum value creation is clear, then the partners should work out the value split. There is no hard-and-fast matrix to which one may refer; each deal is different. Usually, there is a partner (or partners) who supplies the 'go to market' company with technology, IP or other prize assets. They may receive a royalty on sales; a supply agreement; a one-off fee for licence or transfer of ownership; or a fee for service. The quantum of these amounts relates to the size of the contribution, the margin available to divide and the influence of the history and future of the relationship.

Even in heavily asymmetric relationships, where one company is much larger than the other, win–win is fundamental. It is generally accepted that risk should be proportional to return. In asymmetric partnerships, the risks taken by the smaller partner are often very significant in proportion to their size, whereas the impact of the project dying is usually much smaller for the large company. The large company in the relationship should recognize the potential consequences of failure for their partner and seek to manage the projects accordingly.

In the spirit of openness, it is important to understand each other's expectations of value creation. This includes the quantity and the method; how much, and how. After all, this is rarely a philanthropic exercise. Each party should take great care to ensure that the other firm profits significantly from the partnership. If one side does not gain, they are unlikely to remain in the deal and all value is lost, leading both partners to suffer. Thus, win–win is not just a bland statement, it is the crux of any partnership, the foundation for success.

Nestlé's SiW program brings together a number of their core suppliers in a group aimed at driving innovation for Nestlé (Traitler and Saguy, 2009). Around twenty suppliers in areas such as packaging, food ingredients and flavours participate. The title of the program is intended to convey the spirit of the partnership without implying a naïve sacrifice of proprietary territory. Partners involved in SiW, with Nestlé as the hub, include Barrie Callebaut, BASF, Cargill, Cognis, DSM, DuPont, Firmenich, Fonterra, Givaudan, IFF, Kerry, Mane, Symrise and Tetra Pak.

The wish to protect confidential information of competitive advantage is vital in any situation. It is underlined in a context such as SiW. It was endorsed by Chesbrough (2006), who commented that 'companies that keep their IP too close to the vest risk missing out on critical business innovations that idea sharing could generate'. It is therefore important to have so-called 'open business models' that provide a framework within which new value can be created through the sharing of ideas and IP.

13.3.4 Intellectual property (IP)

Given that companies wish to provide as much protection for innovation as possible, it is not surprising that IP can be one of the most difficult areas to address in an open innovation partnership. Each partner may have different corporate policies on ownership and exploitation. There may be disagreements on definition of field, licensing, indemnities, exploitation outside of the field and many other areas.

In general, the stronger the relationship and the deeper the partnership, the more likely it is that issues of IP are resolved to the benefit of all partners. On the other hand, a transactional approach is adversarial, focused on buying or paying for a licence to IP and trying to 'win'.

The principle used for IP in Nestlé's SiW is that the competency-providing partner owns every physical solution, such as ingredients or technologies. The receiving party owns the applications of this solution, which in this case is usually Nestlé (Traitler and Saguy, 2009). The latter ownership is in specified areas. In this way, the partnership avoids unnecessary and repeated negotiations and potential conflicts. Even though complex situations may still arise, the existence of a founding principle enables them to be kept to a minimum.

As partnerships evolve away from the transactional end of the spectrum towards the community, and partners understand each other in more detail, the approach to IP can also be prepared jointly. In this way the IP strategy and future protection can be planned together, resulting in a stronger 'fortress', in a more natural and harmonious way (Traitler and Saguy 2009).

There may be applications of one partner's IP in fields unrelated to the other partner's, and in which there is no competitive conflict. A transactional approach seeks to gain from every aspect of the deal. Where one partner's interests are considered, it makes sense to allow them to exploit what is after all their IP, in areas which do not have influence on the other partner's business. The wider application of IP tends to strengthen it against future challenge, and involves more parties who have a vested interest in its protection.

13.3.5 Structuring the open-innovation organization

Organizing for successful partnering in open innovation is an important part of the process. There are three basic ways of structuring both strategic alliances and open innovation in large companies:

1. a central group acting as a centre of expertise and best practice, working with business units on a matrix basis;
2. a devolved structure with responsible managers based in each business unit; or
3. a hybrid of the previous two with defined responsibilities.

Both Kraft Foods (Goers 2011) and General Mills have a core central group with devolved business unit managers. The advantages of this approach are many. The company can retain a view of both strategic goals and short-term imperatives. Learning is efficient. A common approach can be taken to aspects such as legal, IP, process and negotiating. Partnership management principles can be embedded more easily. Last but certainly not least, the company has the ability to be brave, try new things and work out what should be rolled out further, at relatively low risk.

In addition to the dedicated open innovation team, the General Mills Worldwide Innovation Network (G-WIN), General Mills has a dedicated central group called XPD (External Partner Development). This multidisciplinary team has dedicated team members with expertise in R&D, Supply Chain, Sales, Marketing and Legal. Led by a senior procurement executive, it has the power and diversity to integrate all aspects of an open innovation project to the company, rather than go through a convoluted and lengthy alignment process.

13.3.6 The small company perspective

The vast majority of studies on open innovation are from the perspective of large companies, in a context where the other partner in the open innovation relationship is usually smaller, often considerably so. Small to medium-sized enterprises (SMEs) approach partnerships from a different perspective. They are often in a weaker financial position, have a different corporate culture, and may be privately held. Smaller companies tend to make decisions more quickly and may have an overly positive view of their technology.

Successful open innovation companies such as General Mills understand these differences, and ensure that smaller companies appreciate the realities of working with larger companies. The advantages to the SME of working with a large open innovation partner are clear. The large company has access to a much larger customer base, and has the ability to roll out any product developed from the partnership often around the world, and in all the major retail entities. They can take technology that can be barely beyond proof of principle and transform it into a profitable product manufactured very efficiently at high volume.

It is therefore important at the early stage of the partnership that each company manages the expectations of the other. The large company should explain the process and timescales, justify the money needed to invest to successfully get to market, and identify the key contact people who are there to help. The SME should give an honest assessment of the positive and negative aspects of their technology and be prepared to spend more time than they probably expect on the translation from technology to product.

Each company should look at the partnership through the eyes of their partner. As the relationship develops, and more projects follow, each one becomes easier as less time is spent establishing ground rules and ways of working. As the relationship evolves and expands involving elements of community, the SME and large company uncover more opportunities and new ways of accruing mutual value.

13.4 Ecosystems

Ecosystems are the next stage of development within which open innovation relationships can flourish. These can be developed around existing relationships such as those between a core company and its suppliers; or in some cases involve shared physical infrastructure.

13.4.1 Building an ecosystem

When a company has reached the stage of acquiring many different partners with complementary competencies and abilities, there is the question of how to manage them. The management may continue in a purely bilateral manner, with just the company and one partner in the room at any one time. Alternatively, multiple touch points and interactions with diverse activity may be achieved throughout a connected ecosystem.

There are examples of the creation of innovation ecosystems in which open innovation can thrive. An ecosystem can be defined biologically as an environment consisting of all the organisms living there as well as the physical components with which they interact. It implies interaction between those organisms, a symbiosis.

13.4.2 Supplier networks

Suppliers are often the richest source of open innovation opportunities. There are many advantages. The companies know each other; their enterprise systems are connected; there is often a strong understanding and awareness of each other's business; and personal relationships exist. There may be direct contact between the relevant R&D groups. In such a situation it is easier to move to a relationship built on innovation as well as commercial supply.

If the core company retains a traditional approach to procurement, with the 'cards held close to the chest' and the focus purely on price and performance, the innovation potential is limited. It is only when companies start to share information on strategic direction and future priorities that suppliers can direct their efforts to harmonize with those of their customer. This is demonstrated by the work done by Kraft Foods and General Mills.

The first stage of supplier engagement is to prioritize those companies within the supply base using relevant criteria. These may include the value of current business; technical capability; match of competency with future trends; and last but not least the status of the working relationship. Once the key suppliers are selected, they must be brought on board.

The second stage is engagement. Suppliers can not be reluctant, passive or suspicious if the initiative is to succeed. The corollary of any win–win situation is mutual commitment to the goals of the program. Once approached by the customer, the supplier needs to ensure full internal support for participation, including, crucially the allocation of the required resource.

The third stage, briefing, depends on whether the challenge is a single project or a longer-term innovation program. In some projects, the solution needed is very specific and it is a question of supply. In more difficult situations, it is preferable to provide as much information as possible on the background of the problem.

Confidentiality can be a concern here, but as General Mills' leaders state, the problem is rarely a secret within the industry; the solution should be.

The approach can produce strong results. For example, at a Supplier Summit in 2009, General Mills presented seven specific challenges to its supplier base. As a result they received over 200 submissions, 76 of which were considered, resulting in 15 active projects. An important internal action is to ensure that all the critical stakeholders take part in the assessment of the external opportunity, as is done by Kraft Foods in a face-to-face 'showcase' meeting.

If the third stage involves participation in a longer-term innovation program, the briefing involves much higher-level information. This includes a clear understanding of company strategy; trends in the market and consumer needs; consumer insights; and specific technical challenges.

Supplier summits are also important to send a cultural message to suppliers. For example, between the 2009 and 2011 supplier summits, General Mills' suppliers felt that the company was more 'open for business', was sharing information more frequently and much earlier in the process, and had become more transparent.

The fourth stage is sustaining the momentum. As supplier innovation networks evolve, the emphasis shifts from requesting specific solutions to unprompted submission of opportunities by suppliers, based on information and understanding of their customer's business. These often give the hub company a pleasant surprise, and provide a source of new, non-obvious value for product development.

The final stage is the building of supplier communities. The development of this stage may be somewhat in the future, but elements of it can be detected today. It is partly reflected in the prediction by Jeff Bellairs of General Mills that open innovation in the food and drink industry will evolve from 'one to one' to 'many to many'. Examples exist of suppliers and other external innovation partners connecting spontaneously and producing opportunities based on the synergy of their competencies, to the benefit of the customer or hub company.

13.4.3 Physical ecosystems

Another approach to the creation of an open innovation ecosystem is to dedicate physical infrastructure, as in the case of Chemelot and the High Tech Campus (HTC) Eindhoven, both in the Netherlands. These campuses rely on a large company as the driving force behind the initiative; DSM with Chemelot and Philips with HTC Eindhoven. They provide the vision, a large part of the residency, and a coherent thread running through the business interests of the tenants.

Open innovation principles apply quite strongly. There is a belief that colocation of small and large companies with common interests ultimately results in new and shared value. At the start, there does not appear to be clarity on what shape that value will take.

Chemelot describes itself as 'a unique chemical and materials community that ensures accelerated business growth through the open exchange of ideas. Chemelot has been planned around one central idea: to bring together the knowledge and skills normally found only in major organizations, and to apply these within a flexible community of small and large chemical businesses, radically changing the view of the chemical industry'. Even though the focus is on chemicals, much of the output has direct relevance to the food and drink industry through suppliers like DSM. More details can be found at www.chemelot.nl.

13.5 Human factors

Successful open innovation relationships are underpinned by a strong consideration of human factors. These include a focus on the success of all partners, involving the right people, valuing communication, recognising that conflict is not necessarily bad and working hard at maintaining the partnership.

13.5.1 I win vs. we win

The traditional transactional approach to business deals has many metaphors, including war and sport, with the overwhelming objective to win, to defeat an opponent. If you are constantly defeated, do you want to work with the opponent again? I hesitate to use the word 'partner' in the context of a purely transactional business deal. The company on the receiving end also feels it is in a battle, and constantly seeks to defend its position, building metaphorical walls, wearing armour, and being prepared to retaliate.

Such an approach does nobody any good. It is a waste of mental and emotional energy. It therefore makes so much more sense to enter into an open innovation relationship with the clear intent that your partner's interests and objectives should also be met. If they are, both parties will want to work on the next project and time is not wasted plotting political manoeuvres intended to outflank the other partner. The bulk of mental energy and management time is applied to working out the best way to meet the technical and business challenges, tapping into the full capability of the partner.

The phrase 'win–win' is often used in almost a throwaway fashion. In 2010 Procter & Gamble set two objectives for their Connect & Develop open innovation program. The first was a very tough financial objective. The second aims for P&G to be the 'win–win' partner of choice in their industry. This will be harder to measure but shows the importance that open innovation leaders place on the partnership. Increasingly open innovation leaders in the food and drink industry recognize that innovators have choices. Each wants to become the partner of choice and, as Dave Wagner of General Mills says, 'we're willing to work hard to earn that'.

Clorox, the US-based household products company are strongly competitive but not dominant. In order to be as competitive as possible in open innovation, they consciously take an open innovation team approach that treats partners fairly and well; ensure the internal teams behave well; bear the 'three alliances in one' in mind; and build a co-creation community to bring more ideas in house in collaboration with internal teams.

The objectives of open innovation leaders such as General Mills, Kraft Foods, Nestlé and Clorox include looking after the interests of their partners, both short and long term. This is the best way to build sustainable partnerships.

13.5.2 Choosing and training the right people

Emotions are key to relationships. It is important to realise that even though the strategic and business rationale should never be over-ruled by emotion, whether positive or negative, it can influence the way an open innovation partnership is approached. In a classic transactional arrangement, those emotions often present as aggressive type A macho behaviour. This is clearly not conducive to a successful partnership, but perhaps, historically has been viewed as positive for the financial results of a single project.

Open innovation managers are prime candidates for the classic 'T' profile (Guest, 1991). Such people have a breadth of knowledge, experience and internal networks. They also have depth of knowledge in a small number of fields, where they can be considered experts. Their knowledge allows them to rapidly assess the potential of new technologies, understand potential connectivity and work out how external innovation can be implemented in internal product development. The T-profile experience gives them a deep understanding of how the company works, how it approaches collaborations and how people work together. It strengthens the sixth sense, intuition.

People who have long experience within a company usually have good networks. It is important to choose people who not only have that asset but who are continual and strong networkers. Formal mechanisms are necessary to get things done, but they are often lubricated by the networks, and by people working closely together in informal ways. Open innovation people are the 'integrators', the ones who connect the external and internal.

Unilever have recognized the importance of training open innovation managers in the so-called 'soft skills'. They formally train their open innovation managers in seven of these skills: intrapreneurial orientation; strategic influencing; communication; relationship building and maintenance; quick study; tolerance for uncertainty; and passion and optimism (Martino and Bartolone, 2011). They recognise the essential part played by soft human skills, ones that are more difficult to assess.

One company has established an 'open innovation academy' (Mortara *et al.*, 2009) to train employees in key aspects of the discipline, particularly those in R&D and the supply chain. Training is delivered in a variety of formats, and multiple interactions also strengthen internal networks.

13.5.3 Communication and conflict

The resource and time needed to establish successful relationships should not be underestimated. In addition it also takes time to build the external networks from which relationships spring, complemented by the research that in today's web 2.0 environment is not only readily available but more easily connected.

Open innovation also needs an attitude of openness. General Mills has made good use of existing information such as consumer insights and technology, in a transparent way with partners, to add more value to the relationship. Not everything a company knows is a really deep secret, and often the more a partner knows the better able they are to help.

The human and relationship side is also built with regular meetings, whether that is supplier summits, 'town hall' meetings and in depth co-creation workshops. As Jeff Bellairs of General Mills says – 'with the right opportunity, the right partner and the right relationship, magic can happen'.

All relationships have expectations, whether personal or business. Relationships fail when expectations are not met. Usually we expect something good to happen, and for promises to be met. If events turn out to be more positive than anticipated, those surprises are much easier to accept. However, if the predicted future does not materialize, trust in the relationship starts to erode and recovery can be difficult, if not impossible. It is therefore essential to any relationship to manage expectations appropriately, particularly in the context of innovation, a process that can be inherently difficult to predict.

Misunderstandings in communication are probably the biggest source of friction and conflict in partnerships. In classic transactional deals, the adversarial stance of winning versus losing can consider conflict as a challenge to score more points. Healthy partnership can value conflict if presented and handled in the right manner. Conflict and disagreement about issues are healthy, particularly if personal conflict is not part of the discussion.

A lack of conflict is not always healthy in advanced partnerships. Disagreements should be 'on the table', important ones resolved and little time spent on unimportant ones. There are lessons for open innovation communities from open crowdsourced community sites, where open online discussion can be quite strong, but is driven from a common passion and a strong belief in the opinions expressed. The end result is the general view of the community prevailing. It is important that this happens, because passive agreement can lead to the mediocrity of the lowest common denominator.

Communication is also central to one of the understated benefits of open innovation: learning, both individual and corporate. Participation brings exposure to new technologies, new ways of working and experience of new and sometimes difficult situations. The deeper and more committed the partnership, the deeper such learning becomes. This is because companies communicate more about strategy and technology; there are more frequent and deeper personal interactions; there are often more partners involved and, consequently, more diversity.

Additional learning does not necessarily mean acquisition of competency to a level sufficient to practice. Unless the open innovation agreement specifically includes technology and/or competency transfer, the original rationale for the partnership is likely to remain.

Internal communication and alignment is also important. Clorox believe that the internal open innovation team should be the optimistic realists. They must be strongly networked both internally and externally; and take a strong view that each deal must be fair and right for the partner as well as the company. Each business category has a Business Connection Leader, who is embedded in that group. They usually have deep experience of the brands and products. One of their primary roles is stakeholder navigation. To do this, they need to be able to choose the right moment to present the right idea; to be able to navigate their colleagues through quite far out ideas; in effect, to constantly socialize ideas.

13.5.4 Maintaining the relationship

As stated previously, attractiveness as an open innovation partner involves both hard and soft factors. If you are the biggest company in a particular sector, it is easier to attract and retain open innovation partners, as your size and reach are unarguable hard facts. If your company is smaller, it is less likely to have the internal resource of a larger competitor, so that open innovation becomes potentially more important as it can provide complementary access to incremental resource and capability.

For these reasons, both General Mills and Clorox approach ongoing partnerships with more consideration. Both have actively constructed systems to follow the best interests of existing and potential partners, not just themselves. One feature on the Clorox Connects website (www.cloroxconnects.com) provides a community-based co-creation facility. Members can construct concepts that may, or indeed may not, be directly relevant to Clorox's business. They can receive monetary awards; find partners to work with and network with the whole community for feedback. Some of the concepts also receive consumer feedback. The response to this initiative is very positive, as it gives something else to the members of the Clorox Connects community, other than the ability to contribute to Clorox's future innovation pipeline.

Regular relationship audits and open communication enable open innovation managers to keep a close eye on the areas to celebrate and those that may give cause for concern. This will not only improve the existing partnerships but also enhance the approach and management of future deals. Often the feedback contains unpalatable information that may be difficult to change. Not every problem needs to be, or even can be, solved. They should, however, be discussed openly between partners and prioritized for attention into those that can be addressed, will deliver the largest commercial opportunity and reduce any 'pain points'.

This approach should not be interpreted as a philanthropic endeavour, even though we may all hold generous charitable principles. In addition to being the best way to meet objectives, there needs to be a hard-edged approach to negotiations. This is made easier in an open and clear relationship.

13.6 Building a community

Crowdsourcing is a term first coined by Jeff Howe of Wired magazine. There are two basic definitions. The first or 'white paper' one describes crowdsourcing as the act of taking a job traditionally performed by a designated agent (usually an employee) and outsourcing it to an undefined, generally large group of people in the form of an open call. The second, or 'soundbite' version defines it as the application of open source principles to fields outside of software (Howe, 2006).

The relevance of crowdsourcing to this text is that it relies on online communities. These communities are

- self policing;
- motivated by persuasion and collaboration;
- pooling resources;
- motivated by recognition and respect; and
- motivated by a competitive element.

Open innovation communities have very clear vested business interests as the objective of their work. They are motivated by the achievement of business objectives and adding value to their brands and products. Even though there is no direct relationship between crowdsourcing and open innovation communities, there are lessons to be learned, particularly on the human and relationship aspects.

Firstly, participants feel part of the community. There is an involvement and motivation that goes beyond a mere subscription. Secondly, the motivation includes a strong desire to see a positive result for the community that may not accrue an immediate benefit for each member. This is driven by a belief that what is good for the community in the longer term will also be good for them. Thirdly, although the element of self-policing is not relevant, successful open innovation communities are self-directing in the sense of seeking to create value without being specifically asked to do so. They create their own momentum and links.

Open innovation communities of companies are unlikely to get to the intensity of community created by crowdsourcing companies such as Lego. Community members, although motivated by many of the same objectives, ultimately have the commercial goals of their company as the primary criterion by which actions and commitments are judged. They are not like the individuals who write code for Lego Mindstorms for free; who in addition spend thousands of dollars per year on Lego products.

Communities do not need to be exclusive. Companies and individuals can have the sense of commitment, involvement and passion for a number of different communities. The personal analogy is a good one to draw in this situation. Each of us has a mix of communities in which we are involved, for example company, local community, church, sports team, hobby club and online. They are rarely incompatible, and often complementary.

In order to reach the community stage, there are several steps to be followed. Firstly, there should be a statement of intent from the hub company. Suppliers should be encouraged to talk to each other when they possess complementary portfolios. For example, flavour and sweetener suppliers may overlap in their areas of interest and create new value for the customer. Secondly, the core company should pay attention to the cultural aspects of innovation communities. Finally, the structure and organization should be put in place to manage and nurture the community. This does not just mean people and cash; it also includes ensuring the right networking tools, meeting structure and frequency are in place.

A cursory glance at the list of companies in Nestlé's SiW shows that several fierce competitors are contained in the list. As long as the appropriate safeguards are in place, then competitive conflicts can also be managed. In the case of a flavour supplier to the food and drink industry, the prominent companies supply similar ingredients to customers who compete, at the same time as competing with other flavour suppliers. They can be part of several open innovation communities, often sitting in the same room as their competitors. The safeguards include confidentiality, respect for competition rules, and totally ethical behaviour. The first sign of betrayal of confidential information results in a massive decrease in credibility and trust, and damages any future relationship.

The opportunity for hub companies is to create as much of a spirit of community as possible. The degree of involvement and commitment benefits all involved and, based on the lessons of crowdsourcing, strategic alliances and open problem solving, leads to a greater likelihood of producing a stronger flow of superior innovation.

Expert communities such as suppliers may have a higher degree of control, but often need to be invisible to each other to preserve unprotected IP or commercial confidentiality. This may lose a lot of the synergy of communities, so the logical next step is to move to a collaborative model that can keep track of the collaboration and help to allocate IP accordingly.

Clorox are a good example of a company who have made the evolution from a transactional network to a strong partnership model and are now moving further to a community-oriented approach. They have users with a passion and common interest in household cleaning, leading to fairly advanced ideas that have received a lot of thought and consideration. Often those users have a common capability. Even those ideas with much less potential can be easily filtered, an important process requirement for any crowdsourced or community application.

All projects finish, eventually. The deep integration that two or more partners feel during a project inevitably wanes. In a transactional situation, the end of a project means 'goodbye'. In a deeper partnership, where both parties may wish to work together on a future project, links are maintained. In deeper, more embedded communities of innovation, it is highly likely that the partners will want to remain actively engaged to look for new synergies and potential value creation opportunities together.

The commitment engendered by a sense of community allows greater sharing and a wish to exploit the potential more fully. Companies must never lose the business rationale and hard-edged approach to value creation that is the foundation of open innovation communities. It is likely that such relationships are more mature and open, certainly than transactional ones. In that context, it should be easier for partners to close down projects and disengage from immediate work, while at the same time remaining committed to future success.

13.7 Conclusion

An evolution has been described in this chapter, where open innovation is not just viewed as a simple transaction where a classic aggressive negotiation is conducted to achieve the best one-off deal. In the eyes of leading companies in the food and beverage industry, open innovation is now seen as a major strategic platform which requires substantial and sustainable long term relationships in order to deliver its full potential. Consequently, paying strong attention to the identification, selection and management of partners in open innovation is recognized as an essential part of implementation. Increasingly the human factors in that partnership are also being acknowledged.

As open innovation becomes a stronger part of the innovation strategy of many companies, the relationships and interactions become more complicated to design and manage. In addition to the more commonly addressed challenges such as IP, the management of partnerships is also more complex. Some companies are taking an ecosystem approach, whether or not this involves physical infrastructure. The next logical step, supported by observations in the food industry and others, is to build communities organized around the interests of the hub company.

Communities of open innovation are the ultimate destination for leading companies, pulling together diverse and complementary people and companies in a mutually beneficial and dynamic network. These communities consider human interaction as much as contracts. The will to be involved and to see your partners do well is characteristic of the best open innovation, and is fundamental to successful open innovation communities. These communities still have the achievement of business objectives as the number one priority; they simply recognize a better way to get there.

Finally, the leaders in open innovation all believe they have a long way to go in the partnership journey. Their attitude is characterized by a lack of arrogance that comes from a deep ability to develop partnerships, and to be aware of the needs of their partners. As stated previously, this is not a philanthropic exercise. If a company's partners do well, they will want to continue the relationship, which will benefit the company. Ultimately, strengthening partnering ability with community elements, increases the flow of winning innovation and strengthens the ability of a company's teams to find and develop.

13.8 Acknowledgements

I am very grateful for the comments and input of Miles Eddowes of Kraft Foods; Jeff Bellairs, Dave Wagner and Seema Shah of General Mills; and Navin Kunde, Greg Piche and Ed Rinker of Clorox.

13.9 References

CHESBROUGH H (2006) *Open business models*, Harvard Business School Press, Boston (MA).

DARBY M (2006) *Alliance brand*, John Wiley & Sons Ltd, Chichester.

DOZ Y L AND HAMEL G (1998), *Alliance advantages*, Harvard Business School Press, Boston (MA).

GOERS S (2011) 'Institutionalizing open innovation'. In *A guide to open innovation and crowdsourcing*, ed P Sloane, Kogan Page, London.

GUEST D (1991) 'The hunt is on for the Renaissance Man of computing'. The Independent, September 17 1991.

HOWE J (2006) *Crowdsourcing: why the power of the crowd is driving the future of business*, Crown Business Publishing, New York.

MARTINO G AND BARTOLONE J (2011) 'Soft skills for open innovation success'. In *A guide to open innovation and crowdsourcing*, ed P Sloane, Kogan Page, London.

MCFARTHING K G (2011) 'Attracting open innovation partners – portals, publicity and people'. In *A guide to open innovation and crowdsourcing*, ed P Sloane, Kogan Page, London.

MORTARA L, SLACIK I, NAPP J J AND MINSHALL T H W (2009), *How to implement open innovation*, Institute for Manufacturing, Cambridge University.

SLOWINSKI G AND SAGAL M W (2003) *The strongest link*, AMACOM, New York.

SPEKMAN R E, ISABELLA L A AND MACAVOY T (2000) *Alliance Competence*, John Wiley & Sons, New York.

TRAITLER H AND SAGUY I S (2009) Creating successful innovation partnerships. *Food Technology*, **63**(3), 23–32.

14

Managing co-innovation partnerships: the case of Unilever and its preferred flavour suppliers

M. Tepic and S. W. F. Omta, Wageningen University and Research Centre, The Netherlands, F. T. J. M. Fortuin, Food Valley Organization, The Netherlands and Wageningen University and Research Centre, The Netherlands and A. Saris, Wageningen University and Research Centre, The Netherlands

DOI: 10.1533/9780857097248.3.254

Abstract: Although the trend towards more collaboration for innovatiion is growing, there is limited knowledge about the managerial capabilities needed in each phase of this process, such as partner selection, governance of knowledge sharing, avoiding opportunistic behaviour or shirking, as well as innovation appropriation; ways of addressing this knowledge gap are identified by specifically focusing on the implementation of a long-term co-innovation partnership between Unilever and three of its preferred flavour suppliers. This case study allows the key determinants for success in each co-innovation phase to be identified and the managerial implications for co-innovation in the food industry are discussed.

Key words: co-innovation, food industry, management, flavour suppliers, Unilever.

14.1 Introduction

In the food industry, a company's long term survival on the market has become increasingly dependent on the continued improvement and introduction of new products and processes. The fast changes in consumer demands put increasingly more pressure on food companies to engage in innovation to safeguard their profitability (Sarkar and Costa, 2008), e.g. by developing innovative food products with higher quality and nutritional content (Senker and Managematin, 2008). In order to realize these innovations, companies can no longer only rely on their internal skills and capabilities. Therefore, open innovation (Chesbrough, 2003), collaboration with different types of partners, with complementary resources, skills and capabilities, is increasingly used to increase the number of ideas and innovations. Accordingly, inter-firm collaboration is a trend which is increasingly observed in the food industry. However, a number of characteristics specific to the food industry remain to differentiate this sector from other industries. In the food industry, end-consumers are hesitant towards too radically new products (Sarkar and Costa, 2008). Food companies have to be careful with regard to the kind of innovations they introduce (Senker and Managematin, 2008). Therefore, consumer preferences, e.g. with regard to flavour, need to be carefully investigated.

For a long time, the main focus of large, multinational, food companies has been on enlarging production capacity and on cost reduction (Vanhaverbeke *et al.*, 2007; Duysters *et al*, 2006; Costa and Jongen, 2006; Senker and Managematin, 2008). Food companies' R&D budgets (Unilever's 2010 R&D budget was 2.1% (€928 million) of its 2010 sales (€44 billion)) are relatively low compared with those of high tech industries (Avermaete and Vienne, 2002; Christensen *et al.*, 1996; Galizzi and Venturini, 2008; Stewart-Knox and Mitchell, 2003; Wilkinson, 2002). Although limited budgets create all the more reason for collaboration, the food industry, as a relatively mature and slow-growing area of business (Costa and Jongen, 2006), has been marked by limited collaboration. However, as Ziggers and Henseler (2009) found, food companies can engender sustainable competitive advantage by fostering close working relationships with a limited number of partners, building effective network structures and developing a long-term orientation. Although the trend towards more collaboration is growing, there is still limited knowledge about the problems food companies encounter in establishing innovation collaborations (Sarkar and Costa, 2008). More knowledge is needed about the way in which this industry deals with the problems inherent to co-innovation issues, such as partner selection, governance of information sharing, opportunistic behaviour or shirking, as well as innovation appropriation (Omta and Van Rossum, 1999). This chapter tackles this knowledge gap by focusing on a case study of the implementation of a long-term co-innovation partnership between Unilever and three of its preferred flavour suppliers.

In Section 14.2, the theoretical background is portrayed by discussing the various phases in a co-innovation partnership: the initiation, partner-selection, formalization, implementation and evaluation phases. In Section 14.3, a description of the case of the co-innovation partnership between Unilever and its preferred flavour suppliers is provided. In Section 14.4, an in-depth description of the different phases of the co-innovation partnership is given from the perspective of Unilever and the perspectives of the respective flavour houses. In Section 14.5, the key determinants per co-innovation phase are presented with reference to the Unilever case study, and the managerial implications for co-innovation in the food industry are discussed.

14.2 Co-innovation

An open approach to innovation is developed over several years, where, in addition to internal resources, competences and capabilities, external ones are also sought and deployed (Chesbrough, 2003). In this way, in open innovation both internal and external knowledge, capabilities and paths to the market can be considered (Chesbrough, 2003; Sarkar and Costa, 2008). Depending on the goal of the company, it may choose to collaborate with universities, research centres, firms within their value chain or even with customers (Vanhaverbeke, 2006). Although collaborative innovation can be pursued through a variety of organizational forms (e.g. R&D consortia, joint ventures, or learning networks), an alliance is considered to be the most efficient and effective form and is therefore often the preferred instrument (Bossink, 2002; Lord *et al.*, 2005). More specifically defined, a co-innovation alliance is 'a business relationship, in which two or more independent firms or research institutes work co-operatively on a specific project, which is aimed at the development and commercialization of new products or services that are clearly defined in terms of activities, geographic location and time. Although partners remain independent, they also share rewards and risks' (Stel, 2011: 16).

Co-operating with third parties involves risks and uncertainties. Most firms are used to considering their environment as an exogenous variable. However, in co-innovation alliances firms create value together; the success is therefore dependent on the alignment and commitment of all partners (Vanhaverbeke, 2006). Therefore, developing, implementing and maintaining a co-innovation strategy requires new competences. For instance, firms are likely to face paradoxes such as trust versus control, co-operation versus competition, and flexibility versus efficiency (Stel, 2011). During the lifecycle of a co-innovation partnership, these challenges become apparent and need to be dealt with. Co-innovation partnerships and the accompanied challenges can be divided into a number of stages. On the basis of previous studies, a number of these stages are distinguished: initiation, partner selection, formalization, implementation and evaluation phase (Das and Teng, 2002; Kale and Singh, 2009; Ring and van de Ven, 1994). In the following, these stages are used to explore how food companies manage the co-innovation processes.

14.2.1 Initiation phase

Previous research has identified numerous factors that lead firms to collaborate and form partnerships. From a transaction cost perspective, the risks and uncertainties that are present in innovation processes can be reduced by forming partnerships, such as strategic alliances and joint ventures (Glaister and Buckley, 1996; Hagedoorn, 1993). Creating economies of scale has also been described as an important driver to collaborate (Glaister and Buckley, 1996). In line with the transaction cost perspective, organizational scholars have described the use of alliances to gain access to valuable resources belonging to other firms; such resources could be specific skills or more abstract resources such as knowledge (Das and Teng, 2000; Eisenhardt and Schoonhoven, 1996; Kogut, 1988; Porter and Fuller, 1986). Lastly, strategic considerations are suggested as rationales to form partnerships, e.g. to improve the market position or to gain access to new markets (Gulati, 1995; Hagedoorn, 1993; Kogut, 1988).

In practice, firms choose to collaborate because of a number of the above mentioned factors. In open-innovation partnerships, value is created through the linkage of internal and external resources (Chesbrough, 2003; Slowinski and Sagal, 2010), thus firms should have a clear understanding of the resources or assets that are required to meet their goals (Slowinski and Sagal, 2010). Through interorganizational relationships, firms are able to gain access to resources that they are, to some extent, dependent on, rather than acquiring ownership of all critical resources (Hillman *et al.*, 2009). Besides access, interorganizational partnerships also enable firms to retain and develop their own resources by combining them with others' resources (Das and Teng, 2000). Both rationales are key to an open-innovation strategy. In order to initiate an open-innovation trajectory, the firms first need to recognize the opportunities and added value of opening up to external knowledge, capabilities and other resources.

The shift towards an open-innovation strategy then starts with the internal alignment within each of the participating firms; this is essential for the success of the new strategy. Because such a partnership often involves several functional groups (e.g. R&D, marketing, and finance), individual business units, and senior leadership, all of them must thoroughly understand and be committed to the objectives and terms of the alliance (Slowinski and Sagal, 2010).

14.2.2 Partner-selection phase

There has been relatively little study of the second phase of the process: the partner selection phase (e.g. Dollinger *et al.*, 1997; Hitt *et al.*, 2000; Lambe *et al.*, 2002; Saxton, 1997). This has been described as one of the most crucial factors of alliance success (Lambe and Spekman, 1997). Potential partners are supply chain actors such as suppliers and customers, but also competitors or knowledge institutions, such as universities and research centres (Miotti and Sachwald, 2003; Batterink, 2009).

Three main factors have been identified as being essential in the partner-selection phase: technological, strategic and relational alignment (Emden *et al.*, 2006). Technological alignment includes the recognition of a firm's technological capabilities, complementary resources and overlapping knowledge bases. The latter is necessary because it allows a firm to see the value in the potential partners' competences (Emden *et al.*, 2006). However, this has been described as very costly and difficult (Dyer and Singh, 1998). Previous alliance experience might ease this because such firms have a better understanding of potential partners and/or resources and this allows them to create a competitive advantage (Gulati, 1995; Dyer and Singh, 1998). In addition, the ability to acquire information from a potential partner enables firms to better evaluate a potential partner's skills and resources (Dyer and Singh, 1998). The rationale behind the decision to incorporate external resources influences the selection of partners. When a firm primarily aims at reducing costs and risks through economies of scale and rationalized innovation processes, similar resources should be pooled. However, if partners aim at managing technological convergence or interdependence among innovation processes, they are likely to combine complementary resources (Miotti and Sachwald, 2003).

According to Emden *et al.* (2006) strategic alignment refers to the degree of motivation and goal correspondence. Although the motivations of each partner may differ, they need to correspond in order to indicate whether partners have mutually beneficial intentions. Furthermore, such an alignment reduces the likelihood that the partners act opportunistically in the partnership. Finally, to reach relational alignment, firms should look for partners that have a compatible culture or fit organizationally, are willing to adapt in line with the requirements for collaboration change, and have a long-term orientation (Emden *et al.*, 2006; Saxton, 1997; Shah and Swaminathan, 2008). The latter has also been referred to as commitment and, as with trust, reduces the risk of opportunism (Shah and Swaminathan, 2008).

14.2.3 Formalization phase

The formalization phase of the partnership has also been reported as important to its success (Kale and Singh, 2009). Open-innovation collaborations might take a number of forms, such as licencing-in, outsourcing and collaborations (Batterink, 2009). Within collaborations, forms can range from contractual agreements to joint ventures (Sampson, 2007). Contractual agreements have a number of functions to manage the risks of a partnership. First, firms use a contract to describe the mutual rights and obligations, such as each firm's contribution to the partnership, as well as the organizational processes necessary to solve problems and to divide the expected outcomes of the alliance. Furthermore, a contract specifies how each partner interacts with third parties and outlines what happens if one of the partners wants to or has to end the alliance. Claims on intellectual property have also been reported (Kale and Singh, 2009; Reuer and Ariño, 2007). These aspects must be carefully planned, negotiated, and captured in an alliance agreement to avoid misunderstandings during the later phases of the partnership. Being a bilateral process, it is important that the negotiated terms must be acceptable to both parties (Slowinski and Sagal, 2010). However, various negotiation strategies have been described which differ in their concern for the outcomes of the own and partner firm. Firms with a long-term orientation in their alliance also take the needs and wishes of the partnering firms into account. The firms can also choose to compromise, resulting in some gains, but the firms may not be willing to put in the necessary effort to truly obtain an integrative solution (Das and Kumar, 2011).

When negotiation is successful, the firms are able to reach an agreement, often in the form of a contract. The number of provisions in a contract and their stringency are influenced by several factors. First, transaction-specific investments increase the need for complex contracts. In particular, when high levels of asset specificity are present, which means that the investments are difficult to redeploy in alternative usages, firms attempt to negotiate and formalize the consequences of agreement violation and termination (Dyer, 1997; Reuer and Ariño, 2007). In contrast, the presence of social mechanisms mitigates the need for a large number of provisions in a contract. Social mechanisms such as trust, goodwill and mutual understanding of each other's culture and organizational routines can arise from prior experience with the partner, and a partner's reputation (Granovetter, 1985; Gulati, 1995; Oliver, 1988; Reuer and Ariño, 2007; Uzzi, 1997). Consequently, the presence of trust may already have a significant impact in this phase, because it shapes the institutional arrangements of the partnership.

14.2.4 Implementation phase

The implementation phase has also been referred to as operation phase (e.g. Das and Teng, 2002), execution phase (e.g. Ring and Van de Ven, 1994) and postformation management phase (Kale and Singh, 2009). In this phase, the partner firms collaborate and implement all agreements of the alliance (Das and Teng, 2002). One of the critical factors in this phase is trust. Although the building of trust starts in the earlier phases, the process has to be accelerated, because the actual collaboration begins in this phase. Firms can facilitate trustful relationships by placing themselves in a vulnerable position, showing commitment and living up to expectations, enabling the development of interpersonal trust (Kale and Singh, 2009). Because operational level staff are involved in the partnership, the development of interpersonal trust is essential because it impacts the day-to-day efficiency of alliance operations (Zaheer *et al.*, 2002). Training programs, workshops and network meetings enable employees to meet each other, develop relationships and share (tacit) knowledge (Dyer *et al.*, 2001; Ring and Van de Ven, 1994). Furthermore, these meetings serve as a mechanism for (operational) team members to obtain a clear and consistent insight into the objectives of the alliance and how the agreement terms should guide their actions (Slowinski and Sagal, 2010).

Another key issue in the implementation phase is the co-ordination of activities and the creation of mutual benefits (Kale and Singh, 2009; Todeva and Knoke, 2005). Partnerships require ongoing co-ordination of activities that are completed jointly or individually across organizational boundaries. Tasks need to be decomposed, assigned and monitored, and this requires internal and external communication, decision-making and conflict resolution (Gulati and Singh, 1998; Child *et al.*, 2005). In addition, firms may have to bridge cultures. Cultural diversity can arise from differences in national context, corporate culture (ideologies and values that characterize an organization) and management practices. Developing an understanding of the partners' beliefs and behaviour and setting up unitary management processes might overcome cultural diversity problems (Parkhe, 1991).

14.2.5 Evaluation phase

In the evaluation phase, the outcomes of the alliance become tangible and can be evaluated (Das and Teng, 2002). A distinction should be made between the 'process performance' and the 'goal performance'. Where the first refers to the extent to which an alliance performs well as a (virtual) organizational unit, the latter refers to the extent to which the objectives of each partner are realized in practice (Child *et al.*, 2005). This evaluation merely encompasses perceptual measures, such as how well the alliance partners interacted with each other; how flexible each partner was to requests made by the other partner; the satisfaction with the governance structure; and the overall satisfaction with the alliance (Kumar and Nti, 1998; Monczka *et al.*, 1998; Reuer and Ariño, 2002). The evaluation is a dynamic process, the pattern of interaction between the partners evolving as the relationship unfolds, consequently affecting the partners' psychological attachment to the relationship and as a result influencing the pattern of interaction positively or negatively (Kumar and Nti, 1998).

More objective measures have been described to evaluate whether the goals of the alliance have been achieved. Mostly, these are financial measures such as joint profits (Lambe *et al.*, 2002). For (open) innovation collaborations, firms could evaluate their partnership on the share of sales derived from new products developed in the last three years (Batterink, 2009). Chesbrough (2004) also proposes metrics on time-to-market and project failures.

After evaluation, firms have to decide on the future of the alliance. Several outcomes are possible, such as stabilization, reformation and termination (Das and Teng, 2002). When the alliance is operating successfully the firms may try to maintain the collaborative relationship and invest more resources and capabilities in order to retain its added value in the future (Jiang *et al.*, 2008). However, the evaluation phase might also result in an alteration of the initial partnership agreement or even its termination (Das and Teng, 2002). Dissatisfaction can arise from system performance (e.g. interpartner conflicts), goal performance (e.g. discrepancy between expected and actual outcomes) or both. Unsatisfactory outcomes require the firms to carefully assess the origins of the discontent and attempt to make the necessary adjustments (Kumar and Nti, 1998). When firms fail to make these adjustments, the partnership is likely to be terminated.

14.3 The co-innovation partnership between Unilever and flavour suppliers

In order to be able to extract lessons for successful management of co-innovation in the food industry, the study of the different phases of the co-innovation partnership between Unilever and three of its preferred flavour suppliers (houses) is valuable because of the relatively high level of innovation in this partnership. Unilever and its partners engage in both incremental and radical innovation, thus increasing the spectrum of lessons that can be learned from them.

The partnership is called the Flavour Operating Framework (FLOF) and aims at developing better flavours for tastier foods, and to speed up flavour development for Unilever. This partnership is selected because it provides an in-depth view on the process towards openness, briefing processes and co-operation, as well as the aims, concerns and incentives of both types of partners to get involved in a co-innovation partnership. It represents one of Unilever's most successful partnerships and has been used by Unilever as an exemplar case for other collaborations. Not only a more efficient working method was envisaged in this partnership, but also joint capacity building including mutual learning and joining forces to strengthen each other's position in the market. The sharing of insights, expertise and co-ordinating efforts has resulted in better applications of existing flavours, but also in radically new flavours.

Unilever, being the second largest food company in the world, has 400 brands in the field of food, household and personal care, including both global and local brands. The worldwide turnover in 2010 was over €44 billion. Unilever has around 167 000 employees, with production locations in 100 countries and a market coverage of 180 countries. The company invests almost €1 billion per year in R&D. More than 6000 employees work in Unilever's R&D departments, applying for 250 to 350 patents each year and the company has around 20 000 registered patents and patent applications. In 2010, the revenues from innovations accounted for one third of the corporate revenues.

The three flavour houses are all world leaders. The first one is a company which has over 6000 employees, 30 flavour expertise centres and 18 flavour manufacturing sites, and is present in more than 60 countries. Its sales amounted €2.3 billion of which they invest 10% in research, mainly on tonality, shelf-life and consumer preference of flavours. Their scientific research focuses on identification and manufacturing of the key flavour raw materials that have a direct positive effect on food quality. They possess a large number of patents on new flavour ingredients and technologies and in the fiscal year of 2011 they filed 36 patent applications. The second flavour house is a global supplier of fragrances, flavours, aroma chemicals and cosmetic ingredients for the fragrance, cosmetics and food industries. They employ 5288 employees. In 2010, their sales amounted to €1.6 billion, spread over 130 countries. In terms of sales, this flavour house is the world's fourth largest flavour supplier in the flavours and fragrances market with an estimated market share of around 11%. R&D expenditures amounted to €108 million in 2010, comprizing 7% of total sales. It offers a number of unique technologies and, with 42 patent applications in 2004, it is one of the most innovative manufacturers in the marketplace. The third flavour house focuses on the creation of unique, superior and economically competitive products through integration of flavour/fragrance expertise, consumer insights and brand understanding. It operates in 60 different countries with headquarters in New York. From the total of 5500 employees, approximately 1100 are employed in R&D. Its annual sales exceeds €2 billion and 8% of its annual sales is spent on R&D. Sales of flavour products accounted for almost 50% of total sales during the last three years.

The data collection consists of semistructured interviews and reviews of documents such as the FLOF partnership agreement, project contracts, and brochures about the FLOF partnership, company websites and business publications. At Unilever, four semistructured interviews with senior management of Unilever and the R&D and Procurement Department were conducted. At the flavour houses, three interviews were conducted with the global account directors. The interviewees elaborated on the different stages of the co-innovation partnership. For example, the Senior Vice President of Unilever provided information about the initiation and strategy formulation phase, as well as the formalization phase. The R&D and Procurement management provided information on all phases of the co-innovation partnership.

The interviews were 60 to 90 min in length. Interviews began with background information, followed by the chronology of events, activities and decisions which took place during the formation and implementation of the partnership. On the basis of the identification of important events and milestones, topics such as trust building and information exchange were addressed. For example, the procedures used to select the partners, the negotiation processes, and which types of agreements were made were described by the interviewees. All interviews were audiotaped and transcribed. The transcriptions were sent to the interviewees giving them the opportunity to comment and verify the information.

14.4 Implementation and development of the Flavour Operating Framework partnership

14.4.1 Initiation phase

At Unilever, the starting point of the partnership was the recognition of the need for a change in strategy. The traditional co-operation between Unilever and the flavour houses acted as two 'black boxes'. On the one hand, Unilever requested a flavour without extensively specifying for which product it would be used and, on the other hand, the flavour house delivered the flavour without much detailed information on its contents. This closed co-operation led to time-consuming product development at Unilever and a constant feedback to the flavour houses that the flavour needed to be tastier and cheaper.

Unilever realized that their innovation processes needed to be more effective in order to be a truly innovative company. They became aware of the fact that it would not be the most beneficial to do everything in-house; an external network could provide more. As one of the Unilever respondents stated: 'Unilever thought that they can do a lot themselves and were very wary about the actual added value or contribution of the flavour houses'. At some point, Unilever realized that through the use of the expertise of the flavour houses it might be able to increase efficiency, reduce the innovation cycle time, and speed up the time-to-market of new products. In other words, Unilever considered that an open-innovation trajectory would be a possibly advantageous strategy to meet its goals. The flavour houses felt the urgency to become more efficient, because they were confronted with an increased market demand for natural products which tend to be more expensive for flavour houses.

Because flavours are an important ingredient in the final (consumer) product, knowledge of the context of its use enables the flavour companies to tailor the flavour in order to improve the total quality of the product. In the traditional co-operation set-up, this context was not extensively communicated by Unilever, so the flavour houses were not able to fully predict the effect of the specific flavour in the end product. As one respondent stated: 'Because we did not know in which product/context the flavour was intended to be used, it was not possible to know exactly how the taste would work out in the final product'. Therefore, a closer partnership with more information exchange was expected to result in cost reductions at the flavour houses, because goal-directed development of flavours would be possible, thus leading to fewer unexploited products. In addition, a firm such as Unilever with a strong position in the global food market is an attractive partner. The urgency to change and the recognition of the potential that an open-innovation strategy could bring, led the firms to explore the options and to initiate organizational changes that support the implementation of an open-innovation strategy.

The initiation phase was characterized by the internal alignment of the organization. Supported by top management commitment, Unilever managed to overcome internal disagreements between the R&D and Procurement Departments. In the traditional relationships with the flavour houses, no co-operation took place between R&D and Procurement. To permit information exchange with the external partners, first an increase of internal information exchange at Unilever needed to be achieved. Marketing, R&D and Procurement started to share their insights with the intention of integrating the internal innovation processes more effectively. For example, R&D shared information on recipes with the marketing department, something they were not accustomed to. It took management quite some effort to convince all employees of the added value of the open-innovation strategy and to make them acquainted with the new working routines. In addition, global compliance was assured. The new policy of preferred suppliers needed to function within all Unilever departments across the world. Acceptance and support was required from all local Procurement and R&D departments. Without their support, excluded flavour houses would try to find other ways to continue their relationship.

14.4.2 Partner-selection phase

To reduce the level of uncertainty that a strategic partnership encompasses and to select the best partners, Unilever undertook a careful partner assessment. The total process took nine months, and as one of the selected flavour houses stated: 'It was one of the longest and most professional set-up phases we have ever experienced, everything was checked and re-checked'. Unilever only evaluated global suppliers, because local suppliers would not be able to serve Unilever on a global scale. Based on conversations with and the reputation of the flavour houses, six were selected as potential partners. Resulting from several debates at Unilever, a list of approximately 20 criteria was developed on which the potential partners would be assessed. It included aspects such as company strength; strategic fit; competences; property rights ownership; ability to serve Unilever global market needs; and their willingness to assign the best employees to Unilever's projects.

The second phase of the selection process was characterized by the disclosure of strategic and critical information by the selected flavour houses. 'Unilever asked all kinds of questions and required to reveal a lot of information'. Unilever aimed to get insight into the competences of the flavour houses, in order to assess the level of complementarity to their own competences. The potential partners were invited to discuss these criteria on a senior management level, with as much transparency as possible. This was positively interpreted by the flavour houses as is illustrated by the following statement: 'We liked the fact that Unilever chose the approach of seeking consent from the highest level at our firm. They went for the president of the company, which means that commitment is assured'. For Unilever, it was important to make clear that this was meant to be a long term endeavour. These meetings were used to exchange more information and to find out to what extent willingness and strategic fit was present. A major concern of the flavour houses in this early stage was that information about their company or products would leak to their competitors. Confidentiality agreements became crucial elements in this stage, because they assured the non-disclosure of vital information. In addition, openness from both sides was essential at this stage. The reaction of one of the flavour houses was: 'If you want us to reveal information, then we want to acquire information from you as well. Complete transparency or not'. For the flavour houses, the meetings represented milestones, because, as one of the interviewees indicated, they realized that 'They have to start talking to the girl which they fancy but never dared to ask for a dance'. Furthermore, the flavour houses had to consider whether other clients, being competitors of Unilever, would be reluctant to continue their relationship when they would hear about the partnership. The capability to create and organize an environment in which all parties felt comfortable to reveal information on their competences and resources was regarded as one of the key success factors of this partnership by all interviewees. Finally, three flavour houses were selected for the FLOF partnership.

14.4.3 Partnership-formalization phase

In the partnership-formalization phase, the firms had to negotiate and balance their demands. 'It took a long time and a lot of effort', according to one of the respondents. The distribution of the intellectual property rights (IPR) was an important issue during this phase. Traditionally, Unilever tended to be rather possessive with respect to the ownership of knowledge and technologies. In order to stay a first mover in the market, it acknowledged that time-to-market was crucial in the innovation process and acquiring all the property rights would not benefit this. Therefore, during the establishment of the FLOF partnership, the firm relaxed its attitude and agreed to a fair distribution of the property rights. As one of the flavour houses commented: 'It is one of the best relationship agreements I have ever seen, especially the part on the IPR'. It was decided that all firms retain ownership of their specific input into the partnership. Thus, the flavour houses acquired all inventions related to flavour materials and the manufacturing thereof and Unilever acquired all the innovations related to packaging, use and/or application of flavour materials and manufacturing of the final products. The flavour houses, therefore, kept ownership of the flavour technology and Unilever got the right to be the first user of the new flavours. Unilever had time to indicate whether they wanted to use the flavour, after which the flavour houses were allowed to exploit it commercially through other (non-competing) customers.

Another point of attention during the formalization phase was the question of what, and to what extent, agreements had to be safeguarded by means of a contract. 'It includes everything that is important for a partnership, with the exception of purchasing clauses which are of a commercial character'. The partnership agreement includes that confidentiality is maintained during the project and five years after the contract. Another aspect that is dealt with in the partnership agreement is a growth model of the sharing of short and long-term costs and benefits.

It was decided to keep the partnership agreement rather standardized and add detailed contracts for the individual projects. Although a project contract was drafted for all projects, it was especially important for the radical innovation projects. The contracts consist of the objectives, time schedule and task division, possible deviations from the partnership agreement and the procedures for an unanticipated termination of the project. Concerning the co-ordination of the project, the project contracts also describe the frequency of communication (e.g. progress meetings). Working with a general partnership agreement and specific project contracts enabled the firms to shorten the negotiation period and to tailor the contracts to the specific project requirements. The formulation of the partnership agreement took up to three months in total.

14.4.4 Implementation phase

One of the main motivators of the set-up of the partnership was the increase in information exchange which would tackle the uncertainty problems, in particular for radical innovations. The increase in information exchange would make the innovation process more focused, enabling more informed decision-making and reducing the risk of failure. To facilitate information exchange and open communication, trust has to be present, as according to one of the flavour house respondents: 'The partnership encourages trust to communicate; it provides employees with a solid basis and clear frame which inhibits the situation that they start creating chaos'. Building trust has not been an easy process. According to one of the flavour houses, it started with an insight into Unilever's strategy: 'We all knew what the partnership would entail, the framework was set up and the agreement was made, but the most important element, trust, started to increase when Unilever started to share information about its internal strategy'.

The new working routines resulted in an intense relationship. The input of the suppliers and Unilever was included in a project briefing. In this manner, both parties were able to share their knowledge on the flavours and products to be developed, which resulted in an effective innovation process: 'The briefing process was an excellent tool for the advancement of the relationship'. Besides the briefings, Unilever also involved the suppliers in its internal workshops and working groups, even before the official briefing process had started. These meetings consisted of employees from marketing, procurement and R&D brainstorming about the innovations. The positive effect is expressed by one of the flavour houses: 'These kind of sessions have been important for the building of trust and the creation of openness'.

Open communication has proven to be vital in this integrated project process. Sharing vital information with the supplier or client was not customary and therefore the employees at both Unilever and the flavour houses had to make big changes in their mind-set. At first, employees were double checking whether they 'really should communicate all this information'. In this stage, the commitment of senior management was of great importance to assure the employees that this was the new way of working. Now, within the partnership, the employees from the flavour houses and Unilever act as a team with the possibility of picking up the phone, speaking to each other about the project and asking for additional information. 'The contact with Unilever is now very different from before. In the past when Unilever asked for something, it was first the question whether it was interesting for us to start working on this and spend time on it. Now, because we have more insight in the strategy of Unilever, it is easier to assess whether something is relevant and when unclear a telephone call takes place much easier, asking Unilever what they exactly envisage'. Furthermore, expectations and concerns are openly discussed at the start of each project, which saves a lot of resources.

During the first three years, the firms strengthened their relationship further by sharing strategic information, informing each other on new developments and involvement of cross-sectional teams in an early stage of the innovation process. This enabled the different departments to contribute and share in the strategy development. At the same time, the flavour houses developed a trusting relationship amongst each other. Although it is a competitive market, the flavour houses came to realize that they were all there for the same purpose: to grow their businesses, develop innovations with Unilever and to deliver value together. Their joint participation in meetings with Unilever facilitated the development of trust. At some point they started to feel comfortable in discussing their vision of industry developments with each other. As one of the flavour houses expressed: 'There is no point in competing with each other when it comes to certain aspects; that's what comes with trust'. Over the years, the firms have been able to develop a philosophy that each participation contributes to a collective strategy. The involvement of the flavour houses in the early stages of the innovation process and the organized meetings with the partaking of all partners has led to trust and the understanding of the shared goals of the partnership.

Another point of attention was the bridging of different corporate cultures. Unilever is more a marketing-oriented company, whereas the flavour houses are more technology-oriented. This results in culture differences and complicates the integrated working routines at the operational level. Unilever employees experienced difficulties with creating a thorough understanding at the flavour houses, which they assigned to their non-marketing background. However, flavour houses' employees commented: 'They have difficulty accepting that someone who never developed soup before would now be interfering and contributing to its development'. The firms realized that these kinds of bottlenecks are serious threats for an open-innovation partnership, and they managed to bridge the different cultures over the years.

However, changing routines and cultures in a firm is not enough; the changes also need to be institutionalized. Because people are the foundation of a trustful relationship, changes in the project teams or departments proved to complicate the process. As some of the respondents stated, 'It remained difficult to transport values and trust to other people' and 'You have to make sure that the new people also go through that process of trust development'. In the FLOF partnership, team members were strongly encouraged to pass on the values and understanding of the new working routines to new team members to ensure a continuing process. In addition, the account managers of Unilever who normally would be relocated to other projects or accounts after a while were intentionally not relocated during the FLOF partnership, in order to sustain the routines and culture that they had become familiar with.

14.4.5 Partnership-evaluation phase

The partnership-evaluation phase plays an important role in the continuation and the future shape of the partnership. The partnership brings different tangible and intangible benefits to the partners. Tangible benefits for Unilever were mainly related to the reduction of the innovation cycle time. Flavour-related innovation projects that would normally have taken 1.5 years have now been shortened to 3 to 6 months. In this way, Unilever gets the opportunity to introduce flavour innovations into the market more rapidly, heading to the first-mover advantage. This enabled Unilever to increase its market share in a highly competitive market. Owing to the increased and more transparent information exchange with the flavour houses, the quality of the innovative products also increased as did the success rate of the projects. In terms of intangible benefits, the employees at Unilever went through a process of change and their motivation has grown. They participated in several change management and leading-by-example meetings. Unilever obtained an exemplary role in the area of open innovation, the Global Procurement Manager visited various R&D departments of other companies to share his experiences about open innovation and the open way of working. He learned that a lot of companies still have difficulty with an open way of working and are eager to learn from Unilever's example.

In terms of tangible benefits for the flavour houses, it is particularly in the flavour houses that a significant development time reduction took place. From a period of 7 years needed to get from molecule to revenue generation, this period was shortened to about 3 years. This has led to saving up to 33% of the resources. Also, it involves shorter lines of action. Instead of having to acquire internal approval first, direct interaction with the client is now allowed. One of the most important intangible benefits for the flavour houses is the employees' preference for working on Unilever projects, because innovations developed in co-operation with Unilever are three to four times more often introduced into the market.

14.5 Conclusion

The aim of this chapter is to contribute to the empirical research on open innovation by focusing on co-innovation partnerships in the food industry. The apparent success of the Unilever co-innovation partnership model makes it a good example to learn from. Several factors distinguished as key determinants of success in the different phases of Unilever's co-innovation partnership are therefore discussed here.

Through inter-firm collaboration, firms are able to acquire access to resources which they need, but do not possess themselves (Hillman *et al.*, 2009). The case study clearly shows that, in the initiation phase, the capability of a firm to recognize the potential added value of the complementary resources and skills of a partner, is one of the most important managerial aspects (Slowinski and Sagal, 2010). In particular, in food innovation, where a high level of secrecy with regards to recipes and ingredients is the norm, the ability to recognize the value of sharing this secret information is the main determinant of success. To overcome internal hurdles, firms first have to incorporate external thinking in the strategic planning process and to have a clear understanding of the resources that are needed to meet these goals (Slowinski and Sagal, 2010). Then, management has to initiate the necessary organizational changes to align the different departments (e.g. R&D, Procurement, Marketing and Sales) to work closely together and with external partners in the co-innovation process.

According to Emden *et al.* (2006) technological, strategic and relational alignment are the three most important aspects during the partner-selection phase. This is confirmed by the Unilever case which shows that developing the assessment criteria to identify the relevant partners and executing an open procedure offering the secrecy needed for transparency from both sides are among the most important capabilities needed in this phase. The ability to provide a safe and open environment where the partners are willing to share company-specific and critical information is important. Only in this way it is possible to actually assess the potential complementary character of the partners' capabilities (technological alignment) and the possibility of reaching strategic and relational alignment. The Unilever case study also showed that the ability to ensure commitment of senior management of all the firms in the co-innovation partnership is an important element in this phase.

Previous studies recognize the importance of the partnership-formalization phase as one of the key determinants of success (Kale and Singh, 2009). Contractual agreements are considered important to manage the risks of a partnership and avoid misunderstandings and conflict during the collaboration (Slowinski and Sagal, 2010). Mutual rights and obligations, organizational processes for decision-making and problem-solving and property rights can be the subject of the contractual agreements in order to safeguard the collaboration process. However, as well as the types of agreements which can be made, the capabilities of the firm to negotiate on these different agreements are also important for a successful formalization phase. Different negotiation strategies can be followed. Partners need to engage in the analysis of their own position in the network of suppliers and buyers and the availability of other options to be able to assess to what extent they can balance their own demands with the partners. The case study also shows that, particularly if long-term collaboration is envisioned, uncertainty plays an important role. For a long-term co-innovation partnership, it is therefore very important to be able to find and to emphasize shared interests. As it is impossible to make all-encompassing agreements to cover all the aspects of collaboration beforehand, successful management of the formalization phase includes the ability to determine the extent to which the agreements have to be formalized. The key to this capability is the attribution of time and effort to assessing possible future disagreements or points of difficulty in the collaboration.

The implementation phase is one of the most important and longest phases of the co-innovation partnership. It is the phase where the partners actually start to collaborate and, if they were not engaged in earlier collaborations, get to know each other and each other's competences. The building of trust among the partners is one of the most important aspects in this phase, because it impacts on the efficiency and effectiveness of the daily operations and tasks (Zaheer *et al*, 2002). As the case study shows, this can be done by institutionalising routines and processes which stimulate the employees to share information and knowledge (Kale and Singh, 2009; Todeva and Knoke, 2005). This may require a change in the existing internal routines and practices. In order to ensure this, management needs to facilitate open communication and trust among the employees, and especially among employees who are not involved in the initial set-up of the partnership. The creation of mutual benefits needs to be emphasized (Kale and Singh, 2009; Todeva and Knoke, 2005) and possible frictions because of different corporate cultures (e.g. a more technology versus a more marketing oriented culture as in the present case) need to be bridged (Parkhe, 1991), in order to make the implementation phase of the co-innovation process successful.

As described in the theoretical background section, co-innovation partnerships can be evaluated on the basis of different process and performance criteria (Child *et al.*, 2005). Both subjective, perception-based evaluations (Kumar and Nti, 1998; Monczka *et al.*, 1998; Reuer and Ariño, 2002) and objective measures (Batterink, 2009; Chesbrough, 2004; Todeva and Knoke, 2005) can be used. Evaluation of a partnership is dynamic process that can develop over time, resulting either in stabilization, reformation or termination of the partnership (Das and Teng, 2002). The case study shows that one of the most important managerial aspects in this phase is the enabling of open reviews and the development of a flexible attitude to adjust the initial agreements, if needed.

14.5.1 Key factors for a successful co-innovation partnership

The managerial capabilities found in the Unilever case study can be considered as key determinants for the success of a co-innovation partnership. They demonstrate the essentials of managing the different phases of such a partnership. In line with earlier case study research on open innovation (e.g. Dodgson *et al.*, 2006), the shift towards an open-innovation strategy entailed a significant cultural and organizational change within all the partnering firms. The capabilities to manage and execute the co-innovation partnership have developed over time and require continuous management attention because of their dynamic character. Their relevance goes beyond this particular case, proven by the fact that this partnership serves as an exemplar case for all other co-innovation collaborations of Unilever. For example, Unilever learned to set up the formalization phase in a better way. The thorough attention given to this phase was important to ensure commitment to the partnership. On the basis of their current experience with the set-up stage and the formulation of the partnership agreement, they are able to approach this process in a more efficient way in their other partnerships. The construction of a general partnership agreement, which is legally binding, and separate supplementary project contracts which can be agreed upon for each new project where this is needed, is a new formula that has been used successfully in other co-innovation partnerships.

The success of the partnership does not mean that the partners did not experience any hurdles in the set-up and implementation of the partnership. However, these hurdles were used to learn and improve. One of the biggest challenges related to attaining a higher information exchange level in the partnership. In the first place, this had to be achieved in the partner-selection and partnership-formalization phases. As described above, this was done on the basis of indicating and

Table 14.1 Key factors for a successful co-innovation partnership

Initiation phase

- Incorporation of thinking about possible external linkages in strategic planning
- Recognition of the potential value of complementary external resources and skills

Partner selection phase

- Partner selection based on assessment criteria and open transparent information exchange
- Getting commitment of senior management of all the co-innovation partners

Formalization phase

- Consideration of the own network position and that of possible partner suppliers and buyers
- Determination of possible points of (future) disagreement and attempt to avoid these through mutual agreements on e.g. property rights or organizational processes for decision-making and problem-solving

Implementation phase

- Trust-building through institutionalization of routines and processes which stimulate information sharing and bridge differences in corporate cultures
- Organizational changes within the company to assure that cross-functional collaboration works also in the context of co-innovation
- Emphasizing the potential mutual tangible and intangible benefits of the co-innovation process for both partners

Evaluation phase

- Open reviews of the co-innovation process and the initial agreements, if flexible adjustments are needed crystallizing the complementarities of resources and skills, as well as the shared interests and goals. Also, assurance of top-management commitment and the pin-pointing of key points of possible disagreement, as well as ensuring that formal agreements were made on these issues, helped the partners to overcome their fear and led to increased information exchange. At the operational level, time and effort was needed to achieve the cultural and organizational change to ensure that fear of information exchange was eliminated.

Next to the difficulty related to sharing product and ingredients information, also the sharing of critique towards each other's conduct or approaches was one of the hurdles they experienced. Despite the partnership, the client–customer relationship remains which makes it complicated to express critique. However, the trust that was built up during the collaboration contributed to increase the confidence among the partners to also share vital information in order to reduce the chance of failure in the more radical innovation projects. The lessons learned are summarized in Table 14.1.

14.5.2 Recommendations

First of all, for a co-innovation partnership to be successful, long-term and short-term co-innovation objectives have to be balanced. Partnerships usually start with short-term co-innovation projects that are revenue-oriented and move quickly to market-driven projects. Fast-cycle-ideas-pushing can be used to focus on the efforts and knowledge of the co-innovating partners on a specific challenge, and to quickly generate usable ideas. This involves, for instance, that, instead of having loose review processes every couple of months or once a year, short-term events are held where employees from different organizational levels and co-innovation partners can participate. Participation is eased through the use of mobile phones and skype, decreasing the hurdle of ensuring the presence of all participants. In addition, knowledge brokers can be assigned. These are highly visible employees who generate and disseminate innovative ideas throughout the company and with co-innovation partners. Idea hubs can also be created where innovative ideas are posted, voted on or discussed. Another possibility to speed up the co-innovation process is to use pre-existing solutions more effectively. Knowledge management and open-innovation processes can be created which help to find and re-use the internal and external solutions, instead of re-inventing them. Finally, all co-innovation partners need to be aware of the fact that they are engaged in a process, instead of a tightly planned project, in which different organizational levels are involved in multilevel project teams. Therefore, differences of opinion have to be accepted, and a good understanding has to be created on how innovation benefits can be captured and risks can be divided in a balanced way, while generating alternatives and new research possibilities and effective actions are taken to translate these possibilities into new products and processes.

14.6 Future trends

The multipartners perspective on the management of co-innovation processes taken in this study, contributes to the understanding of successful co-innovation in the food industry and might serve as a point of departure for larger scale research. In addition, the inclusion of more co-innovation trajectories and research into the capabilities of an unsuccessful co-innovation partnership can result in interesting insights that can enhance the theory on co-innovation. Other possible studies may be dedicated to the long term dynamics of co-innovation partnerships, to investigate whether, and to what extent, the partners are able to improve their co-operative capabilities over time. Finally, comparative case studies with other industries, for example from the pharmaceutical industry, might increase our understanding about general and food-industry specific co-innovation management capabilities.

14.6.1 Trends in co-innovation

The goal of innovation is to increase the speed with which new products or services with high-quality and/or unique features are introduced into the market. We expect that co-innovation will increasingly be tested to share risks, resources, skills and knowledge to aid companies in the reduction of costs and speed up development time. However, co-innovation is thwarted by challenges, such as finding appropriate partners and the time needed to align to each other's working methods and processes, and confidentiality aspects and openness to share knowledge while also providing critiques, if needed. These are all aspects that have the potential to slow down the innovation process and the initial goal of engagement in co-innovation. Therefore, future co-innovation partnerships have to draw lessons from the innovation management approaches in highly innovative co-innovation partnerships, such as the FLOF partnership between Unilever and the flavour houses.

14.7 References

AVERMAETE, T. AND VIAENE, J. (2002). On innovation and meeting regulations: the case of the Belgian food industry. *DRUID summer conference*, Copenhagen and Elsinore.

BATTERINK, M.H. (2009). *Profiting from external knowledge: how firms use different knowledge acquisition strategies to improve their innovation performance.* Wageningen Academic Publishers: Wageningen.

BOSSINK, B.A.G. (2002). The development of co-innovation strategies: stages and interaction patterns in interfirm innovation. *R&D Management*, **32**(4), 311–320.

CHESBROUGH, H. (2003). *Open innovation: the new imperative for creating and profiting from technology*. Boston, MA: Harvard Business School Press.

CHESBROUGH, H. (2004). Managing open innovation. *Research Technology Management*, **47**(1), 23–26.

CHILD, J. FAULKNER, D. AND TALLMAN, S. (2005). *Cooperative strategy*. Oxford: Oxford University Press.

CHRISTENSEN, J.L., RAMA, R. AND VON TUNZELMAN, N. (1996). *Study on innovation in the European food products and beverages industry*. Brussels, European Commission Report, EIMS/SPRINT.

COSTA, A.I.A. AND JONGEN, W.M.F. (2006). New insights into consumer-led food product

development. *Trends in Food Science and Technology*, **17**(8), 457–465.
DAS, T.K. AND KUMAR, R. (2011). Interpartner negotiations in alliances: a strategic framework. *Management Decision*, **48**(8), 1235–1256.
DAS, T.K. AND TENG, B.S. (2000). A resource-based theory of strategic alliances. *Journal of Management*, **26**(1), 31–61.
DAS, T.K. AND TENG, B.S. (2002). The dynamics of alliance conditions in the alliance development process. *Journal of Management Studies*, **39**(5), 725–746.
DODGSON, M., GANN, D. AND SALTER, A. (2006). The role of technology in the shift towards open innovation: the case of Procter and Gamble. *R&D Management*, **36**(3), 333–346.
DOLLINGER, M.J., GOLDEN, P.A. AND SAXTON, T. (1997). The effect of reputation on the decision to joint venture. *Strategic Management Journal*, **18**(2), 127–140.
DUYSTERS, G.M., GILSING, V.A. AND BIJMAND, J. (2006). Innovation in agriculture and food: the paradox of many cooperatives, yet few collaborations. An essay commissioned by Transforum.
DYER, J.H. (1997). Effective interfirm collaboration: how firms minimize transaction costs and maximize transaction value. *Strategic Management Journal*, 18(7), 535–556.
DYER, J.H. AND SINGH, H. (1998). The relational view: cooperative strategy and sources of interorganizational competitive advantage. *Academy of Management Review*, **23**(4), 660–679.
DYER, J.H., KALE, P. AND SINGH, H. (2001). How to make strategic alliances work. *MIT Sloan Management Review*, **42**(4), 37–43.
EISENHARDT, K.M. AND SCHOONHOVEN, C.B. (1996). Resource-based view of strategic alliance formation: strategic and social effects in entrepreneurial firms. *Organization Science*, **7**(2), 136–150.
EMDEN, Z., CALANTONE, R.J. AND DROGE, C. (2006). Collaborating for new product development: selecting the partner with maximum potential to create value. *Journal of Product Innovation Management*, **23**(4), 330–341.
GALIZZI, G. AND VENTURINI, L. (2008). Nature and determinants of product innovation in a competitive environment of changing vertical relationships In: (Ed.) Rama, R. *Handbook of innovation in the food and drink industry*, The Haworth Press: New York.
GLAISTER, K.W. AND BUCKLEY, P.J. (1996). Strategic motives for international alliance formation. *Journal of Management Studies*, **33**(3), 301–332.
GRANOVETTER, M.S. (1985). Economic action and social structure. *American Journal of Sociology*, **91**(3), 481–510.
GULATI, R. (1995). Social structure and alliance formation patterns: a longitudinal analysis. *Administrative Science Quarterly*, **40**(4), 619–652.
GULATI, R. AND SINGH, H. (1998). The architecture of cooperation: managing coordination costs and appropriation concerns in strategic alliances. *Administrative Science Quarterly*, **43**(4), 781–814.
HAGEDOORN, J. (1993). Understanding the rationale of strategic technology partnering: interorganizational modes of cooperation and sectoral differences. *Strategic Management Journal*, **14**, 371–385.
HILLMAN, A.J., WITHERS, M.C. AND COLLINS, B.J. (2009). Resource dependence theory: a review. *Journal of Management*, **35**(6), 1404–1427.
HITT, M.A., DACIN, M.T., LEVITAS, E., ARREGLE, J.L. AND BORZA, A. (2000). Partner selection in emerging and developed market contexts: resource-based and organizational learning perspectives. *Academy of Management Journal*, **43**(3), 449–467.
JIANG, X., LI, Y. AND GAO, S. (2008). The stability of strategic alliances: characteristics, factors and stages. *Journal of International Management*, **14**(2), 173–189.
KALE, P. AND SINGH, H. (2009). Managing strategic alliances: what do we know now, and where do we go from here? *The Academy of Management Perspectives*, August 2009, 45–62.
KOGUT, B. (1988). Joint ventures: theoretical and empirical perspectives. *Strategic Management Journal*, **9**, 319–332.

KUMAR, R. AND NTI, K.O. (1998). Differential learning and interaction in alliance dynamics: a process and outcome discrepancy model. *Organization Science*, **9**(3), 356–367.

LAMBE, C.J. AND SPEKMAN, R.E. (1997). Alliances, external technology acquisition, and discontinuous technological change. *Journal of Product Innovation Management*, **14**(2), 102–116.

LAMBE, C.J., SPEKMAN, R.E. AND HUNT, S.D. (2002). Alliance competence, resources, and alliance success: conceptualization, measurement, and initial test. *Journal of the Academy of Marketing Science*, **30**(2), 141–158.

LORD, M., DEBETHIZY, D. AND WAGER, J. (2005). *Innovation that fits, moving beyond the fads to choose the right innovation strategy for your business*. New Jersey: Pearson Prentice Hall.

MIOTTI, L. AND SACHWALD, F. (2003). Co-operative R&D: why and with whom? an integrated framework of analysis. *Research Policy*, **32**(8), 1481–1499.

MONCZKA, R.M., PETERSEN, K.J., HANDFIELD, R.B. AND RAGATZ, G.L. (1998). Success factors in strategic supplier alliances: the buying company perspective. *Decision Sciences*, **29**(3), 553–577.

OLIVER, C. (1988). The collective strategy framework: an application to competing predictions of isomorphism. *Administrative Science Quarterly*, **33**(4), 543–561.

OMTA, S.W.F. AND VAN ROSSUM, W. (1999). 'The management of social capital in R&D collaborations', In: *Corporate social capital and liability*, Dordrecht: Kluwer Academic Publishers, pp. 356–376.

PARKHE, A. (1991). Interfirm diversity, organizational learning, and longevity in global strategic alliances. *Journal of International Business Studies*, **22**(4), 579–601.

PORTER, M.E. AND FULLER, M.B. (1986). 'Coalitions and global strategies'. In M.E. Porter (ed.), *Competition in global industries*. Harvard Business School Press, Boston, MA, pp. 315–344.

REUER, J.J. AND ARIÑO, A. (2002). Contractual renegotiations in strategic alliances. *Journal of Management*, **28**(1), 47–68.

REUER, J.J. AND ARIÑO, A. (2007). Strategic alliance contracts: dimensions and determinants of contractual complexity. *Strategic Management Journal*, **28**(3), 313–330.

RING, P.S AND VAN DE VEN, A.H. (1994). Developmental processes of cooperative interorganizational relationships. *The Academy of Management Review*, **19**(1), 90–118.

SAMPSON, R.C. (2007). R&D alliances and firm performance: the impact of technological diversity and alliance organization on innovation. *Academy of Management Journal*, **50**(2), 364–386.

SARKAR, S. AND COSTA, A.I.A. (2008). Dynamics of open innovation in the food industry. *Trends in Food Science & Technology*, **19**(11), 574–580.

SAXTON, T. (1997). The effects of partner and relationship characteristics on alliance outcomes. *Academy of Management Journal*, **40**(2), 443–461.

SENKER, J. AND MANGEMATIN, V. (2008). Biotech innovation in Europe's food and drink processing industry: promise, barriers and exploitation. In: Rama, R. (Ed.). *Handbook of innovation in the food and drink industry*, The Haworth Press: New York.

SHAH, R.H. AND SWAMINATHAN, V. (2008). Factors influencing partner selection in strategic alliances: the moderating role of alliance context. *Strategic Management Journal*, **29**(5), 471–494.

SLOWINSKI, G. AND SAGAL, M.W. (2010). Good practices in open innovation. *Research–Technology Management*, September–October, 38–45.

STEL, F.G. (2011). Improving the performance of co-innovation alliances. Thesis Tilburg University.

STEWARD-KNOX, B., AND MITCHELL, P. (2003). What separates the winners from the losers in new product development? *Trends in Food Science and Technology*, **14**, 58–64.

TODEVA, E. AND KNOKE, D. (2005). Strategic alliances and models of cooperation. *Management Decsion*, **43**(1), 123–148.

UZZI, B. (1997). Social structure and competition in interfirm networks: the paradox of embeddedness. *Administrative Science Quarterly*, **42**(1), 35–67.

VANHAVERBEKE, W. (2006). 'The inter-organizational context of open innovation'. In: Chesbrough, H., Vanhaverbeke, W. and West, J. (Eds.). *Open innovation: researching a new paradigm*. New York: Oxford University Press.

VANHAVERBEKE, W.P.M., DE ROCHEMONT, M.H., MEIJER, E. AND ROIJAKKERS, A.H.W.M. (2007). Open innovation in the agri-food sector. Research paper commissioned by TransForum.

WILKINSON, J. (2002). The final foods industry and the changing face of the global agro-food system. *Sociologia Ruralis*, **42**(4), 329–346.

ZAHEER, A., LOFSTROM, S. AND GEORGE, V.P. (2002). 'Interpersonal and interorganizational trust in alliances'. In: Contractor, F.J. and Lorange, P. (Eds.), *Cooperative strategies and alliances*. London, Pergamon.

ZIGGERS, G.W. AND HENSELER, J. (2009). Inter firm network capability: how it affects buyer–supplier performance. *British Food Journal*, **111**(8), 794 810.

15

Managing asymmetric relationships in open innovation: lessons from multinational companies and SMEs

D. Oughton, University of Cambridge, UK and L. Mortara and T. Minshall, IfM Centre for Technology Management, UK

DOI: 10.1533/9780857097248.3.276

Abstract: The challenges of partnerships between large and small companies are explored and suggestions from case studies that might be of help in overcoming them are made. The difficulties are explored from both the large and small companies perspectives. Culture, complexity and communication problems are explored. Case studies of open-innovation partnerships involving Nestlé, Tate & Lyle, Unilever and Ben and Jerry's are presented.

Key words: asymmetric partnerships, alliances, company collaboration, partner of choice, culture, trust and complexity, open innovation.

15.1 Introduction: the importance of large and small company partnerships in the food industry

As large companies adopt open-innovation (OI) models (Chesbrough, 2003) and look beyond their boundaries for ideas, small companies are one obvious source. Small companies, and start-ups in particular, typically lack the strategic and operational rigidities that can stifle innovation in established firms (Minshall *et al.*, 2008). Moreover, as argued by King *et al.* (2009): 'Small businesses share a number of characteristics that enable their ability to innovate and provide competitive advantage relative to large corporations. They are also affected by innovation amplifiers, which are factors that augment the ability to innovate and the success of their innovation'. Small and early-stage firms are significant generators of innovation but are typically resource-constrained and struggle to access the complementary assets they need to get their ideas to market.

Larger companies, on the other hand, need access to new ideas and technologies to feed their innovation processes. They need to source them from wherever they are generated. This is particularly true for firms competing in the food sector where, as described by Chris Anderson (2008), 'in 2003 alone, 26 893 new food and household products were introduced, including [...] 187 breakfast cereals [...]'. These numbers are significant indicators of the explosion in product variety to serve niche markets across the globe and of the intent of firms to increasingly serve individual consumers' needs.

Bringing together small and large companies in mutually beneficial partnerships seems the obvious way forward: harnessing the speed, entrepreneurship and innovative capacity of small firms to feed the channels, brands and resources of the large company should create the new value for consumers that neither firms could deliver alone.

Specifically in the food and, more generically, in the fast moving consumer goods (FMCG) sectors, major corporations have embraced OI models consciously and, as a result of top managers' encouragement, they are implementing OI, co-ordinating their efforts through dedicated teams (Mortara *et al.*, 2009; Mortara and Minshall, 2011). Such 'conscious adopters' of the OI model have been attracted by Chesbrough's approach and have contributed to its popularity (Mortara and Minshall, 2011). Firms in this sector seem to have a homogeneous view of what OI can achieve and many intermediary organizations are concentrating on these firms setting up activities to support OI in FMCG. For instance a major FMCG organization, Proctor & Gamble (P&G), has substantially reviewed and redesigned their innovation processes in the light of the OI paradigm (Dodgson *et al.*, 2006; Huston and Sakkab, 2006). Firms in FMCG have traditionally relied on internal resources to innovate, and now OI is seen as an opportunity to accelerate innovation and to promote growth. However, in this sector, radical innovation is very hard to achieve (Zairi, 1995), competition is very high and the market is very demanding. As Mortara and Minshall (2011) found in one of their case studies, this sector recognizes the potential power of OI: ''The 'outperformers' in the food industry use external sources of innovation. [..] OI seems a successful approach''. Thus, a trend is noticeable whereby, as a result of the difficulty in achieving sustained high growth and in containing the innovation, costs are forcing many firms to adopt OI. Furthermore, FMGC's innovation is strongly dominated by brands and the adoption of OI contributes to the reinforcement of branding message (Mortara and Minshall, 2011).

In understanding 'Why food startups are getting hot', Klein (2011) found that a number of current drivers are encouraging an environment where entrepreneurial start-ups can deliver vital innovation. These include: the rise of the demand for organic and sustainable products; the need to respond to issues such as obesity, diabetes, and allergies in the population and the discovery and demand for ethnic foods by the mainstream.

When investigating the 'Food startup landscape' (500 Startups, 2011), William

Rosenzweig from Physic Ventures, which partners Unilever and Pepsico to help fund new innovators in the field, identified two of the primary obstacles for food start-ups as:

1. the entrenched nature of food distribution channels and systems; and
2. the lack of seasoned entrepreneurs in this segment of the industry.

Clearly these are two areas where small companies can leverage the resources of large, established players to great effect.

15.2 The difficulties of open innovation

Though the benefits of the OI model in food and FMCG are clearly evidenced, research shows that making such partnerships work can be problematic (Minshall *et al.*, 2008, 2010). To understand why, it is important to look at the challenges faced by these partnerships from both ends of the 'small company – large company telescope', which reveals two markedly different perspectives in terms of business culture, management style and expectations. These are reported in the following subsections.

15.2.1 The large company perspective

Five aspects are of concern for large companies when dealing with small and, in particular, inexperienced firms (Minshall *et al.*, 2008, 2010).

Managing intellectual property

Start-ups are often reluctant to reveal details of their technology without a non-disclosure agreement (NDA), fearing their intellectual property may be appropriated. However, large companies want to avoid contaminating their own pre-existing IP, especially without evidence to suggest the small company has anything of value to propose. The result is a circular discussion: 'I don't want to sign an NDA until I know what you've got…' and 'I don't tell you anything until you sign an NDA', leading to frustration and mistrust at the very outset of a relationship.

Brand protection

Large companies are fiercely protective of their brands, and rightly so considering their value. According to Interbrand for example, in 2011, the 125 year-old Coca-Cola brand was worth US$71 861 million (Interbrand, 2011). Associations with such recognized brands could be extremely important for the future of small companies and could represent assets to be leveraged both in the consumer and investor communities. However, large companies recognize that there is potential for abuse in such relationships, with start-ups using the partner's brand (perhaps naively) in inappropriate ways in pursuit of commercial credibility.

Technology readiness
Large companies pursue OI models to go 'bigger, better, faster' (e.g. Unilever, 2010). In pursuit of 'faster' the large firm is likely to want a ready-for-market solution (ideally fully proven, documented and fit for launch), Start-ups often perceive their role as to provide an enabling technology to be incorporated into a large firm's product. This gap in expectations can be quite significant and large companies claim that start-ups often do not appreciate the time and cost involved in moving from technology demonstrator to fully-supported product (Mortara and Ford, 2012). Vital regulatory or third-party approvals, essential for the large company to go to market, might be seen as jumping through unnecessary hoops by small firms.

Financial stability
Start-ups sometimes fail to understand a large firm's need for due-diligence checks to give potential partners confidence in their viability. These can seem intrusive, especially when they are undertaken by professional advisors, but large companies cannot afford to leave any stone unturned before entering a partnership. Again, these 'due diligence' checks are necessary at the front-end of a relationship, where early bridges of trust and understanding have yet to be built.

Culture
Ironically, the very cultural traits that make small companies so innovative, i.e. entrepreneurial drive and passion; a lack of rules and bureaucracy; speed and agility, also make them very challenging collaborators for large companies. Start-ups may be run by individuals impatient for progress but unwilling to be governed by schedule and discipline. 'We ask for simple things like a business case or cash-flow projections or reports and they get resentful. They don't see why they should have to justify everything!', said a large company manager (Minshall *et al.*, 2008). We discuss culture as a central theme in Section 15.3.

15.2.2 The small-company perspective

At the same time, small companies also suffer from biases and challenges when facing collaborations with the large firms (Minshall *et al.*, 2008, 2010), including the aspects discussed in the following subsections.

How to engage
Increasingly, large companies who want to access the benefits of OI have established very clear points of contact, with dedicated teams of OI professionals and well signposted web portals and contact points. However, many do not and, in any event, the complexity and scale of large-company operations mean that even their own staff are sometimes unable to help a start-up contact the right people. One start-up CEO commented: 'The [large company] people would start every meeting with us looking at their organization charts to try and work out where they fitted into the company. If they didn't know who did what, what chance did we

have?'(Minshall *et al.*, 2008). Finding the right entry-point is the first of many challenges faced by the early-stage firm.

Understanding company roles

It is very hard for the small companies to work out the different roles of people in a large company. Who is the decision maker? Who influences them? Who will be working on implementing the partnership? Who will be affected by its outcome? It is likely that external people might assume that they are 'in discussion with XYZ plc' whereas they are actually 'having a chat with Fred Bloggs of XYZ plc'.

Changing points of contact

The previous point gets even more troubling when start-ups begin by talking to the large company's technologists who are likely to be enthusiastic and speak the same language. These individuals may be someone met at a conference or a trade fair, but as they move towards formalizing a deal, the start-up will have to talk to the procurement and legal teams, who may treat them quite differently. Moreover, as soon as a contact point is established and developed there is every likelihood of them being moved along by the 'big company merry-go-round'.

Slow decision cycles

Small start-ups are usually able to make decisions very quickly. Large firms, owing to their complexity and multiple layers of management often find it very hard to make decisions at 'start-up speed'. This can be very frustrating for the start-up, undermining the very trait of agility that the large company was hoping to achieve through partnering, i.e. getting to innovation 'faster'. The slowness in decision can also have repercussions on the relationships with others, for instance when the small company contact person gets squeezed in the middle between their anxious senior management and investors and the unresponsive large partner.

Power imbalance

The large firm may intentionally or unintentionally 'abuse' its position by drawing out negotiations and attempting to prevent discussions with competitors. Although this might be perceived as cynical manipulation by the small company, it may just be a manifestation of the different context of the relationship in the two organizations. For the large company, this may be just one of many technology positions within a portfolio whereas for the start-up, the partnership with a large company is likely to be a survival issue. This can push the cash-strapped start-ups towards accepting less lucrative deals or simply serves to undermine confidence in the partnership.

Absorptive capacity

Even without intent, the inevitable imbalance in resources can present a sizeable barrier for small companies to participate in OI. Despite OI's widespread applications, numerous small and medium-size enterprises (SMEs) and others operating in traditional sectors are struggling with its implementation owing to their relatively low level of absorptive capacity (Spithoven *et al.*, 2010). Many SMEs perceive the OI model as prohibitively expensive. Some of this cost is associated with the substantial upfront investment in human resources required for assessing, selecting, and negotiating with the external innovation contributors, and/or the paramount organizational changes required for implementation.

Ignorance of start-ups

Demands made of start-ups by large firms can sometimes show the lack of awareness of how a start-up operates. 'They would ring us up and ask to speak to our Latin America sales director or ask us to train 20 000 of their consultants. Our whole business was six people in one room!' Start-up CEO (Minshall *et al.*, 2008). This example typifies how incorrect assumptions, lack of thoughtful preparation for the partnerships and lack of thorough evaluation of resources required might lead to paradoxical situations that rapidly compromise the relationship.

15.3 Culture, complexity and communication problems

Overall the scenarios outlined above describe problems linked to culture, complexity and communications.

15.3.1 Culture

As we've already seen, the markedly different cultures that prevail in large firms and early-stage companies are at the root of many of the barriers to successful OI collaboration. Doz and Hamel (1998) found that this extended to the stereotypical perceptions that each party may hold of the other.

A large firm (as perceived by a small one) may be ponderous, slow and stupid; preoccupied with reviewing everything to death; awash in mindless procedures; risk averse, procrastinating; characterized by paralysis through analysis; divided, fragmented.

The large firm's perception of the people in small companies is not much more flattering as they are described as a bunch of cowboys, shooting from the hip; disorganized, slippery; going off in all directions, unfocused; characterized by sloppy work; exclusive, clannish, hostile (Doz and Hamel, 1998).

In the light of these observations, it is remarkable that any collaborations succeed. Of course these caricatures actually mask some of the very attributes that attract the parties in the first place. For 'risk averse' one could read 'professional'; for 'going off in all directions', 'experimental' could be the meaning. What is clear is that these cultural differences, stemming from the different operational contexts and the typical backgrounds and outlooks of the protagonists, need to be bridged to deliver success.

15.3.2 Complexity and individual perspectives

So far we have identified the internal challenges that need to be overcome to deliver value from collaborations: the different cultures which exist; the unmatched 'clock-speeds'; and the difficulty of navigating into and through the organizational jungle that is the large company. However, that is just the start of the complexity. Each party exists within an even more complex web of stakeholders and influences which further complicates any relationship (Minshall *et al.*, 2010).

Typically, the small company has investors or owners who have more than a distant financial interest in their operations. Company founders, venture capital (VC) providers, business angels and family members all bring expectations and agendas that are unlikely to be aligned with each other, let alone with the large-company partner. This becomes potentially more distracting when the funding backing the start-up is from a VC provider (whether associated with the potential large-company collaborator or with one of its competitors, though the tensions are obviously different in each case). As the nature of a possible technology acquisition route moves along the spectrum from licensing to equity participation and outright takeover, these tensions are only likely to increase.

Pre-existing relationships between the small company and other outside organizations also add to complexity. There may be technology ties to a university (perhaps the former parent of a spin-out) or to funding programmes (perhaps from national or EU governmental sources) which need to be understood, if not unraveled, before a deal can be concluded. There are also a number of professional advisers in the company (each matched by an army of counterparts in the large company) including lawyers (perhaps covering both IP and commercial aspects separately); consultants (perhaps a contract R&D house or expert technologist) and financial advisers (their own and potentially those of their investors).

The complexity is also a challenge for the large company, even if many of these functions and roles are performed by 'internal' people. Interfaces between different departments, with different reporting lines and often divergent metrics, may offer at least as much potential for friction as external relationships. Large companies also have their own set of relationships with parent and sister companies to manage.

Finally, as if these complexities were not enough to challenge the collaboration between a given pairing of large and small companies, it needs to be recognized that this will only be one partnership of many being explored. The small company is likely to be offering its technology to as many potential partners as possible, many of whom will compete directly with each other. Similarly, the large company will be scouting to identify other technology and partners probably pursuing a number of options at the same time. All in all, this is a very crowded and complex field indeed.

15.3.3 Communication

Given the differences in culture between large and small companies, and the numerous different actors surrounding the relationship, there are inevitable challenges in communication. Some of these challenges relate to the language: what each partner means by, for example, 'plug and play' technology is likely to be very different, as exemplified by a request for the Mandarin version of the installation manual being made to the five-person early-stage company that barely has a technology demonstrator on the lab bench.

Inevitably the differences in culture, context and perception can result in misunderstanding. Delays in communication (for example awaiting the next large company stage-gate meeting) can easily be misinterpreted as 'stalling' by an entrepreneur focused on his cash burn-rate. Delivery of a prototype without associated test documentation (because it was more expedient) could easily be seen as 'sloppy' by the receiving large company. In the pressured environment of innovation and new product development, such misunderstandings can quickly lead to mistrust. Navigating a way through this complex cultural maze requires good understanding between individuals, and in 15.5.3 the importance of people having the right skills and backgrounds at these crucial interface points is discussed.

15.4 The importance for companies of focusing on risk, reward and balance

Roberts (2007) defined innovation as being a combination of invention and exploitation (innovation = invention + exploitation). If this is the case, then open (or collaborative) innovation is about exploiting (and therefore deriving value from) shared efforts and by implication sharing this value. One of the challenges of the OI model is that this value-share takes place between two separate profit and loss accounts and, therefore, the mechanism for sharing needs to be established. A common feature of early discussions and deal-making can be that this contentious issue can become very time-consuming and overwhelming. Indeed, it sometimes seems that more energy is invested in negotiating how to share the value of the collaboration than in actually creating the value in the first place!

Whilst this focus is understandable from both perspectives, in many ways the start of a relationship is not the best time to settle these issues: There is little evidence for what value might be created, or certainty of success, and the partnership is at its most fragile. Importantly, such negotiations risk being overly focused on just one type of 'value': monetary. This can result in the negotiation being based on the assumption of a zero sum game (for me to win, you've got to lose). In reality, much of what is valued by the small company (for example access to market knowledge, brand association and a route to market) can be provided by the large company at little cost. Similarly a large company's principal concern may be with achieving a position of exclusivity (at least for a time), rather than with squeezing the acquisition cost to the minimum. The opportunities for a win–win situation are significant as long as some creativity and flexibility can be exercised by both sides.

Another challenge around deal-making is the fundamental inequality that exists between the partners, in terms of size, resources and, in particular, the implications of failure. A large company with a well developed OI programme might generate 10 000 contacts during a year that result in 1000 discussions, 100 NDAs, 10 investments and only one commercial success, but the scale of that success could well make the rest of the investment worthwhile. Seen from the small company perspective the same story looks like 9 999 disappointments of which perhaps nine involved significant investment of scarce resources and, in the extreme, the difference between survival and business failure.

15.5 Overcoming obstacles to achieve successful company partnerships

There are a number of lessons to be drawn from how successful companies manage these asymmetric relationships that can help to mitigate these challenges. There are lessons for both the small and large partner in these collaborations, involving strategy, process and (perhaps most important) people (Minshall *et al.*, 2010).

15.5.1 Strategy: shared vision and agreed business model

Spending time in developing a mutual understanding of strategy and business models, before diving into discussions about technology and its application, can greatly facilitate the development of successful win–win partnerships. This requires a degree of self-awareness by each party and then a candid exchange, which builds a platform for context and trust.

The start-up is likely to consider multiple possible application areas for its technology. It can greatly assist negotiations if these can be captured and shared to highlight the various opportunity areas and show the resources needed for implementation. The start-up should also be aware of three possible outcomes of a partnership: it may help to implement the intended business model(s); it may open new opportunity areas; but it may also restrict future opportunities.

The large firm should try to create a 'wants' list or portfolio that can be shared with start-ups. This should position the large firm's technology capabilities and needs (including an assessment of the level of criticality) and indicate different opportunity areas. Depending on the level of criticality, the large firm may decide to spread risk by having parallel technology acquisition routes. If this is the case, then openness and transparency are likely to build trust and underpin a long-term relationship (or at least reveal any conflicts at an early stage).

15.5.2 Process: technology readiness and shared roadmaps

Clear communication of the technology that the small company has to offer, the need that the large company has to fill, and the necessary effort required to bridge the gap is essential to evaluate and successfully launch a collaboration. Failure to manage expectations and clearly define these at the outset is a common cause of later difficulties. Technology readiness levels (TRL) (Mankins, 1995) provide an unambiguous common language in which these discussions can take place, helping to eliminate subjectivity (and overoptimism). The independent use of technology roadmaps (Farrukh *et al.*, 2003) by both parties (and ideally the consolidation of these into a joint roadmap for the collaboration) provides an excellent framework for documenting: 'Where are we now? Where do we want to get to? and What do we need to do to bridge the gap?' This will also provide an invaluable monitoring and management tool for the on-going collaboration.

Through the roadmapping process, a start-up could be able to investigate more deeply and make a realistic assessment of the readiness level of its technology and identify tasks and costs associated with preparing it for manufacture, including finding out who owns any complementary resources required. The start-up should also evaluate whether specific developments required to meet a specific partner need are applicable to opportunities in other application areas (and therefore part of its core roadmap) or bespoke. Small companies demonstrate a tendency for using scarce resources to pursue bespoke developments for niche applications, and run the risk of inadvertently moving towards a contract R&D business model.

The large firm should use its roadmap to position the start-up's technology within the broader range of its activities. It should show what complementary resources are needed to bring the technology to market and how this may change over time. It should assess the readiness levels of the start-up's technology and how much of the technology is tacit (undocumented) versus explicit. The commercial viability of the start-up needs to be monitored bearing in mind how critical the technology is to the large firm, and, if necessary, contingencies put in place to secure access to critical technology in the event of business failure.

15.5.3 People: company organization and culture

Many of the challenges in collaborating are rooted in the differences in culture and context between the parties. To be aware of this, and to put in place measures to compensate, are some of the most valuable lessons for success. Specifically, it is essential to recognize that communication takes place between people rather than companies, and to ensure that those people acting at the interfaces between the two companies have appropriate skills and personal attributes.

Start-ups find it useful to check whether the large company has ever worked with a start-up before. It is helpful to have team members who have experience of working in large companies involved in the discussions, and if the start-up lacks such experience themselves they should seek advice from non-executive directors, mentors or investors. Talking to the large firm's suppliers as part of a due-diligence process can also help to develop a sense of how the larger company works. It is also a good idea to encourage informal interaction between the teams so that the large firm gets a better sense of start-up culture. Investing time in building the joint team pays dividends in the long run.

The large firm should spend as much time as possible helping the start-up to understand the needs, internal processes and culture of the large firm. Process maps can be used to show start-ups how decisions are made, and used to communicate progress through the evolution of the collaboration (a source of reassurance when 'things seem to have gone quiet'). Most of the large FMCG firms use a dedicated team or individual champion to act as first point of contact (Mortara and Minshall, 2011). These OI deal-makers will ideally have a mix of well-developed interpersonal skills, and a strong technology and commercial understanding, in tandem with typical negotiation experience, and are as likely to have an innovation background as one in procurement. This can help shield start-ups from unnecessary bureaucracy and smooth communication in both directions (Mortara *et al.*, 2009).

15.5.4 Setting up the deal

Having established the need and technology readiness, aligned roadmaps, shared an understanding of culture and context and appointed appropriate people as interface points, the collaborators are in a good position to move forward to make a deal. In constructing the deal it is important to focus on maximising the value of the collaboration rather than trying to carve out the best share of the spoils. It is also helpful to remember that this is not generally a zero-sum negotiation and there are aspects which each party value that are of little cost or consequence to the other. Coming away from the negotiating table with both parties feeling comfortable provides a strong foundation for an enduring partnership; resentments have a habit of resurfacing when inevitable rocky patches are encountered in the collaboration.

The start-up should find out who is likely to influence and authorize the decision to form a partnership. The start-up should have a clear idea what is really expected from the partnership on both sides, what realistically can be delivered, how things may change over time and what the possible direct and indirect benefits might be. Legal advice should be sought at the outset. Though costs will be incurred, they are likely to be less cumbersome than fixing problems later. As decisions relating to the partnership are likely to be made in the start-up's absence, the start-up should make sure their large company 'champion' is armed with the start-up's viewpoint.

The large firm should ensure that overarching principles concerning the deal are agreed first before moving on to detailed issues. It is essential to be as open as possible with the start-up about any concerns and to be aware of the start-up's cash flow position. Working with the start-up on a small-scale, cash-generating project first can be very useful. It gives both sides a feel for how the other operates and may reveal ways the partnership could develop in the future. An incremental approach to the depth of the commercial relationship (for example moving from underwriting agreed joint revenues before moving to IP ownership or an equity position) can de-risk the collaboration for both parties. Above all, the large company should be mindful of the asymmetry in the power relationships and avoid abusing this natural position of power.

15.5.6 On-going management of the relationship

The common roadmap or plan for the collaboration, and agreed deliverables, provide an invaluable touchstone for the ongoing management of the collaboration. Using these as the basis for continuous monitoring and dialogue provides a good basis for early warnings and avoiding surprises. During implementation it is natural that different team members in each organization become involved in the relationship, but it is important that the original key players in making the deal continue to play a role as changing faces at this critical stage can lead to miscommunication and a loss of trust. Any new faces need to be integrated carefully into the team and fully briefed on the background and culture.

The start-up needs to keep in regular contact with its larger partner, not just when there is a problem. Assigning members of the management team to 'mark' key contacts at the large firm is one way to receive early warning of any emerging problem areas. Documenting all interactions should be a standard part of any partnership management process in case there are later disagreements. Staff in the large firm who are key to the partnership may change roles and strategies and business models can change. Regular reviews of the partnership help ensure the relationship continues on the best footing.

The large firm should ensure time is devoted to managing communication between the partners. The start-up should be kept informed of developments, for example by attending internal conferences, and should be told of impending milestones and their relative importance. If underperformance is noted, the start-up should be informed as soon as possible and given help to address the problem.

15.6 Collaborations between companies: case studies

The following case studies illustrate a number of different collaborations between large and small companies, providing some examples of the motivations for each party and the value that can be derived from such relationships.

15.6.1 Case 1: Nestlé innovation partnerships (INP) and sharing-is-winning (SiW) model

Nestlé's adoption of OI through its innovation partnerships (INP) and sharing-is-winning model (SiW) (Traitler *et al.*, 2011) has been utilized as a paradigm-shifting driving force for the reinvention of R&D. For more than 140 years in business, Nestlé's exceptional successes (Nestlé is the largest food company in the world: it was ranked 36 of the 50 most innovative companies in 2009), and most of its innovations came from within. This has created an understandable natural internal resistance to accessing externally developed intellectual property. In 2009, its R&D department included approximately 5000 people with a spend of almost 4.4 times more than Kraft Foods and ranking second after P&G among all consumer companies. However, no relationship was found between R&D spending and the primary measures of economic or corporate success, such as growth, enterprise profitability, and shareholder return (Jaruzelski *et al.*, 2005). It appears that excess resources may be counterproductive, impeding innovation.

These facts highlight perhaps the most important reasons for change, to fight the complacency that stifles innovation. As reported by Traitler 'It took hard work and strong leadership by some very dedicated and committed individuals to change the company's strategy to embrace INP. Adapting the OI philosophy was not easy for Nestlé; nevertheless, the paradigm shift was recognized in 2006 as a strategic change, creating the INP group. INP allowed turning to the outside world for bigger, better, bolder, and faster innovation, and its execution and implementation were prompted by the recognition that universities, academia, small start-ups, biotech companies, and large industrial suppliers are important sources of co-development and partnerships. In particular, academia has been shown to play a very significant role in most breakthroughs and provides a natural and critical partner (Melese *et al.*, 2009).

15.6.2 Case 2: Tate & Lyle: novel salt-reduction technology

In 2011, Tate & Lyle signed an exclusive, worldwide licence agreement with Eminate Ltd, a wholly owned subsidiary of The University of Nottingham, UK, for its novel salt reduction technology. The product, currently known as 'SODA-LO®', enables added salt levels to be reduced by up to 30% in foods such as bread, pizza bases, pastry, savory pie fillings, cheese and baked snacks. SODA-LO® is an innovative 'clean label' salt product that enables salt content to be reduced in food without loss of flavour or structure.

This case clearly illustrates the benefits for both the large and small company partners. Tate & Lyle gain access to a novel technology which provides a solution to a clear and growing consumer need which fits their strategy and routes to market. Karl Kramer, President of Innovation and Commercial Development at Tate & Lyle said: 'The SODA-LO® salt reduction technology is an excellent addition to Tate & Lyle's Health & Wellness offering. Salt reduction is a high-profile priority for many of our customers who are committed to formulating reduced salt products to meet growing consumer demand'.

Meanwhile, the small company benefit from Tate & Lyle's ability to commercialize the technology on a global basis including manufacturing, product development, sales and marketing. Neil Davidson, Chairman of Eminate Ltd said: 'This is a significant deal for Eminate Ltd and we are delighted that Tate & Lyle is working with us and The University of Nottingham to fully commercialize our novel salt reduction technology. Tate & Lyle's relationships with global food companies and its technical and applications development expertise and resources, will help both parties to maximize commercial value from this exciting technology'.

This technology partnership was assessed and developed by Tate & Lyle's OI team, part of its Innovation and Commercial Development group. The OI team utilizes its global network to develop partnerships with universities, customers and start-ups specialising in food science with the goal of successfully bringing their new technologies and products to market.

15.6.3 Case 3: Unilever: an open-innovation infrastructure

Colworth Park is an OI campus in Bedfordshire, UK. It is owned, developed and managed under a joint venture between Goodman and Unilever, having its historical origins in Unilever's global research centre network. The site is host to 17 R&D organisations that benefit from access to a range of specialist services.

This 'innovation campus' model illustrates the motivations and benefits of OI from both the small and large company perspectives: the start-ups benefit from access to equipment and expertise that they otherwise might not have, and Unilever hopes that successful innovative activity of the entrepreneurial start-ups filters out into new Unilever products.

In recognition of the importance of matching cultures, Colworth Park is trying to further support its innovation agenda by arranging its organisational structure to reflect its strategic aims and appointing people with an appropriate background into it. In contrast to some other property-development-owned science parks, Colworth is headed by a board of directors with science and business, as opposed to commercial property, backgrounds.

15.6.4 Case 4: Ben and Jerry's: supporting socially responsible start-ups

Technology is not the only driver for large companies to collaborate with small ones. Brand association and reinforcement, particularly around sustainability and social responsibility, can be important spin-offs from such relationships. In their 'Join Our Core' competition, Ben & Jerry's are calling for socially responsible entrepreneurs to put their business ideas forward. Rewards include the chance of a €10 000 cash prize, but perhaps of greater value is the provision of mentoring and the opportunity to see their name on tubs of Ben & Jerry's ice cream.

Ben & Jerry's characterize the competition as a '... search to uncover the brightest young minds that can create a business embracing the ice cream pioneer's philosophy of giving back to the community.' As co-founder Jerry Greenfield says:

> When we set-up the company in 1978, Ben and I always believed giving back to the community was as important as making great tasting ice-cream. That's why I am really excited about this competition, to find business models with those same values that still lead us today. We would love to see more socially-minded start-ups making a difference and showing that values-led businesses can prosper even in tough economic times.

This example clearly illustrates the value for the large company in extending the search for talent outside of the business and, in turn, delivering the market insights and access to consumers that start-ups crave.

15.7 Conclusion

Large companies are increasingly adopting a systematic and strategic approach to OI, and consider this as a means of generating sustainable advantage in delivering new products and technologies to their customers. This is particularly prevalent in the food and FMCG industries, with numerous examples of best practice emerging in these sectors. The following are a few examples from the OI portals of such companies:

- General Mills – G-WIN: Beyond our walls, innovators like you might have just what we need to bring healthy and flavorful food to the world. At General Mills, we believe that there is a great opportunity for us to enhance and accelerate our innovation efforts by teaming up with world-class innovators from outside of the company. To facilitate this effort, we created the General Mills Worldwide Innovation Network (G-WIN) to actively seek partners who can help us deliver breakthrough innovation in the following categories: ...
- Kraft – Innovate with Kraft: We recognize that valuable solutions can come from anywhere and we invite you to Innovate with Kraft. We have built the capabilities required to integrate your ideas and inventions into winning business propositions that delight our consumers.
- GSK – Open Innovation: At GSK Consumer Healthcare we want to partner with external innovators. All our leading global brands are looking for new technologies to help us grow. There are many reasons to partner with GSK Consumer Healthcare. We provide resources and global reach to potentially develop your innovative technology into a global consumer healthcare product.
- Unilever – Open Innovation: Smart collaboration between ourselves and our partners allows us to leverage a greater mix of technologies and speed up time to market to deliver value none of us could achieve on our own. We welcome approaches from potential collaborators, whether you're a small technology start-up or a major international organisation, in our areas of interest.

Common themes running through each of these is the recognition that innovation comes from many sources: large and small; a proactive call for engagement and a commitment of what the large company can bring to deliver success and value. These clearly underline the importance of establishing long-term, win–win partnerships and sharing success and risk. The battle to become the 'Partner of Choice' is evidently seen as important, with each giant of the food industry vying for the best innovators, in a seeming reversal of the power roles.

This is reflected not only in the sharing of the spoils of successful collaborations with their partners. At least as important are the experiences of those whose ideas are not taken forward, i.e. where a collaboration starts but is not successful. Dealing with failure in an open, transparent and fair fashion leaves a lasting foundation for trust and helps build the brand of trusted partner of choice in the wider innovation community. We explored some of the recognized problems in partnerships and listed some of the potential solutions of these problems (Minshall *et al.*, 2010). Successful large company open innovators are the ones who deliver on these 'partner of choice' aspirations and who back them up through establishing specialist teams equipped with the appropriate skills, culture, processes and motivation to deliver. Having the right people at the interface enables the large company to establish the context for collaboration with the small company; focus on the wider relationship, rather than just the technology; and take the long-term view to deliver a lasting partnership. Transparent processes, with continuous communications and shared roadmaps lead to a foundation of trust. An awareness of the inherent asymmetry in the power relationship of the two parties, and a proactive approach to rebalancing this to deliver shared reward and risk will pay long-term dividends.

According to the Intuit Future of Small Business Report: The Artisan Economy (IFTF, 2008): 'The next decade will see the growth of small business continue, and the social and economic impacts of small business increase.' A bold claim given that small businesses already employ over half of America's private sector workers, produce over half of America's non-farm private GDP, and create roughly 75% of new private sector jobs. The report draws on McKinsey's idea of 'barbell industrial structures' whereby industry consolidation combined with the growing number of small businesses results in a heavily skewed distribution of company sizes at the extremes of the spectrum. At the root of this trend are the information and transportation technologies that have greatly reduced the transaction costs of working with partners and suppliers.

Although these reports are encouraging for the success of small companies in general, winning as an individual small company in this OI environment requires some specific skills. Having an appreciation of how large companies work and their culture is vital, perhaps through enlisting support from advisors or non-executive directors if there is no experience immediately in the team. Forming and communicating a realistic assessment of technology readiness and the effort required to go to market provides a strong platform for managing expectations and delivering success. Maintaining this continuous and open communication throughout the collaboration should sustain this. Being aware of the specific fields of application where a large partner is likely to want to build a defensible position of advantage, and flexible enough to accommodate these without closing other avenues helps to maximize value.

Although there are many lessons to be learnt by both large and small companies, it is inevitably the larger partners who have the advantage of scale to build these ad hoc experiences into a systematic process for successful OI. Building such an efficient 'OI machine' is a major organisational and cultural challenge, while simultaneously managing each collaboration as an individual relationship and respecting the vital importance that it might have for the small company. Achieving the right balance of expediency and humility to carve out a position as 'partner of choice' is a tough road to travel; but projections of the benefit of OI suggest that it is a road worth travelling.

15.8 Acknowledgements

The authors would like to acknowledge the works of colleagues in the Centre for Technology Management, Institute for Manufacturing, University of Cambridge.

This chapter draws heavily on the research of the Technology Enterprise Group, and further detail and case studies can be found at http://www.ifm.eng.cam.ac.uk/ctm/teg/ and www.ifm.eng.cam.ac.uk/ctm/research/projects/alliances.html.

The author would also like to thank members of the Institute for Manufacturing Open Innovation Practitioner Forum in the Food and FMCG Value Streams for all of the insights they have shared. The OI Forum is a private membership group, so no case studies are attributed to these sources. Details can be found at http://www.ifm.eng.cam.ac.uk/ctm/teg/oi_forum.html.

15.9 References

500 STARTUPS (2011). The food startup landscape: investing lessons learned. http://500.co/2011/08/23/the-food-startup-landscape-investing-lessons-learned/. Retrieved 24 April 2012.

ANDERSON, C. (2008). *The long tail: why the future of business is selling less of more*. New York, Hyperion.

CHESBROUGH, H. (2003). *Open innovation: the new imperative for creating and profiting from technology*. Boston, Harvard Business School Press.

DODGSON, M., D. GANN AND A. SALTER (2006). 'The role of technology in the shift towards open innovation: the case of Procter & Gamble'. *R&D Management*, **36**(3): 333–346, DOI: 10.1111/j.1467-9310.2006.00429.x.

DOZ, Y. L. AND G. HAMEL (1998). *Alliance advantage: the art of creating value through partnering*. Boston, Mass., Harvard Business School Press.

FARRUKH, C., R. PHAAL AND D. R. PROBERT (2003). 'Technology roadmapping: linking technology resources into business planning'. *International Journal of Technology Management*, **26**(1): 2–19. DOI: 10.1504/IJTM.2003.003140.

HUSTON, L. AND N. SAKKAB (2006). 'Connect and develop: inside Procter & Gamble's new model for innovation'. *Harvard Business Review*, **84**(3): 58–66.

IFTF (2008). 'The new artisan economy'. *Intuit Future of Small Business Report*, http://http-download.intuit.com/http.intuit/CMO/intuit/futureofsmallbusiness/SR-1037C_intuit_future_sm_bus.pdf.

INTERBRAND (2011). 'Best global brands 2011' http://www.interbrand.com/en/best-global-brands/Best-Global-Brands-2011.aspx.

JARUZELSKI, B., K. DEHOFF AND R. BORDIA (2005). Money isn't everything. *Resilience Report*, No 05406, New York, Booz Allen Hamilton.

KING, S., A. TOWNSEND AND C. OCKELS (2009) Intuit future of small business report: defining small business innovation, www.intuit.com/futureofsmallbusiness, retrieved 24 April 2012.

KLEIN, K. E. (2011). Why food startups are getting hot. *Bloomberg BusinessWeek*, http://www.businessweek.com/small-business/why-food-startups-are-getting-hot-07112011.html, retrieved 24 April 2012

MANKINS, J. (1995). Technology readiness levels, http://www.hq.nasa.gov/office/codeq/trl/, retrieved 24 April 2012.

MELESE, T., S. M. LIN, J. L. CHANG AND N. H. COHEN (2009). 'Open innovation networks between academia and industry: an imperative for breakthrough therapies'. *Nature*, **15**, 502–507.

MINSHALL, T. H. W., L. MORTARA, S. ELIA AND D. PROBERT (2008). 'Development of practitioner guidelines for partnerships between start-ups and large firms'. *Journal of Manufacturing Technology Management*, **19**(3): 391–406, DOI: 10.1108/17410380810853803.

MINSHALL, T. H. W., L. MORTARA, D. R. PROBERT AND R. VALLI (2010). 'Making 'asymmetric' partnerships work'. *Research Technology Management*, **53**(3): 53–63.

MORTARA, L. AND S. FORD (2012). *Technology acquisitions. A guided approach to technology acquisition and protection decisions*, http://www.ifm.eng.cam.ac.uk/ctm/estap.html, retrieved 24 April 2012, ISBN: 978-1-902546-39-1.

MORTARA, L. AND T. MINSHALL (2011). 'How do multinational corporations implement Open Innovation?' *Technovation* **31**(10–11): 586–597. DOI: dx.doi.org/10.1016/j.technovation.2011.05.002

MORTARA, L., J. J. NAPP, I. SLACIK AND T. W. MINSHALL (2009) *How to implement open innovation. Lessons from studying multinational companies.* http://www.ifm.eng.cam.ac.uk/service/books/form_oi09.html, retrieved 24 April 2012, ISBN: 978-1-902546-75-9.

ROBERTS, E. (2007). 'Managing invention and innovation'. *Research-Technology Management*, **50**(1): 35–54.

SPITHOVEN, A., B. CLARYSSE AND M. KNOCKAERT (2010). 'Building absorptive capacity to organise inbound open innovation in traditional industries'. *Technovation*, **30**(2): 130–141, DOI: 10.1016/j.technovation.2009.08.004.

TRAITLER, H., H. J. WATZKE AND I. S. SAGUY (2011). Reinventing R&D in an open innovation ecosystem, *Journal of Food Science*, **76**(2), R62–R68.

UNILEVER (2010). 'Innovation: bigger, better, faster,' http://www.unilever.com/images/mc_innovation-fact-sheet_tcm13-269251.pdf.

ZAIRI, M. (1995). 'Moving from continuous to discontinuous innovation in FMCG: a re-engineering perspective'. *World Class Design to Manufacture*, **2**(5): 32–37.

16

Challenges faced by multinational food and beverage corporations when forming strategic external networks for open innovation

L. Tolhurst and J. Brown, Oakland Innovation, UK

DOI: 10.1533/9780857097248.3.294

Abstract: Current and future trends in strategic external network formation in the food and beverage industry are explored and key challenges identified. Oakland has developed a ten-step guide to help companies navigate the process of establishing strategic external networks for open innovation. They can take the form of strategic advisory boards, interconnected ecosystems or 'hub and spoke' relationships, and can be used in a variety of ways. Such networks for open innovation are being established to effectively share and leverage technologies, knowledge and ideas.

Key words: open innovation, networks, external collaboration, ecosystem, food and beverages.

16.1 Introduction

Across sectors open innovation has challenged the historical route of sourcing innovation internally by opening doors to external resources and providing a spectrum of collaboration opportunities. In today's open-innovation era, a rich group of actors, including consumers, lead users, parallel industries and even competitors, can be found working together to improve innovative performance (Laursen and Salter, 2006). This is the case even in traditional and mature industries such as the food and beverage industry (Sakar and Costa, 2008).

In order to enable this new way of working, networks and new relationships are being established to effectively share and leverage technologies, knowledge and ideas. Although various studies address the importance of internal networks (Cross *et al.*, 2010; Johansson, 2006; Meister and Willyerd, 2009; McKenzie and van Winkelen, 2004) and tactical external networks (Ballinger *et al.*, 2011; Birkinshaw *et al.*, 2007), there is a remarkable lack of studies on longer-term strategic external networks for open innovation, particularly in the food and beverage sector.

The aim of this chapter is to explore current and future trends in strategic external network formation in the food and beverage industry and identify any key challenges. The study focuses on formally recognised strategic external networks formed by companies for longer-term benefits, in which network members are engaged in the business in an on-going manner under some form of contract.

16.2 Strategic external networks for open innovation

According to Cross *et al.* (2010), the key to delivering both operational excellence and innovation is to allow innovative solutions to emerge through interactions among individuals who see problems from different perspectives. Cross *et al.* (2010) describes internal connectivity, but the principle equally applies to connections with the external world. Strategic external networks are one way to facilitate such connections.

16.2.1 Definition

The Oxford English Dictionary defines a network as 'a group of people who exchange information, contacts and experience for professional or social purposes'. Oakland Innovation is using the working definition of an external, or 'open-innovation' network as a group of people with a related knowledge set, whether geographic, thematic or by competence area, and may cut across different industries, which is then managed to leverage added value, such as:

- ideation,
- technical or market insight,
- project strategy or design advice,
- policy or regulatory guidance, and
- strategic insight.

It is important to distinguish between tactical and strategic open-innovation networks, the latter of which is the subject of this chapter. Strategic networks are implemented with a broader remit and with longer-term significance, whereas tactical, open-innovation networks are formed to solve a specific short-term problem or task. Typically, strategic networks are based on establishing relationships and in Oakland's experience need to be in place for at least two to five years to make them worthwhile. There may be a renewal process where some members are replaced, and eventually the network may be disbanded if it no longer serves a purpose.

16.2.2 Purpose

Oakland considers that there are three broad purposes for establishing a strategic external network for open innovation:

Strengthening a core competency that will remain core over the medium to long term

The aim of the network is to bring in new thinking and knowledge in a strategically important area that will remain relevant to the business over time. The network may bring in experts from adjacent areas or new geographies to build on and complement established in-house expertise e.g. on a core ingredient or processing technology.

Expanding knowledge of generic technology topics that span through the portfolio of products

Knowledge of nutrients or health-related issues that are relevant across a portfolio of products can be expanded by external knowledge of technology and/or market developments applicable to an intrinsic need across the different product types. There is likely to be less internal expertise than in the core-competence areas of the company, but it is an area recognised as strategically significant over at least three to five years.

Exploring and experimenting in new technology or business areas

As an alternative to internal recruitment or training, which can be risky and costly, some companies have used extended external partnerships to scope and build knowledge in new areas.

16.2.3 Types

In Oakland's experience there are two main models for strategic external networks for open innovation (see Fig. 16.1).

Hub and spokes network

A hub and spokes network involves a series of one-to-one relationships between a company (the hub) and various external partners that complement the company's capability (the spokes). Interactions are with individual external partners, and the network members are not brought together. This is the less resource-intense system, and has less risk in terms of IP management arising from the connections than in the ecosystems described below. However, there may be less opportunity for synergies between different topics or insights. This type of network is most appropriate for areas that are close to the core competence of the company, and can also be used for expanding generic knowledge or exploring new business opportunities. These one-on-one interactions help to 'gap fill' the internal knowledge, problem-solve on specific current issues, and can be vital to keep abreast of developments in the external world. The most appropriate experts within the network are targeted to source knowledge and insight as required.

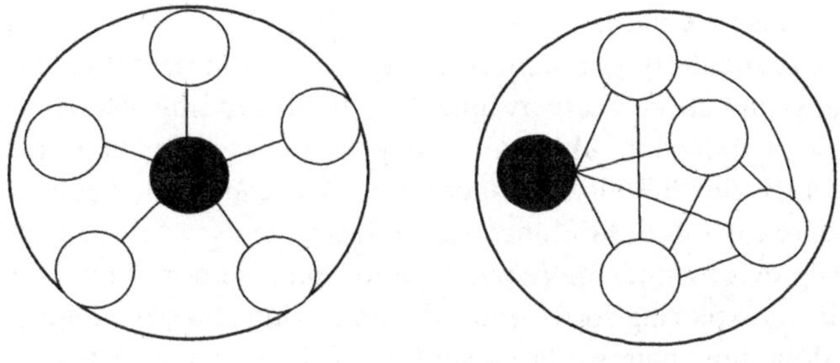

Fig. 16.1 The two main models for strategic external networks for open innovation: (a) hub and spokes network; (b) ecosystem network. *Source:* Oakland Innovation.

Collaborative, or ecosystem, network

A collaborative, or ecosystem network involves orchestrating a community of various external partners, with physical or virtual interactions between network members. This model is appropriate where one-to-one relationships do not address the identified needs, and a more holistic perspective of the issues is desired. The combined diversity can be particularly helpful for more complicated areas, that require multiple elements to be combined from different partners and for areas that are further away from core competencies, where internal stakeholders are less likely to already possess great depth and/or breadth of understanding. Although generally requiring greater input from the company than the hub and spoke model, the ecosystem's value derives from bringing together partners who would not normally interact, giving rise to new, and potentially more disruptive, thinking.

The most important factor in both models is that the company plays a central and active role in order not to 'drop out of the middle'. This means that the company does not just derive value from the network, but also gives value back to the partners or ecosystem to remain an active part of the system.

16.2.4 Benefits and drawbacks

Strategic external networks for open innovation can offer a number of benefits to the innovation process. Bringing in alternative, and broader, perspectives and knowledge from multiple directions can provide both a horizon-scanning function and new solutions to identified problems. In particular, external partners can offer value when they have a leading edge over the company in a competence area, or when they can provide extra capacity, facilities or capabilities (e.g. external suppliers). Lateral knowledge flow in an ecosystem-type network can also enable a company to put together different pieces of a puzzle to show the bigger picture. Furthermore, the company can benefit from the reputation of its external partners, and the network can give a higher level of credibility to products and services. This can be especially relevant where there is a need for evidence-based proof of activity and benefit, and in areas where it is desirable to be seen as moving to positive messaging, for example in health, safety or environmental areas.

Using a strategic external network to explore new areas can also help mitigate against risk, particularly in a rapidly changing environment, by enabling a company to change direction relatively quickly without building in too many assets that may then not be relevant. Although it may be more expensive in the short term, there is less overall liability. Furthermore, as it can be difficult to predict the necessary core capabilities of longer-term R&D, the use of floating resources and capabilities provides much more potential to 'future-proof' without hindering the company from exploring ideas. Indeed, some companies have turned down the option of doubling their R&D capability, believing it is better to keep more flexibility through partnering.

Many relationships exist between employees and external partners, but with the increasing trend in employee mobility, personal connections can easily be lost to a company (Ballinger *et al.*, 2011). A formal network can help guard against this. It also provides a facilitated and established route to external-partner views on a particular area or problem as the need arises.

One perceived drawback relates to the time and resource input required to set up a network and maintain momentum. One of the key differences between personal and corporate networks is the visible cost. It is possible to leverage more from personal networks without cost, but where a company is leveraging a network in a more strategic way, there is a contractual element with some form of remuneration or commercial expectation, making the cost more tangible.

When setting up external networks with multiple partners, it is important to consider confidentiality concerns. Many food and beverage products are not highly technically differentiated, so that, in many areas competitors can move in relatively quickly. The risk of sharing internal knowledge and strategy with external partners needs to be managed without overly restricting open-innovation processes.

Overall, the company must assess the benefits and drawbacks, to determine whether a strategic external network provides sufficient added benefits to the innovation process to make the resource dedication worthwhile.

16.3 Research methodology

Oakland Innovation has helped many leading, global food and beverage companies to define, evaluate and set up various types of networks, from scientific advisory boards and technology platform networks through to innovation ecosystems. In order to conduct the current research, interviews were undertaken in November 2011 with four of Oakland's internal Principal Consultants, with over 50 years' experience working in the food and beverage industry between them. Each Principal Consultant was asked about:

- the definition of a strategic external network for open innovation (or different types of network);
- the key challenges in forming such networks, including issues surrounding:
  - defining the need, particularly issues around changing corporate strategies

  resulting in discontinuity in the competencies required from external networks;
  - the natural tendency towards the most 'comfortable' relationship rather than the best relationships for a network; and
  - challenges around tacit versus explicit relationships;
- how companies have overcome these challenges;
- a future vision of strategic external network formation in the food and beverage industry.

The interview recordings were studied and data collated and categorised against the key challenges highlighted by the interviewees. Within these, subcategories were created for themes emerging from the data. Quantitative analysis was not appropriate due to the limited number of interviews, but emphasis has been given to recurring themes. Oakland's contractual and code of conduct obligations mean that we do not divulge our clients' identities.

16.4 Findings

Findings regarding the current status of strategic external networks are discussed in the following subsections and challenges are identified.

16.4.1 Current status of strategic external networks for open innovation in the food and beverage industry

In Oakland's experience, the food and beverage industry appears to be shifting its emphasis much more towards exploring ideas with externals. This interest is demonstrated by the success of the annual Institute of Food Technologists (IFT) conferences, in which hub spots facilitate the connecting of companies with technology providers.

Many companies in the industry already have a strategic advisory board (SAB) comprising experts with a breadth of knowledge from many years' experience within a sector or across different sectors. Traditionally, such boards have been relatively local and insular; however, Oakland has seen a shift in some companies now seeking a broader remit from their SAB in order to obtain a more global perspective and find experts who are more widely connected.

In many instances there are no formally-recognised networks beyond an SAB, but companies systematically use personal relationships, or seek external experts, for specific tactical requirements. In some cases, specific business units or technical teams within food and beverage companies have established formal networks. Most discipline-related networks have followed the hub and spokes model described above. Forming these networks around a particular technology or geographical need is an emerging area of interest.

Certain food and beverage companies have successfully run discipline-specific ecosystems with connected members, although this not common. This may be owing to the highly-regulated environment and safety implications for innovation, making disruptive or radical changes more challenging to implement.

Many companies are still trying to establish how best to set up and run strategic external networks. Oakland has conducted a number of studies to identify and evaluate leading experts as potential members of a network around specific topic areas, for the purpose of building on-going relationships. However, some companies have struggled to convert their efforts to form established networks owing to a number of challenges around their set-up, which are discussed in the following sections.

16.4.2 Challenges in establishing strategic external networks

Defining the need

In Oakland's experience, the greatest challenge when creating a strategic external network is to define its purpose adequately. A great deal of excitement is often generated over the idea of setting up a network, and considerable time and budget may be dedicated to map people out, sometimes generating huge lists of centres of excellence, stakeholders, and individual experts. However, difficulties can be experienced if the network is considered a 'fringe' activity whilst the company is trying to manage the core business, and this relates to how well the need has been defined to drive the effort required for its implementation, facilitation and management.

There are three key reasons why defining the need is essential:

1. Time and resources: networks take time and effort to generate and then maintain, particularly when they involve building relationships over several years. If the need is not perceived to be sufficient, then adequate resources may not be allocated and the network struggles to get off the ground.
2. Senior buy-in: owing to the commitment required by staff and external partners, there is a need for senior management to support these networks and make participants feel their efforts and contributions are worthwhile and valued. Such a buy-in is unlikely without a clear purpose for the network.
3. Staff turnover: internal stakeholders may leave the company and if the value of the network is not recognised then the network may disintegrate. However, if the need is clear then efforts are made to ensure handover to a new stakeholder to maintain the network.

It is easier to define a longer-term need for a strategic external network when operating on a long-term horizon strategy where the direction is clear and does not change frequently. Companies working on renovation, where the strategy direction can change every six months, find it more difficult to identify a longer-term purpose for such networks that will not become redundant within a year.

Along with the need, the remit of the network must also be defined. One issue in the food and beverage industry is that timelines to commercialisation are typically relatively short, and there can be conflict between teams looking for the longer-term stretch and those that want results now. Some people want the network to bring disruptive/radical changes in technology; however, since most innovation in the fast-moving consumer goods (FMCG) sector activities is incremental, there may be tension between what is desired from the network and what the company can actually cope with.

A related issue is defining the right balance of breadth and depth of expertise: the network must have enough breadth to maximise its relevance, without becoming too nebulous to manage. The inclination might be to set up a network that is tightly focused on the company's core competencies, but this may not give the necessary breadth to add value to existing internal knowledge. As a result, the network may only lead to incremental developments, missing out on more radical ideas that could lead to more disruptive innovations.

Identifying the need for, and defining the remit of, a network can be started via a road-mapping exercise that reviews potential growth areas and mission statements, and matches this against the company's strategic vision. At present, road-mapping processes to define 'the want' are viewed in food and beverage multinationals as 'patchy… Some do it, some don't, even within one client'.

Making networks explicit

A second challenge is making existing relationships within a company more explicit in order to leverage these when forming a new network. Most employees have external professional relationships, and indeed, there is an expectation amongst many companies that certain employees, such as internal scientists, make strong relationships with external experts. However, such knowledge is often tacit and thus difficult for companies to fully exploit. This is compounded by the trend towards increased workforce mobility and the resulting negative impact departing employees have on informal networks.

The latter point highlights the importance of not confusing personal professional relationships with a corporate network, in particular, it should not be assumed that a company is well networked just because its employees are. This differs from a formalised ongoing network, where the relationships are embedded within the corporation and should be proprietary and protected through contracts. The question is: 'How does the company harness the power of its employee connections?'

Although it is not necessary for a company to make every connection explicit at all times, it can be very valuable to establish a process or system that allows relevant existing connections to be identified according to specific criteria. One obstacle here is human nature: individuals may be protective of their relationships and not wish to share these with others in the company. There is also a company culture element to be considered. In companies with silos and a lack of a sharing or learning culture, it may be more challenging to extract personal relationships from individuals. However, where there is a greater exchange inherent in the company then these networks are more likely to be closer to the surface already, making it easier to identify existing relationships.

Another issue in large organisations is that it can be extremely difficult to collate and share knowledge of a large number of connections in an updated format. Oakland has conducted a number of audits for clients to map relationships, which can be useful to leverage the connections revealed in the short term. Often, however, the resulting databases are not popular with users or content providers, and are left to go out of date. Unless there is a clear purpose and defined area of interest, the database can become too large and nebulous to be seen as valuable. People would rather just send an email around the office or ask a colleague for a recommendation.

It should be noted that personal relationships are owned by employees, and the equity resides in the individual not the organisation. Companies need to be careful not to cross the line of overcontrolling these relationships by being clear on the definition of the network requirements. In addition, the network should not preclude employees working with other people, or having their own connections, but it can be made known explicitly that there is a network in place which the company can leverage across units.

Avoiding attachment

The next challenge relates to the selection of members for the network. Increasing recognition of the visible costs involved in creating a formalised network has resulted in increased pressure to rationalise the selection process and conduct due diligence. Not only does this help make the network more explicit, but a clear selection process provides rationalisation for inclusion in, and exclusion from, the network. At present, however, the implementation of rigorous selection procedures for formal networks is viewed as inconsistent across the sector, with a tendency to select experts that are 'known and loved' by internal stakeholders.

A key requirement is to assess both known and potential new connections against selection criteria in an objective manner in order to ensure that the best partners with demonstrable added value are included in the network. The selection process can be viewed in a similar way to recruitment, with a clear capability/knowledge footprint that is required, whether from an existing relationship or a new potential partner. The following aspects should be considered:

Established versus new

It is vitally important that a company is aware of when it is appropriate to use familiar external experts and when it is not. There is a seductive urge to remain comfortably within an established network; however, this needs to be resisted in order to optimise open-innovation opportunities (Johansson, 2006; Birkinshaw *et al.*, 2007). For example, relationships formed by employees, which are often serendipitous, may not always be the best external connections for the company. Moreover, in core areas there are established long-term connections with a few experts that are believed to be exceptional, but overreliance on such experts runs the risk of the company's world view being based on a narrow set of feeds coming from the most familiar connections, and losing sight of developments beyond these horizons.

Established connections are not necessarily unsuitable, however. Such relationships bring the advantage that the external experts know the business, and the company knows their strengths and weaknesses. Mutual trust has already developed, resulting in better knowledge sharing. Nevertheless, diversity in the network is also important.

To some degree the optimal proportion of new versus established members in the network depends on the potential for radical change. For example, some innovation may be contained or restricted by infrastructure assets (e.g., not stripping out production lines) and the overall absorption capacity of the company in terms of innovation or new thinking. There may still be an argument for being aware of possible opportunities in order to mitigate threats/risks from other directions, but the ability to absorb and use these may be limited.

Global reach

The emergence of economies such as China and India and the resulting wealth of activity now occurring outside traditional horizons add additional impetus to the need for diversity. 'Companies can't afford to hang onto their incumbent relationships for the sake of it anymore … with the globalisation of technology it is complete folly for Western companies to stay with just US or European experts'. Failure to engage with partners in emerging economies runs the risk of being blindsided to major opportunities, in terms of both innovations and markets. Furthermore, it can no longer be assumed that knowledge will flow back from Western partners that themselves are well connected. In Oakland's experience, those companies that have reached out to experts from non-traditional markets such as the BRIC (Brazil, Russia, India, China) countries really appreciate the richness gained from new perspectives and broadened horizons.

Achieving such diversity poses its own challenges, however. Most people historically have long-term relationships with experts from similar backgrounds and in relatively close proximity, and there is the temptation in some companies to look for partners nearby rather than for the best fit. Furthermore, in some geographies, cultural norms in the traditional food and beverage industry are balanced very strongly towards face-to-face interaction, making the transition to remote partnering all the harder. There can also be a slower process of earning respect and trust in emerging economies than in the West, meaning that building relationships takes longer. Initiating such connections remotely, or thinking it is a simple task, is not realistic.

Fundamental issues of language and fluency may additionally come into play. Oakland has seen companies trying to set up multilingual teams with no common language and finding that this did not work.

A further issue that may arise relates to the willingness of potential partners in BRIC countries to participate in networks that feed into another company located elsewhere. China wants to lead the way, so there is less incentive to feed into Western organisations, and, moreover, because academics there have a high level of funding, they have less need for research collaborations.

'Rising star' versus 'wise old sage'

The 'best expert' for partnering might not be the world's leading expert. If a network is too focused on people that have been established in the industry for many years, there is a danger that it does not access cutting-edge developments from younger experts. In addition, 'wise old sages' are in demand by everybody, including competitors.

Establishing new relationships

Establishing a relationship between unknown experts and the company in a network setting is a further challenge. One problem is that if the network is too large initially, it can become unmanageable. Another is that if the interactions are too informal and nebulous to start with, then the relationship is unlikely to gel and may well be lost. Members may need to be managed to optimise the network's performance as personal dynamics, strengths and weaknesses emerge during early interactions.

One company took the pragmatic approach of engaging new experts in transactional research projects before inviting them to become members of the core network. This allowed the experts to be 'tried and tested', with the relationship and mutual respect growing over long-term project work across several years. However, this is not practical in all instances. Where there is a more immediate need for establishing the network, it may be wise to set up a structured series of interactions early on to help fast track the relationship. A relationship manager allocated to each individual external member can also help ensure the new relationship is maintained. (See step 8, Nurture the network in the ten-step guide, 16.7.1).

16.5 Discussion

There are few studies relating to strategic external networks for open innovation available for the food and beverage industry, or even relating to strategic advisory boards. However, from the limited number of interviews conducted with Oakland Innovation Principle Consultants, covering multiple food and beverage companies, it is clear that major challenges remain in formation of these networks, with the key challenges around: defining the need and value; making existing relationships explicit and formalised; and the tendency to use established relationships, rather than seeking out the 'best' partners for the job.

A certain level of consistency within the business over several years is vital to derive the full benefit from longer term strategic open-innovation networks. The aspect of 'defining the need', the principle driver for forming the network, is vital to secure the necessary initial buy-in and resources to make it happen, but the 'need' must remain relevant to the company over the longer term so that value continues to be derived from the network. In addition, some level of consistency in the internal stakeholders is required, although if the value is recognised and the relationships are explicit this will aid any hand-over, and ensure that the network survives changes in personnel.

Different companies are at different stages of evolution in implementing strategic external networks, and many are new and still learning about the process.

Even in the more experienced companies it seems that the process is yet to be perfected and continuous review and improvement is an important aspect of ensuring the long-term success of these networks. Certainly there is no 'one size fits all' and organisations need to adapt the process to fit with their own internal strategic vision, culture, structures, resources and needs. It is likely that mistakes will be made along the way, but with persistence and adaptation to find the best approach to fit the company, the benefits from the contribution from external partners can greatly outweigh the time and effort put in. In an increasingly globalised and fast-paced environment, there may be little option but to open the doors to more external interactions to keep abreast of developments, and it is likely that a shift towards greater emphasis on strategic relationships will accompany this.

16.6 Future trends

In the future there is likely to be much greater clarity around when a strategic external network can add value to the open-innovation process, and greater granularity over what behaviours the network should demonstrate (e.g. ideation, advice, proactively bringing new thought, pure R&D). This will be driven by increasing pressure to develop metrics to measure the value derived from such networks. Metrics are already a major focus for open-innovation managers reporting to senior teams and the topic is a regular theme at conferences, such as CoDev, and discussion forums. Metrics should reflect both the quantity and also the quality of the networks, and the latter is more difficult to measure. Professor Ellen Enkel, director of the Institute of Innovation Management, Zeppelin University, reported (Enkel, 2010) that, although the benefits of open innovation were mainly measured through estimation and gut feeling in the past, concrete cost–value relationships need to be presented in order to decide what to do and when, as well as which downsides are related to those activities.

Globalisation is another important trend that will impact the future of strategic external networks. Traditionally, the largest food and beverage corporations have been predominantly based in Europe and the Midwest and East Coast of the USA; however, current R&D and innovation growth capability is located in Asia, particularly China. Western companies face real challenges in forming relationships and networks in these emerging economies. A key requirement is for companies to start forging foundations in those regions now and to invest heavily in building relationships, particularly in places where networks are very personal. Regional networks developed in BRIC countries will most likely be leveraged locally.

16.7 Conclusions and recommendations

Overall the food and beverage industry is still feeling its way on the formation of strategic external networks for open innovation, particularly to understand more clearly how best to maximise the benefits and value-add against the necessary costs. Where companies are new to these networks, they should be kept small and assessed regularly, with learnings on efficiencies, what works and what does not work fed back into the system.

16.7.1 Ten-step guide

As a result of thinking through the issues to be addressed, Oakland has developed a ten-step guide, outlined below, to help companies navigate the process of establishing strategic networks.

Step 1: Define the need

The first crucial step is to define a clearly-stated need for the network, which is significant enough to achieve senior buy-in and sufficient resource allocation over several years. This can be started via a road-mapping exercise to match potential growth areas and mission statements against the company's strategic vision. This process should carefully consider the value to be derived from the network and how success might be measured. It should also be decided whether a 'hub and spokes' or 'ecosystem' model would be most appropriate, and what types of members are required (including knowledge, geographies, etc.).

The amount of time and resources required to establish a network and maintain momentum should not be underestimated. Starting with a small network of carefully-selected partners and building up the numbers slowly may help 'trial' the process at lower cost, and demonstrate value to obtain further buy-in. A clear plan, adequate budget, and regular innovation cycle make it more likely that such networks will be set up and add value.

Step 2: Develop clear selection criteria

Having determined the scope of the network, precise and rational criteria for selecting members should be compiled. These should include both absolute requirements as well as desirable features. It may be important to cover a range of knowledge, geographies or peer groups; criteria for each category of expert should be clearly defined.

Aspects beyond knowledge or skills, such as personal attitude and values, should also be taken into account when creating the member criteria matrix. For example, to ensure a successful network, it is important to consider an individual's attitude and enthusiasm towards collaboration, their ability to communicate and interface with the company and other network members, and even their availability (Fig. 16.2).

Step 3: Internal audit and gap analysis

An internal audit of existing connections between employees and external partners can be conducted to identify potential network members. Such audits often involve internal interviews across the company; however, unless an extensive audit is to be

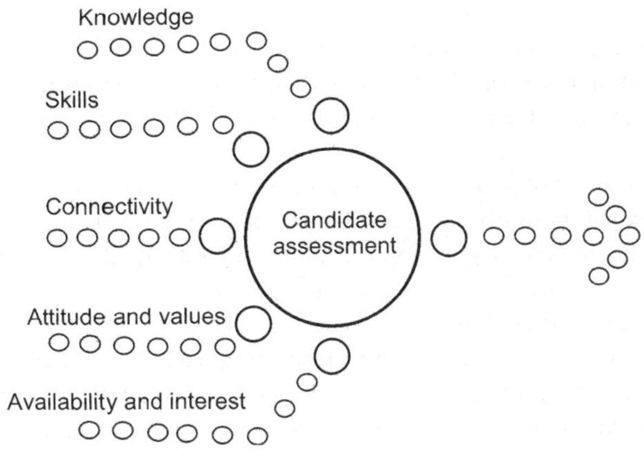

Fig. 16.2 Attributes to consider for partner selection criteria. *Source:* Oakland Innovation.

conducted, it may be more pragmatic to identify and utilise those individuals serving as key nodes of knowledge and connections within the organisation. Such employees are often highly experienced and respected and may have worked in a number of organisations, building up a repertoire of external contacts. They are often people that are approached by other employees if they have a question or need ideas to solve a problem.

Ideally, the audit is an iterative process: people often require some thinking time to dig out all their connections. It works best when an initial discussion is followed up with further exchanges as other relationships are recalled and added to the list. It may also be beneficial to bring groups together to remind each other of different connections and projects. Personal networking tools such as LinkedIn can also be trawled; here, however, care is obviously required to ensure that employees do not feel exploited during this process.

If the company's culture is not conducive to sharing personal connections, then positive or motivational messaging may be required to change attitudes. A reward or incentive, ideally from senior management, that recognises employees who have contributed to sharing external connections can also help change behaviours.

Once the internal audit is complete, existing connections should be assessed against the selection criteria to determine whether they may be suitable network members. Following this, gaps in the network can be identified.

Step 4: External audit

Additional external partners to complete the network can be identified from a number of sources including publications (e.g., scientific literature, trade press), conferences and peer group networks. Key opinion leaders (KOLs) and other commentators can be an excellent source to identify the best candidates to meet specific needs. Known connections can also be leveraged to springboard to new people.

Potential new connections should similarly be assessed against the selection criteria (from step 2) to determine whether they may be suitable network members. It is also essential to assess whether new identified external partners are a better fit for the network than existing connections, especially familiar and comfortable connections.

Ideally, a shortlist of potential new connections are interviewed at an early stage to gain a view on their attitudes, availability and interest, as well as their future direction, to ensure that they are appropriately aligned with the network objectives.

Step 5: Evaluate and rank partners

The shortlisted existing and new connections within each expert category required for the network should next be evaluated against the selection criteria already developed in terms of suitability and disposition, and ranked accordingly, to highlight the most promising individual candidates. This can be done by developing a simple scoring system for each of the criteria, combined with a more qualitative evaluation of the leading player; it may be that a candidate that does not score well overall may still be valuable for the particular expertise or strength that they bring.

Step 6: Selection of candidates

Finally, the overall combination of partners needs to be considered, to optimise not only the best mixture of knowledge and experience but also backgrounds, team dynamics and personality types. For ecosystem models, it is also worth considering whether there are any specific clashes between individuals (e.g., some experts may refuse to work with specific individuals; this can be assessed as part of the early interviews). Figure 16.3 outlines steps 1 to 6 described above, which lead to the selection of external candidates for a strategic network.

Step 7: Engagement process

The inception of the network can be a critical time that sets the tone for future interactions. Ideally, early interactions would be face-to-face, whether in an ecosystem symposium or one-to-one visits. This helps cement a relationship, particularly where there may not yet be any concrete projects or transactions. Another route to build the relationship would be to conduct a transactional piece of work before the expert joins network. Certainly, a more formal agenda of interactions over the first year for newly introduced experts is recommended.

Step 8: Nurture the network

Maintaining the network requires nurturing the connections and ensuring both that the members feel valued and the company is deriving value. This area has not been explored in detail for the purposes of this chapter, but some comments are given below.

Many companies appoint a network facilitator in order to orchestrate and manage the process, as well as to act as a catalyst to make things happen. They do not own the network, but in this way they fulfil the role of maintaining and

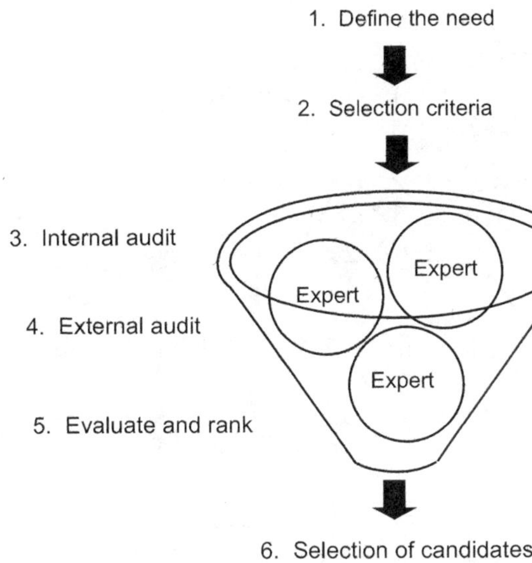

Fig. 16.3 Ten-step guide: definition and candidate selection steps. *Source:* Oakland Innovation.

reviewing the process. In addition, it is generally ideal for the network members to have an allocated relationship manager as their first point of contact. This does not have to be the same person as the network facilitator, and does not have to be the same person for all the members. Also, not all interactions need to go via the relationship manager; the manager's role is to ensure the relationship is maintained and smooth any problems that may occur. Furthermore, although the network should be accessible and visible across units, the workload of individual members should be monitored and managed by the relationship manager, to ensure they are not overexposed to the company. Conversely, it is also important to ensure that the relationship is maintained by some regular contact, and this can be mediated by phone, face to face, or in virtual working spaces.

Open-innovation managers may be appointed to these roles, or it could be an R&D manager, depending on the priority and maturity of open innovation in the company. As well as these network managers, it is important that the contact points within the business units also play a role as relationship managers, while they are interacting with the external partners in execution of any work. Balancing technical and relationship skills is not always an easy role to fulfil, but is likely to become more and more important in future within R&D and business units, with the increase in open innovation.

Step 9: Review the network

The network should be reviewed at intervals to ensure it is still serving the original purpose, and that the original need still exists. In addition, a review of what is working, what is not working and what could be changed should be undertaken to

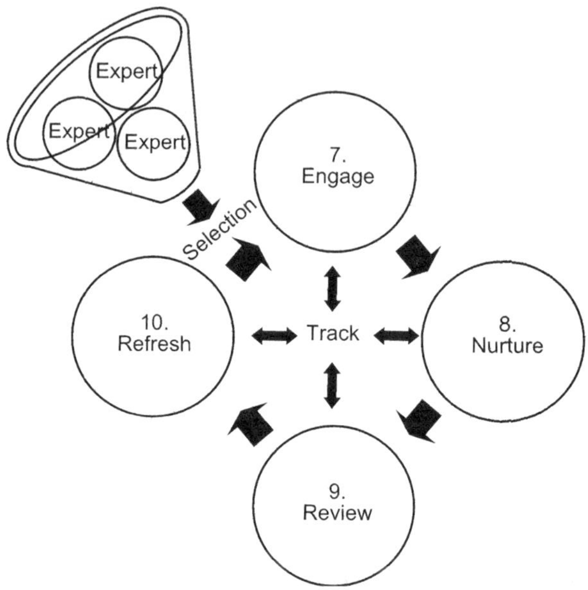

Fig. 16.4 Ten-step guide: running the network steps. *Source:* Oakland Innovation.

ensure that all participants are maximising the benefits of inclusion and the network is running as efficiently and effectively as possible.

Step 10: Refresh the network periodically
Some member turnover is likely to occur: some people may naturally drop out, while others may perhaps not fulfil their expected role. In addition to such obvious replacements, it may be worth periodically undergoing the due-diligence process to ensure there are no additional players that should be included. Figure 16.4 shows that there are four key stages to creating and maintaining the strategic network once the external experts have been selected.

During this research Oakland has collated insights into strategic external networks, and has identified a number of challenges, and possible routes to address these. Very little data appears to be available in the public domain, and further research would be very valuable in this area, particularly case studies from the food and beverage industry.

16.8 References

BALLINGER, G., CRAIG, E., CROSS, R, AND GRAY, P. H. (2011) A stitch in time saves nine: leveraging networks to reduce the costs of turnover, *California Management Review*, **53**(4), 111–133.

BIRKINSHAW, J. BESSANT, J. AND DELBRIDGE, R. (2007) Finding, forming, and performing: creating networks for discontinuous innovation, *California Management Review*, **49**, 67–83.

CROSS, R., GRAY, P., CUNNINGHAM, S., SHOWERS, M. AND THOMAS, R. J. (2010) The collaborative organization: how to make employee networks really work *MIT Sloan Management Review*, **52**(1), 82–90.

ENKEL, E (2010) *The 'state of the art' in open innovation metrics*, www.pure-insight.com.

JOHANSSON, F. (2006) *Break out of your network: ants and truck drivers – making intersectional ideas happe*n, Harvard Business Press Chapters, 14 pp.

LAURSEN, K. AND SALTER, A. (2005) Open for innovation: the role of openness in explaining innovation performance among U.K. manufacturing firms, *Strategic Management Journal*, **27**, 131–150.

MCKENZIE, J. AND VAN WINKELEN, C. (2004) Understanding the knowledgeable organization: nurturing knowledge competence, London, Thomson.

MEISTER, J. C. AND WILLYERD, K. (2009) The über-connected organization: a mandate for 2010, Harvard Business Review Blog.

SARKAR, S. AND COSTA, A. (2008), Dynamics of open innovation in the food industry, *Trends in Food Science and Technology*, **19**(11), 574–580.

Part IV

Open innovation tools, process and managerial frameworks

17

The 'want find get manage' (WFGM) framework for open-innovation management and its use by Mars, Incorporated

M. Garcia Martinez, Kent Business School, University of Kent, UK

DOI: 10.1533/9780857097248.4.315

Abstract: This chapter examines the role of the 'want find get manage' (WFGM) model in assisting firms to execute open innovation effectively. Specifically, the way the WFGM framework has been formulated to meet Mars, Incorporated's particular needs and context is discussed and its application to collaborative knowledge-creation processes is described. The WFGM model has significantly increased the success rate of collaborative projects by forcing business teams to be clear about their resource needs, before they go out to find them and search for potential innovation partners, and before they build the deal and then manage the consequences.

Key words: open innovation, WFGM framework, cultural change, business management.

17.1 Introduction

The benefits of open innovation are increasingly reported in innovation management studies as the trend towards innovation collaboration across organisational boundaries intensifies (for a review of the literature refer to Dahlander and Gann, 2010 and Huizingh, 2011). Today's fast paced business environment requires firms to explore the use of external sources of technology and ideas to augment in-house research and design (R&D) (Wolpert, 2002; Chesbrough, 2006; Gassmann, 2006, Huston and Sakkab, 2006, Hagedoorn, 2002). Advances in information and communication technology have made companies more aware of externally generated scientific knowledge and they are starting to recognise that external R&D can create significant value (Gassmann and Enkel, 2004). Start-up and small and medium-sized enterprises (SMEs) and even individuals now play a far more relevant role in knowledge and technology creation as large firms seek to collaborate with these smaller knowledge creators (Chesbrough, 2006).

The transition towards open innovation, however, confronts firms with considerable managerial challenges, such as the transformation of business models (Chesbrough, 2007), implementing new types of R&D organisations and management structures (Chesbrough, 2006, 2007; Chiaroni *et al.*, 2010; Gassmann and von Zedtwitz, 1999), and cultural change to accommodate a more externally oriented mindset that encourages employees to see the world as their technology base (Huston and Sakkab, 2006). An open-innovation approach takes significant time to implement as deeply engrained organisational mindsets need to be overcome and fears mitigated of losing control over proprietary technology (Chesbrough, 2003, 2006). Harnessing external knowledge then requires the development and implementation of suitable systems to look at this outside-in/inside-out view of the world (Slowinski and Sagal, 2010).

In this chapter, we look at how open innovation is implemented in practice, particularly regarding the tools and processes companies use to frame and communicate their understanding and to assess how value can be created from innovations and which elements have to be sourced internally or externally. In particular, we examine the role of an increasingly adopted framework, the 'want find get manage' (WFGM) model developed by Gene Slowinski (2004), in assisting firms to execute open innovation effectively. The open-innovation approach of Mars, Incorporated, a leading global food company, is described focusing on how the company successfully engages in collaborative innovation agreements to meet its strategic objectives. Key to this success has been leading cultural change and having an efficient process to identify, evaluate, value and contractually acquire external technology: the WFGM model. Having developed new tools and processes, open innovation is gaining momentum within the company with 20% of new product launches having open-innovation elements and a threefold increase in the number of collaborative contracts; an impressive performance considering that open-innovation activity only started in 2008.

First, we discuss the drivers for change at Mars, which, like many other leading global food companies, is dealing with a rapid pace of innovation and consumer change, leading them to consider new business models and innovation approaches. In Section 17.3, we present the company's open-innovation model with particular emphasis on the organisational structure implemented to access and leverage external knowledge and the key role played by the open-innovation team and support network in leading the process of transition from closed to open innovation. In Section 17.4, we present and discuss the WFGM framework and show how it has assisted Mars to frame its open-innovation approach and ensure internal and external alignment in their collaborative knowledge-creation processes. The chapter ends with some concluding remarks.

17.2 History of open innovation at Mars, Incorporated

The open-innovation journey started at Mars in 2007, when Mars' Regional President of the European Business set up a working group to look for ways to increase the efficiency and effectiveness of Mars' innovation processes. The work considered a wide range of issues, in particular, open innovation, recognising that this was a relatively new subject for the company and work was needed to enhance Mars' understanding of open innovation (i.e., what it would mean to the company and the impact of such a new direction on its internal and external stakeholders). Experimenting around open innovation was a significant step change to Mars' traditionally closed and centralised culture where major innovation developments, such as Mars bars, Snickers, M&Ms and Maltesers were all developed over 50 years ago with internal skills. Hence, making open-innovation systematic within the firm has required a cultural change in the way Mars actively seeks for externally generated technologies and ideas, and co-operates with innovation partners to create consumer value. 'Mutuality' is a key principle shaping Mars' culture, where business relationships are developed to generate win–win solutions for all involved.

The drivers for change rest at the heart of open-innovation thinking. First, the locus of technological knowledge; compared with 40–50 years ago, the world has changed in terms of where scientific knowledge is created and the ease of accessing this knowledge (Gassmann and Enkel, 2004). Mars realise that all components of innovation do not need to come from within the company and that they can accelerate their own efforts (or perhaps even broaden the scope of their own efforts) by accessing and leveraging external knowledge to fill the innovation capabilities gap. Second, real breakthroughs in the marketplace require a wider knowledge base. When the Mars bar was originally introduced in 1932, it required mastering nougat, chocolate and caramel technologies for which Mars had extensive in-house expertises. The recently introduced Mars bar for the 21st century has 35–45% less saturated fat per 100 g, a fundamental breakthrough in chocolate technology that comes after five years of development work, an investment of €10 million and over 40 000 man-hours in R&D, working collaboratively with ingredients suppliers. Part of the difficulty of reducing the saturated fats in chocolates is the need to retain the texture of the finished product so consumers do not perceive any change in the taste or quality of the product. This required Mars experimenting with various saturated fat replacers working alongside ingredient suppliers in a more open and collaborative fashion creating synergies around core competencies, away from the traditional closed, transactional oriented innovation model characterised by limited data sharing and trial and error. Open innovation has led to faster, better and more innovative capacity.

17.3 Mars' open-innovation model

The process of transition from close to open innovation has significantly impacted Mars' organisational and management systems. As noted by Christensen (2006, p. 35), 'Open innovation can be considered an organisational innovation'. It requires firms to implement core processes and develop knowledge capacities (Lichtenthaler and Lichtenthaler, 2009) to apply the open-innovation approach effectively (Gassmann and Enkel, 2004). Specifically, Mars faced three managerial challenges in its journey towards open innovation.

17.3.1 Open innovation organisational structure

To kick-start the open-innovation journey, Mars established a dedicated cross-functional team of three within the chocolate business with the mandate to make open-innovation systematic within the company. The open-innovation team is embedded within Mars' innovation structure to ensure that open innovation becomes the responsibility and, eventually, the established innovation model of all innovation teams (rather than exclusive to a small, independent open-innovation business unit). As Mars formalises its open-innovation model, full alignment of organisational units, in particular the operational ones, with open-innovation initiatives has to be executed in full on otherwise well-devised initiatives.

The open-innovation team focuses 100% on open-innovation activities and acts as internal consultants providing support and engaging the entire R&D, procurement and engineering communities in changing their ways of working. A network of open-innovation ambassadors at each of Mars' five business segments (chocolate, food, drinks and gum and sugar) provides support to the open-innovation team ensuring that the open-innovation principles trickle down the global R&D movement (Fig. 17.1). Open-innovation ambassadors spend 20% of their time as open-innovation champions, and combine this with the execution of the innovation pipeline.

Open-innovation understanding, training and skill development of business teams have been gradually implemented across business segments, with most of the open-innovation activity coming from the chocolate business with 15 open-innovation ambassadors actively engaged in the deployment and activation of open-innovation projects. The stage of development in the other segments is lower, but rapidly evolving, as results from the first open-innovation projects show the positive impact of open innovation in Mars' innovation capacity. Over 300 associates have been trained, helping to build an open-innovation culture across the organisation and demystifying the widespread perception that external collaboration requires more time and effort to manage than internal projects. The opportunities and challenges of open innovation are similar to all business segments; its systematic implementation ultimately depends on having open-innovation champions among senior and middle management or project leaders to drive behaviour change. Buying-in and leadership are, therefore, critical to Mars' decentralised management style based on decision-making by consensus and self-implementation. Unlike its major competitors, Mars does not have the critical mass to excel in all technological areas, thus making it more imperative to collaborate with external partners. The recent reorganisation from a strong regional focus to a more globally integrated structure after the acquisition of Wrigley in 2008 has been partly driven in order to increase connectivity and synergies among global R&D hubs.

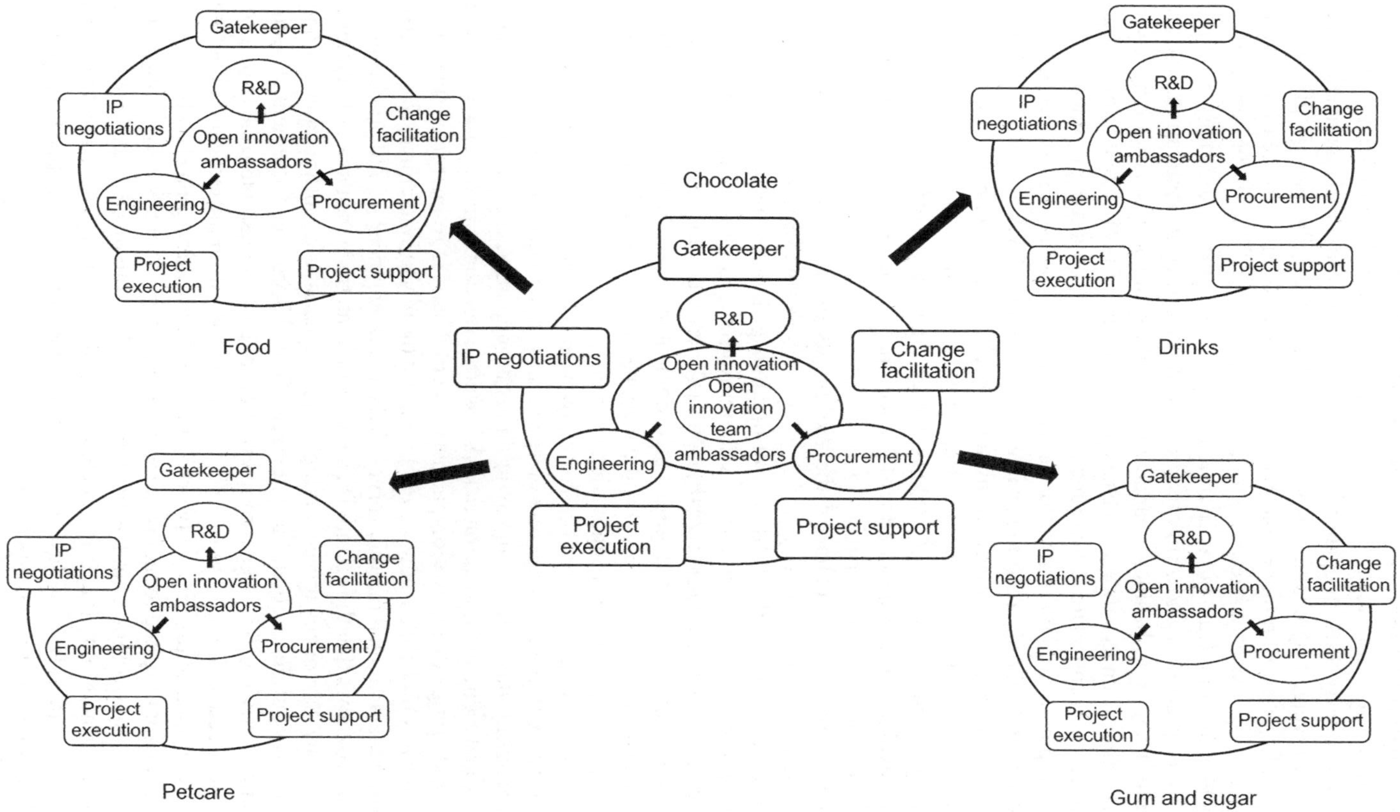

Fig. 17.1 Mars' global open-innovation structure.

17.3.2 Open-innovation team: leading cultural change

Mars' open-innovation team, in its key role of culture change agent, has been instrumental in creating an open-minded culture away from insular concepts such as the 'not-invented-here' and 'not-sold-here' syndromes (Chesbrough, 2003) into an externally oriented mindset that encourages business teams to see the world as their technology base (Witzeman *et al.*, 2006) (Fig. 17.2). One of the key roles of the open-innovation team is facilitating change by developing the understanding, knowledge, processes and skills required to carry out open innovation, and by providing training to business teams. As a result, business teams no longer see external innovation as a threat but as a risk-management vehicle that allows Mars to manage costs and share development risks. Further support is provided during the development of open-innovation projects by guiding business teams through the process and solving emerging problems. This support has been instrumental in overcoming internal resistance from scientists/R&D labs whose skill sets and interests lie in making technical discoveries rather than integrating external technologies (Slowinski *et al.*, 2009). Hence, having an efficient process to identify, evaluate, value and contractually acquire external technology has significantly influenced cultural change.

Equally important to maintain the credentials of the open-innovation team within the firm is the execution and delivery of open-innovation projects by the core team. Recognising the need to carefully manage intellectual assets in external collaborations, the core team includes an innovation contract manager to oversee and manage intellectual property (IP) rights in open-innovation projects. The role involves working closely with global procurement teams to determine the type of non-disclosure agreement (NDA) and the appropriate terms to be proposed given the specificity of each open-innovation project.

Internal resistance to collaborative innovation has been largely overcome by demonstrating with results the role of open innovation in driving revenue growth by accelerating innovation, improving innovation quality, increasing innovation capacity and reducing innovation risk. As external relationships develop over time with a wide range of innovation partners, internal staff is re-energised with new ideas, new resources and a new way of looking at the work (Witzeman *et al.*, 2006). Data from Mars' annual internal survey to the R&D, engineering and procurement community shows that the culture is changing and internal resistance is beginning to disappear. Associates are becoming more enthusiastic about the changes and increasingly confident in their line managers. Global R&D's endorsement of open-innovation activities has triggered internal behaviour change with line managers being more active in identifying potential open-innovation projects and directing staff to the open-innovation network. The open-innovation team has been very active in facilitating and reducing the work required, particularly at early stages of the process, by working with the IP team to put NDAs in place and assessing the potential partners' intent and technical capabilities. The WFGM framework has helped Mars to carry out fast and efficient due diligence on partners and to negotiate high-quality agreements.

Since the creation of the open-innovation team in 2008, 20% of new product

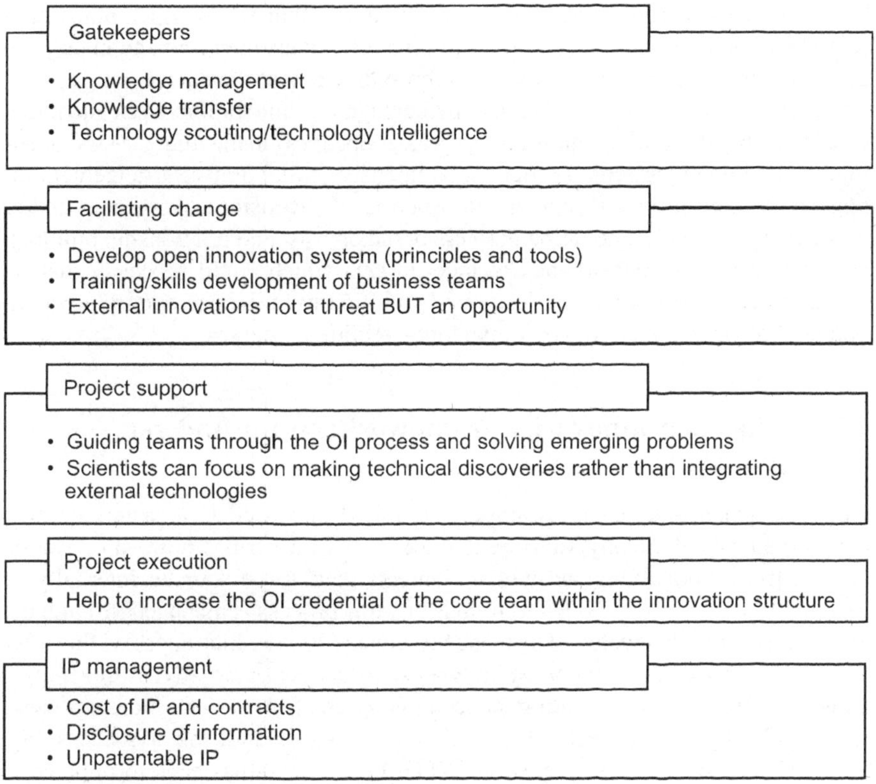

Fig. 17.2 Mars' open innovation team as a culture change agent.

launches have contained open-innovation elements and the number of collaborative contracts has trebled, particularly regarding intensive collaborative agreements (i.e., sharing and building IP). This wealth of knowledge is centrally managed by the open-innovation team so that lessons learnt, experiences and successes from previous open-innovation projects are easily transferred and applied to new external collaborations avoiding typical delays and failures in open-innovation projects owing to lack of knowledge and experience of those involved in collaborative agreements.

17.3.3 Reward and incentive systems

Mars' reward system and corporate communication plans have been equally instrumental in leading cultural change by becoming more appreciative of external capabilities. The 'Magic corner', a collaborative innovation project between Mars and SCA Packaging, has recently received the 'Making a Difference' award, Mars' corporate recognition programme, and has won several top international prizes. These achievements are a great recognition, both internally and externally, of the benefits of collaborative innovation by rewarding external initiatives and specific behaviours, away from some classic reward metrics that oppose open innovation (e.g. R&D labs are now judged on the number of patents they find and manage to get access to rather than exclusively on those they patent themselves).

There are no set targets on the percentage of innovation to be acquired externally, but there is a commitment to greater openness in the future. Mars' open-innovation model particularly focuses on intensive collaborative arrangement (as opposed to transitional IP agreements such as IP licensing) by sharing and/or building new IP with innovation partners based on trust and relationship building. Hence, there is no real urgency to meet targets which could be easily met by buying-in IP, but rather a real desire and ambition to maximise open-innovation potential through collaborative knowledge-creation processes.

17.4 The open-innovation framework: 'want find get manage'

How and when external knowledge is required and used is to a large extent determined by a company's business model, which describes how value can be created from innovations and which elements have to be sourced internally or externally (Chesbrough, 2003). Within Mars, this process is managed through the WFGM framework developed by Gene Slowinski (2004) which suggests that after the firm knows what technology it 'wants' to access externally, it develops a strategy to 'find' it. After finding the technology, the firm has an efficient process to contractually 'get' the technology and 'manage' the relationship to success (Fig. 17.3). This section discusses how the WFGM framework has been formulated to meet Mars' particular needs and context and its application to collaborative knowledge-creation processes.

17.4.1 'Want': creating consumer value

The process of open innovation starts with the identification of resource needs and

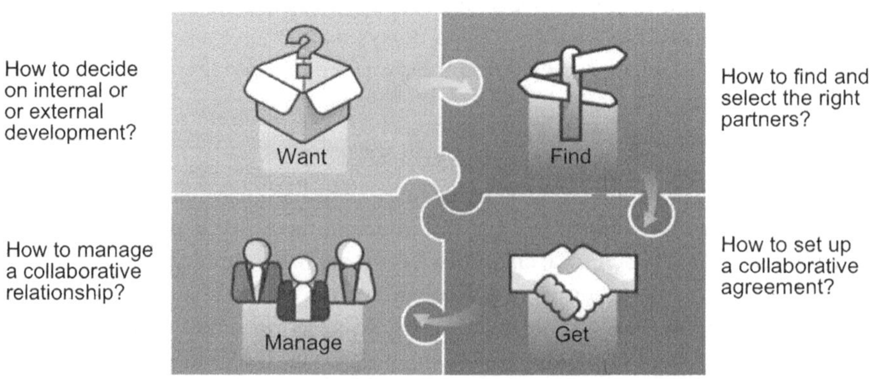

Fig. 17.3 Want – find – get – manage model.

determining which ones should/could be developed internally (in which case there is no need to pursue an open-innovation relationship) and which assets are not available internally and thereby Mars must seek them externally (Slowinski and Sagal, 2010). Identifying the 'wants' is the most challenging part of the process as business teams in general struggle to conceptualise and prioritise their needs; however, this clarity is needed before pursuing the resource intensive 'find', 'get' and 'manage' phases. Desired assets tend to be more permanent than strategic objectives and are likely to prevail after a change in business strategy. For instance, healthy ingredients are a key resource need for Mars leading to various solution areas, such as less saturated fat or sugar-free, whose strategic importance is linked to the prevailing business strategy. Hence, 'want' identification and conceptualisation determines the success of subsequent steps: if the 'want' is ill-defined then the company is looking in the wrong places and it is pointless to sign a contract given the lack of external alignment. Conversely, a well-defined 'want' points the company to the right networks and forms the basis for communication to ensure internal and external alignment.

Mars has developed a suite of tools for assessing the trade-offs between internal development and external acquisition (Fig. 17.4). Key to the discussion is whether the 'want' is part of the core areas the company wants to activate in the next 15 to 20 years or, conversely, if it is a short-term activity (over the next 5 years) for which the company does not need to build in-house expertise but rather acquire them externally. This process was applied to the decision whether to strengthen existing in-house expertise in emulsifiers, an important ingredient in chocolate technology, but not core to Mars' capabilities. Hence, by linking the 'want' to Mars' growth objectives, the company decided not to recruit and build internal expertise but rather to acquire it externally.

Mars' resource needs focus primarily on products, packaging, ingredients and processes. They are linked to a particular brand or product and tied to strategy and growth objectives. There might also be some fundamental enablers to allow future work. The open-innovation team also looks outside to better conceptualise the 'wants'; for instance working alongside technology intermediaries to assess the feasibility of different heating technologies to cook peanuts before selecting the ones offering greater opportunities, which are then scouted in more detail in order to find companies and research institutes with expertise in those areas.

A set of interactive tools supports the decision-making process by first helping the firm to identify 'wants' aimed at creating consumer value, then determining whether they should be developed using internal resources or by adding complementary external resources to Mars' in-place resource base (Fig.17.4). The objective is to find the best existing technology that is relevant to Mars' brands and innovation strategy, and to find the best way to bring it in. Sometimes this technology is found internally, sometimes not. A make/buy/partner decision-making protocol (Slowinski and Sagal, 2010) brings rigour to this determination by benchmarking internal developments against external alternatives to ensure the best option is selected. The planning outcomes are then converted into a set of prioritised 'want briefs' that clearly describe the resource need, thus enabling a

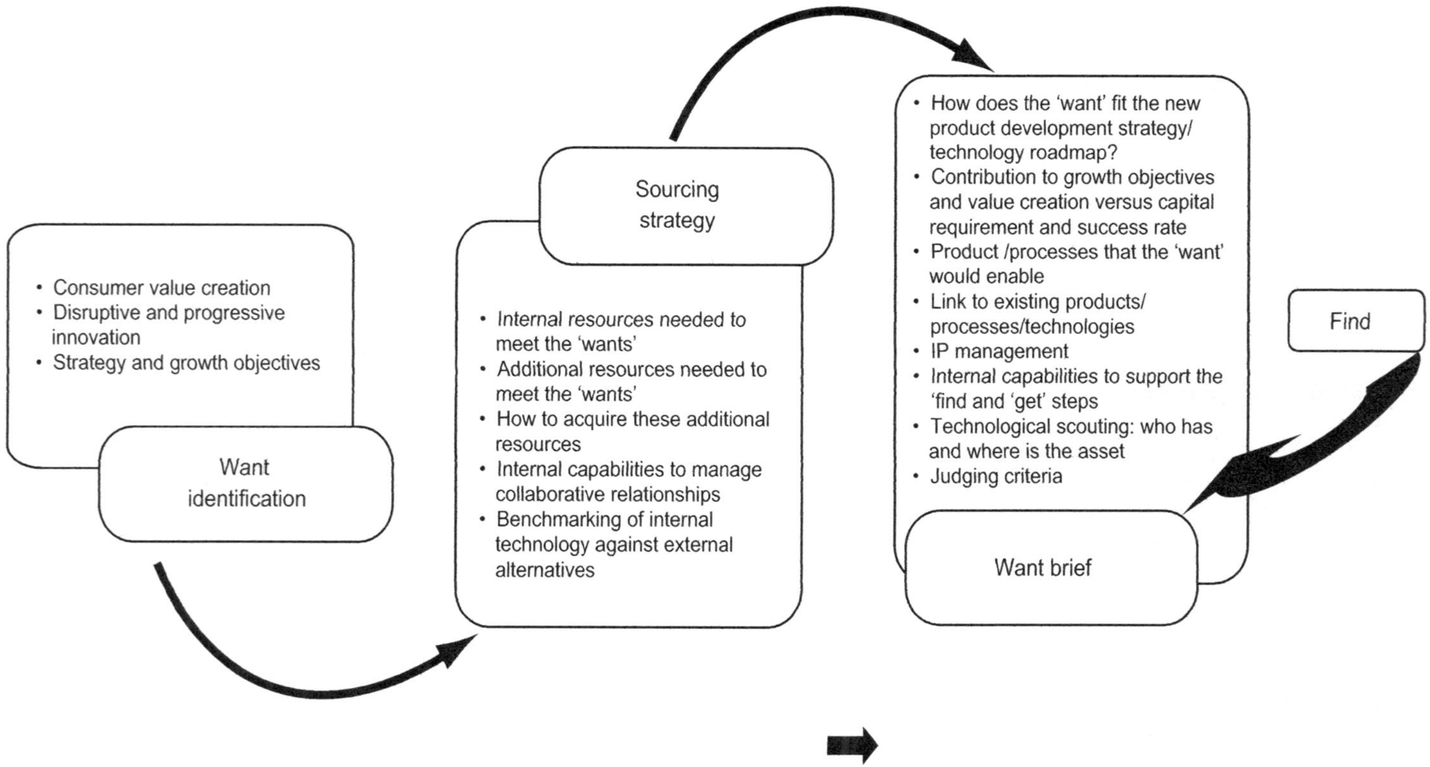

Fig. 17.4 Want decision making process.

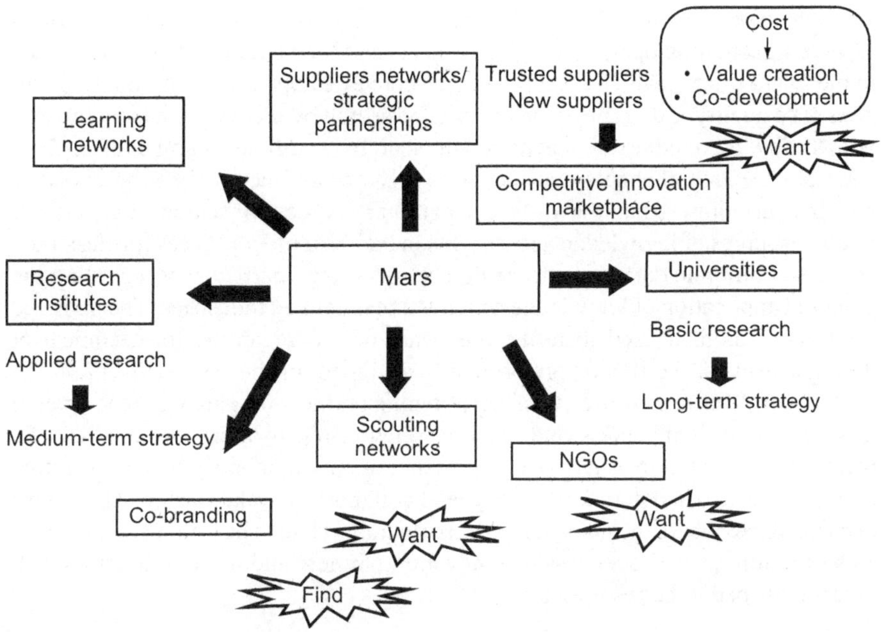

Fig. 17.5 Mars' ecosystem mapping.

decision regarding sourcing (internal or external) and guiding an efficient 'find' effort (Slowinski and Sagal, 2010). The contribution of the assets to growth objectives and value creation are measured against capital requirements and success rate.

17.4.2 'Find': selecting the right partner

The 'find' stage focuses on locating possible sources of external assets based on the information provided by the 'want briefs' (Slowinski and Sagal, 2010). Mars operates within an extensive ecosystem along the innovation funnel with increasing capabilities to connect beyond their supplier base via internal search capabilities or scouting networks (Fig. 17.5). They have recently developed greater internal search capabilities so that scanning and assessment of external knowledge is increasingly performed internally. In addition to cost implications, by relying exclusively on third-party 'find' agents, having internal scouts improves the speed and quality of the 'find' effort because they have free access to sensitive information that might be important to the effort but which Mars might be reluctant to share with open-innovation intermediaries. As a result, intermediaries only receive a narrow brief on the knowledge needs and though the search to locate potential asset sources could be extensive using their proprietary expert networks, the intermediaries unintentionally deliver against the brief with the potential loss of relevant information. Internal scouts, on the contrary, retain acquired information internally for future use. Moreover, internal scouts are in a better position to judge what solutions work best for Mars as they have a better understanding of the implicit aspects surrounding open-innovation projects and the tacit knowledge embedded in the 'want briefs', which is really hard to convey to open-innovation intermediaries. The quality and scope of the knowledge search by internal scouts (compared with that by intermediaries) depends on the search breadth (i.e., number of external sources or search channels that the firms rely on) and depth (i.e., the extent to which firms draw from the different external sources or search channels) of internal sources of knowledge and insights in the form of employees' professional contacts, corporate databases, learning networks, etc., particularly regarding the potential application of knowledge generated in unrelated industries. The gathered information is then used to refine the 'want briefs' regarding the technical or design aspects or the IP strategy originally designed for the assets (Fig.17.4).

A list of 20 criteria is used to select potential partners ranging from the level of trust, expertise, uniqueness and existing relationship to more commercial elements. These selection criteria are important not only in identifying and selecting partners, but also in determining the type of collaborative relationship. There have been instances where the intensity of the partnerships had to be limited owing to the lack of readiness and openness of innovation partners and more transactional IP agreements had to be resorted to.

17.4.3 'Get': setting up win–win solutions

From the 'find' step, a short list of potential external sources and information on each source is generated. Key to a successful 'get' is the internal and external alignment of 'wants' (Slowinski and Sagal, 2010). Internal alignment starts at the 'want' stage by involving organisational units, in particular the operational ones, in open-innovation initiatives. One of the key measures is whether the 'want' fits with the business strategy and technical objectives, what product does it fit with, what is the brand manager perspective and the level of commitment from the business to engage in external collaboration. By having these filters in the 'want' stage, the 'get' step can then focus on external alignment by assessing the potential partner's intent and technical capabilities. A number of supporting tools allow the open-innovation team to plan and negotiate every aspect of the alliance to be captured in a structured alliance agreement to avoid misinterpretation during the 'manage' phase.

By being open from the outset and sharing the 'wants' with innovation partners ('putting your cards on the table'), Mars avoids having different stakeholders working from different understandings on the purpose, terms and priorities of the collaborative agreements. This approach ensures that negotiations focus on 'win–win' solutions for all involved. The challenge is to balance the need for IP disclosure with the desire to avoid IP pollution (Mehlman *et al.*, 2010). Gathering of non-proprietary information on the potential partners' intent and technical capabilities during the 'find' stage contributes to an effective information exchange and, if Mars decides that a valuable collaboration is possible, the IP team puts NDAs in place so that confidential information can be exchanged to make the

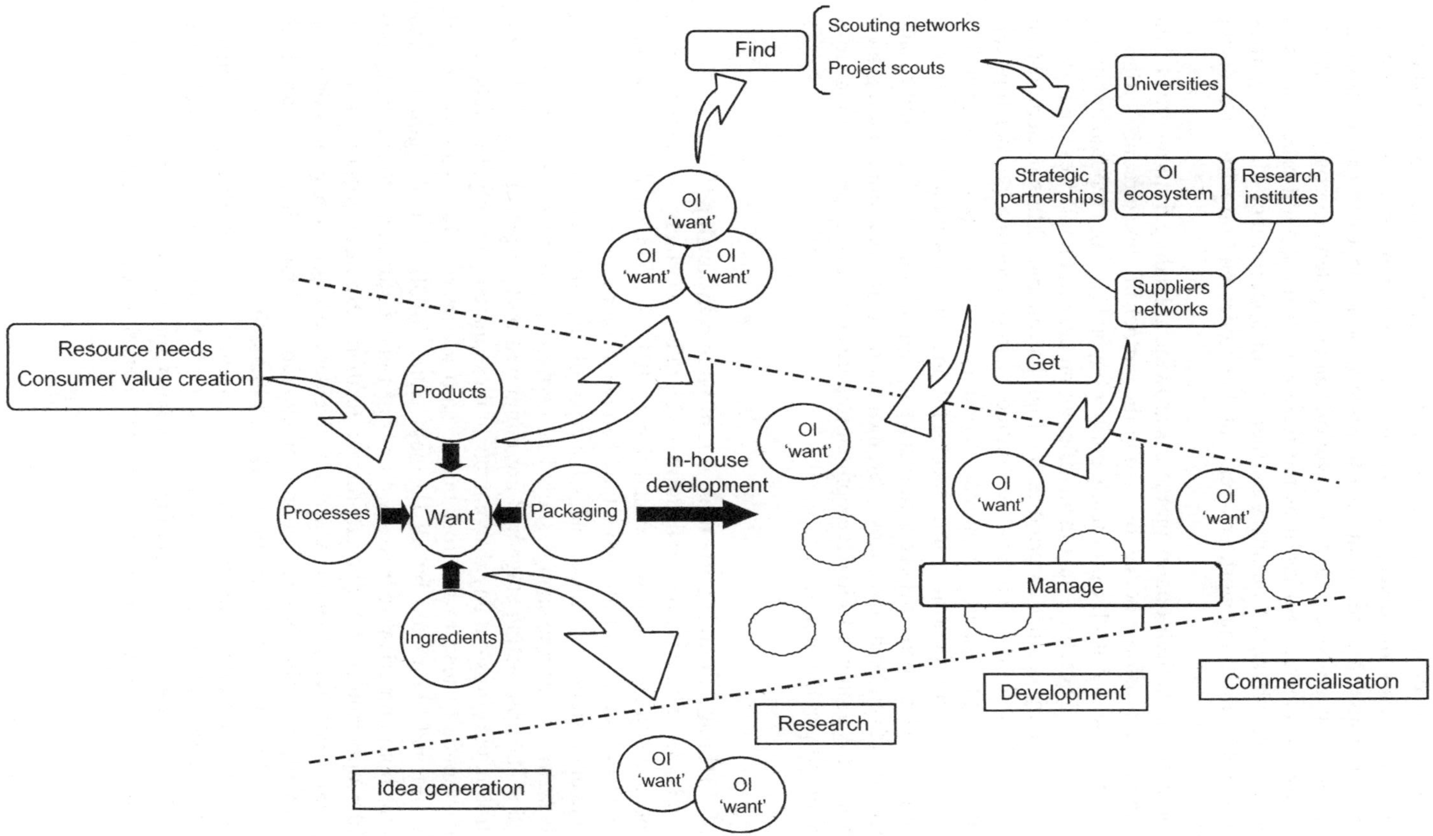

Fig. 17.6 WFGM Framework: getting access to the right competences.

next go/no-go decision. If the decision is to proceed with the collaboration, a steering team involving managers from the partnering companies is set up at the contracting phase to agree both how to review the project and the relationship during the 'manage' stage. Open innovation can be a potent approach to increase innovation capacity, but it requires the firms's own innovation processes to be under control; otherwise, adding a partner is a disappointment for both parties. Using the WFGM framework ensures that companies end up with an enduring alliance (Fig. 17.6).

Competition for high-quality partners is growing as leading firms augment their open-innovation efforts; hence, becoming the partner of choice requires Mars to demonstrate that it is itself a promising partner beyond having a reputable standing in the industry. Therefore, having an efficient process to identify, evaluate, value and contractually acquire external technology has enabled Mars to demonstrate that it is effective, efficient and committed to implementing partnerships and commercialising the results. Mars has started a number of strategic partnerships with core innovation partners after successful open-innovation iterations leading to a high level of trust between partners and full alignment of objectives and expectations from both sides. However, moving from an open-innovation project to open-innovation portfolio development requires time and expertise in creating and managing collaborative agreements.

17.4.4 'Manage': making the most out of your relationships

The goal of this phase is to co-ordinate and integrate the partners' resources to meet the specific objectives (Slowinski and Sagal, 2010). It is not only about meeting technical milestones but also how they behave as partners; that is, ensuring that partners understand who does what, what types of information must be exchanged and how information exchange should be carried out. These parameters are all agreed at the contracting stage and are monitored by the steering team. A short questionnaire brings out any disconnects in these matters quickly, that otherwise could potentially lead to mistrust that is hard to reverse. The questionnaire is addressed and completed by all business teams involved in the alliance and contains both numerical scores as well as open-ended questions to give respondents the opportunity to elaborate further in their responses. Feedback is then reviewed by the steering team which then decides the appropriate course of action to deal with emerging issues. A common challenge is managing expectations in strategic alliances, notably partners asking too much or not being open enough. Managing open-innovation projects presents similar challenges to internal projects with the adding complexity of having to ensure both internal and external alignment. This requires a high level of trust between innovation partners, which Mars takes into account when assessing potential innovation partners and setting up the deal.

17.5 Conclusions

Open innovation is a potent approach to increase firms' innovation capacity, but it requires the development and implementation of suitable systems that are used by well-trained people at every step. The role of R&D management changes completely and new capabilities and competencies are required once firms enter the external sourcing arena (Witzeman *et al.*, 2006). The Mars case study supports previous work highlighting the enabling role of top management in promoting the transition to an open-innovation approach (Vanhaverbeke, 2006; van de Meer, 2007) and the need for champions promoting and leading change (Chesbrough, 2006). Mars' open-innovation team in its key role of culture change agent has been a key enabler to successful open-innovation implementation.

Having a framework was imperative to Mars' inclusive open-innovation model where open innovation is the responsibility of all business teams rather than being exclusive to a small independent open-innovation business unit. The open-innovation team needed a framework to articulate and communicate the impact of such a new direction to its internal and external stakeholders. It provides Mars and partnering firms with a reference point at each stage of the process aided with a set of tools to guide decision making. Its application, however, requires continuous management attention and adequate resources.

The WFGM model has significantly increased the success rate of collaborative projects by forcing business teams to be clear about their resource needs, before they go out to find them and search for potential innovation partners and before they build the deal and then manage the consequences. Collaborative innovation can still fail to deliver its technological objectives, and changes in business strategy might warrant some open-innovation projects irrelevant; however, the framework has given greater structure to collaborative knowledge-creating processes. Lessons learnt, experiences and successes in open-innovation initiatives are easily transferred and applied to new external collaborations avoiding typical delays and failures in open-innovation projects owing to lack of knowledge and experience of those involved in collaborative agreements.

Having proper business reasons ('wants') for engaging in open innovation and sharing them with innovation partners will influence Mars' ability to innovate in new territories. Although open-innovation activity only started in 2008, there is a real desire and ambition to maximise open-innovation potential, and recent open-innovation projects have created the learning experience needed to drive the process further. The application of open-innovation principles has now moved beyond the innovation funnel to include projects looking at operational excellence (i.e., factory and maintenance systems).

Mars is just starting to see the benefits of open innovation with a number of products currently in the market resulting from different types of collaborative relationships and the pipeline is filled with new joint development contracts. The challenge now is the transition from open-innovation activities to accelerate developments to technological platforms leading to paradigm shifts. Operating under an open-innovation context depends on contributions from across a network of partners ranging from suppliers of raw materials and equipment and research institutes to consumers and customers that create value for the end consumer (West, 2006). Therefore, this co-ordinated innovation by network members requires leadership of the network.

17.6 References

CHESBROUGH, H. W. 2003. *Open innovation: the new imperative for creating and profiting from technology*, Boston (MA): Harvard Business School Press.

CHESBROUGH, H. W. 2006. *Open innovation models: how to thrive in the new innovation landscape*, Boston (MA): Harvard Business School Press.

CHESBROUGH, H. W. 2007. Why companies should have open business models. *MIT Sloan Management Review*, **48**, 22–28.

CHIARONI, D., CHIESA, V. AND FRATTINI, F. 2010. Unravelling the process from closed to open innovation: evidence from mature, asset-intensive industries. *R&D Management*, **40**, 222–245.

CHRISTENSEN, J. F. 2006. Wither core competency for the larger corporation in an open innovation world? In: CHESBROUGH, H., VANHAVERBEKE, W. AND WEST, J. (eds.) *Open innovation: researching a new paradigm.* Oxford: Oxford University Press.

DAHLANDER, L. AND GANN, D. M. 2010. How open is innovation? *Research Policy*, **39**, 699–709.

GASSMANN, O. 2006. Opening up the innovation process: towards an agenda. *R&D Management*, **36**, 223–250.

GASSMANN, O. AND ENKEL, E. 2004. Towards a theory of open innovation: three core process archetypes. *Proceedings of the R&D management conference.* Lisbon, Portugal, July 6–9.

GASSMANN, O. AND VON ZEDTWITZ, M. 1999. New concepts and trends in international R&D organisation. *Research Policy*, **28**, 231–250.

HAGEDOORN, J. 2002. Inter-firm R&D partnerships: an overview of major trends and patterns since 1960. *Research Policy*, **31**, 477–492.

HUIZINGH, E. K. R. E. 2011. Open innovation: state of the art and future perspectives. *Technovation*, **31**, 2–9.

HUSTON, L. AND SAKKAB, N. 2006. Connect and develop. *Harvard Business Review*, **84**, 58–66.

LICHTENTHALER, U. AND LICHTENTHALER, E. 2009. A capability-based framework for open innovation: complementing absorptive capacity. *Journal of Management Studies*, **46**, 1315–1338.

MEHLMAN, S. K., URIBE-SAUCEDO, S., TAYLOR, R. P., SLOWINSKI, G., CARRERAS, E. AND ARENA, C. 2010. Better practices for managing intellectual assets in collaborations. *Research Technology Management*, January–February, 55–66.

SLOWINSKI, G. 2004. *Reinventing corporate growth*, Gladstone (NJ): Alliance Management Press.

SLOWINSKI, G. AND SAGAL, M. W. 2010. Good practices in open innovation. *Research Technology Management*, September–October, 38–45.

SLOWINSKI, G., HUMMEL, E., GUPTA, A. AND GILMONT, E. R. 2009. Effective practices for sourcing innovation. *Research Technology Management*, **52**, 27–34.

VAN DE MEER, H. 2007. Open innovation – the Dutch treat: challenges in thinking in business models. *Creativity and Innovation Management*, **6**, 192–202.

VANHAVERBEKE, W. 2006. The interorganisational context of open innovation. In: CHESBROUGH, H., VANHAVERBEKE, W. AND WEST, J. (eds.) *Open innovation: researching a new paradigm.* Oxford: Oxford University Press.

WEST, J. 2006. Does appropriability enable or retard open innovation? In: CHESBROUGH, H., VANHAVERBEKE, W. AND WEST, J. (eds.) *Open innovation: researching a new paradigm.* Oxford: Oxford University Press.

WITZEMAN, S., SLOWINSKI, G., DIRKX, R., GOLLOB, L., TAO, J., WARD, S. AND MIRAGLIA, S. 2006. Sourcing external technology for innovation. *Research Technology Management*, **49**, 19–27.

WOLPERT, J. D. 2002. Breaking out of the innovation box. *Harvard Business Review*, **80**, 77–83.

18

Crowdsourcing: the potential of online communities as a tool for data analysis

M. Garcia Martinez and B. Walton, Kent Business School, University of Kent, UK

DOI: 10.1533/9780857097248.4.332

Abstract: The potential of crowdsourcing is examined as a tool for data analysis in order to address the increasing problems faced by companies in trying to deal with 'big data'. Specifically, crowdsourcing is evaluated as a way of solving predictive modelling problems. Predictive modelling competitions, a particular subset of crowdsourcing competitions with distinct features in terms of design and participants' knowledge and motivation, represent a potential win–win scenario for all involved: contestants get access to real-world data that has been carefully anonymised to eliminate privacy concerns and prize sponsors reap the benefits of the contestants' creativity. The empirical setting of the research is the data predictive modelling start-up Kaggle, which operates as a knowledge broker between companies aiming to outsource predictive modelling competitions and a network of over 31 000 data scientists competing to produce the best solutions.

Key words: crowdsourcing, open innovation, online communities, predictive modelling competition, co-creation, shopper behaviour, data analysis.

18.1 Introduction

The practice of co-creation interactions with consumers and unidentified individuals on the internet in the process of product development and value creation is becoming more common (Howe, 2008; Surowiecki, 2005; Tseng and Piller, 2003; von Hippel, 2005). High-quality interactions that enable consumers to co-create unique experiences with companies are the key to unlocking new sources of competitive advantage (Prahalad and Ramaswamy, 2004, p.7). Co-creation through crowdsourcing in particular, a mode of openness where a firm conducts inbound open innovation (Huizingh, 2011), exploiting knowledge outside of their organisation by outsourcing the function of idea generation to large groups of external contributors who are often unknown or undefined (Howe, 2008), is gaining attention. Advances in the Internet, collaboration tools, and new Web 2.0 technologies have considerably reduced the costs of accessing large number of potential problem solvers (Morgan and Wang, 2010; Terwiesch and Ulrich, 2009; Zwass, 2010).

Crowdsourcing is built on the concept that large groups of people are smarter than an elite few, no matter how brilliant the elite few may be (Surowiecki, 2005, p.1). Howe (2006, p.1) describes this new paradigm 'as everyday people using their 'spare cycles' to create content, solve problems, even do corporate R&D'. These 'crowds' form around particular projects (e.g., innovation or ideas contests) and thus have common objectives. Harnessing the wisdom of the crowd promises many advantages: eliminating biases that lead to an unjustified preference for certain contributors and totally integrating market research and innovation by getting customers to submit their own ideas and voting on each others' ideas (Surowiecki, 2005; Howe, 2006). Crowdsourcing among users of products and services has been found to generate incremental and radical solutions (Bogers *et al.*, 2010; Morrison *et al.*, 2004; von Hippel, 2005). Empirical studies even suggest that innovations developed by a crowd of users may have clear advantages over solutions generated by internal R&D departments (Baldwin and von Hippel, 2009; Bogers *et al.*, 2010). Poetz and Schreier (2009) found that users' ideas exhibit a greater degree of novelty, and promise clearer consumer benefits than ideas generated by firms' professionals which tend to be more focused on flexibility to reduce development time.

One of the ways that businesses are making use of crowdsourcing as a strategic tool is by using the wisdom of the crowd to solve complex business problems. In this chapter, we look at crowdsourcing as a way of solving predictive modelling problems, a particular subset of crowdsourcing competitions with distinct features in terms of design and participants' knowledge and motivation that until now, have received limited study, despite their potential to address the increasing problems faced by companies in trying to deal with 'big data' (Manyika *et al.*, 2011).

In Section 18.2, the idea of predictive modelling competitions and their attributes is introduced. Section 18.3 looks into the effective design of predictive modelling competitions that encourage participation and engagement. Section 18.4 provides a case study, focusing on Kaggle, the world's leading online host of predictive modelling competitions. Finally, Section 18.5 provides a series of conclusions.

18.2 Predictive modelling competitions

Predictive modelling is a process used in predictive analytics to create a statistical model based on future behaviour (Davenport and Harris, 2006). A predictive model is made up of a number of predictors, which are variable factors that are likely to influence future behaviour or results (Hair, 2007). Marketing is among the most frequent applications of the technique. Advances in information technology, data gathering and analytics are enabling companies to manage all phases of the customer life cycle, including acquiring new customers, increasing revenue from existing customers and retaining new customers. Retailing in particular is an area where predictive analytics shows the most promise (Harris and Lowitt, 2007). Using increasingly granular data from detailed demographics and psychographics to consumers' clickstreams on the web, retailers are creating highly customised offers that steer consumers to the right merchandise or service, the so-called 'next big offer', at the right time, at the right price and in the right channel (Davenport *et al.*, 2011). For instance, through its loyalty card programme, Tesco, the leading UK retailer, tracks which stores consumers visit, what they buy and how they pay. As a result, Tesco and dunnhumby have been able to define 267 lifestyle clusters (Humby *et al.*, 2003) allowing them to match offers to specific demographics and target promotional activity based on purchasing history (Rowley, 2007). This demographic and psychographic information about their customers has also allowed Tesco to successfully target coupon offers resulting in the achievement of redemption rates ranging from 8% to 10%, far higher than the 1% to 2% seen elsewhere in the grocery industry (Davenport *et al.*, 2011).

Predictive modelling competitions extract 'core' predictive modelling challenges from their application environment, and these are then presented to participants. They are usually expected to represent real-life predictive modelling challenges. For instance, the Netflix Prize challenged the computer-science communities to develop predictive systems for movies preferences (Bennett and Lanning, 2007). The Knowledge Discovery and Data Mining (KDD) Challenge Cup, the oldest data-mining competition, running since 1997, focuses on analysing different datasets in search, bioinformatics, web mining and medical diagnosis. By extracting a real-life problem from its context, predictive modelling competitions may relax some of the supervised learning assumptions. For example, in the KDD Cup 2007, participants did not see any y_i's (outcomes), instead they had to extrapolate them based on training x_i's (observed data).

Data modelling competitions represent a potential win–win scenario for all involved: contestants get access to real-world data that has been carefully anonymised to eliminate privacy concerns and prize sponsors reap the benefits of the contestants' creativity. By exposing the problem to a large number of participants proficient in different techniques, competitions can very quickly advance the frontier of what is possible using a given dataset. Crowdsourcing allows firms to bypass all five stages in mining external R&D proposed by Porter and Newman (2011). Rather than searching for methodological approaches to be used on the firm's data by conducting literature reviews, research profiling, tech mining, structured knowledge discovery exercises or literature-based discovery exercises, crowdsourcing brings the most effective approaches to the firm. Goldcorp, Inc. was one of the original pioneers to use the crowdsourcing model to analyse geological data (Tapscott and Williams, 2006). From across the world, students, physicists and geologists demonstrated the power of crowdsourcing the task of data interpretation. In all, 1200 people from 50 different countries offered insights

Table 18.1 Distinguishing features of predictive modelling and ideas competitions

	Predictive modelling competition	Ideas/innovation competition
Task	Prototyping and modelling	Ideas and concept development
Crowd	Knowledgeable crowd – data scientists/specialists	'Hobbyists', engaged, loyal customers
Problem specificity	High (searching for a solution to a specific problem)	Low (elaboration of own ideas)
Elaborateness	Complex – code or algorithms	Textual descriptions of rough ideas/concepts
Goal	Community-centric	Individual-centric
Motivation	Mixture of intrinsic and extrinsic motives: intellectual interest, peer recognition and monetary rewards	Intrinsic motives: fun, curiosity
Evaluation	Performance oriented: benchmark score	Participation oriented

to Goldcorp, enabling the company to find substantial gold within its mines. An effort that cost Goldcorp US$575 000 in prize money returned over eight million ounces of gold.

Predictive modelling competitions tend to focus on prototyping and modelling rather than execution. They are viewed by data scientists as the platform to test the robustness of their algorithms and theories. A participant's motivation is driven by the desire to advance the data mining community (community goal) in contrast to ideas competitions which tend to be more participant-centric (individual goal). As such, problems are highly specified requiring the submission of codes or algorithms to improve pre-existing accuracy benchmarks. Table 18.1 provides a synopsis of the key design elements of predictive modelling and ideas competitions and their attributes.

18.3 Design and management of predictive modelling competitions

Online communities let companies draw insights from a deep, diverse knowledge pool and apply them to continuous innovation. The design of effective crowdsourcing competitions that brings the right problem to the table, targets the right crowd and encourages crowd's participation and engagement is central to the success of online communities as a vehicle for innovation.

18.3.1 How to attract the right crowd

Key to the success of crowdsourcing contests are the identification of the right crowd or solver community to a particular problem and crowd participation and motivation. Research has focused on the design of effective crowdsourcing competitions to encourage solvers' participation (Fuller, 2010; Howe, 2008; Leimeister *et al.*, 2009; Morgan and Wang, 2010). Attracting participants to join a crowdsourcing contest is an effective way to improve the contest outcome because the bigger the crowd the higher the quality and diversity of solutions (Terwiesch and Xu, 2008). Moreover, diverse solutions can significantly contribute to the generation of breakthrough innovation and solutions (Terwiesch and Ulrich, 2009; Hargadon, 2003).

Two aspects in particular have been the focus of recent crowdsourcing research. The first aspect is participants' motivation examining the role of extrinsic and intrinsic motivation in participation and contribution performance. The major source of intrinsic motivation is the sheer fun and enjoyment of developing innovative solutions to challenging problems (Franke and Shah, 2003; Fuller, 2006). For data scientists, participating in predictive modelling competitions serve to test the robustness of their algorithms and theories and satisfy their intellectual interest by sharing knowledge more openly and effectively. Trying to contribute to the creative discovery of solutions seems to be a source of positive feelings of competence, autonomy and self-expression (Shah and Kruglanski, 2000).

Regarding extrinsic motives, scholars document three key factors: career aspirations; monetary rewards; and the personal need for innovations. Developing concepts for new products and services or writing lines of codes or algorithms allows individuals to demonstrate their talent to potential employers (Jeppesen and Fredericksen, 2006). Companies such as Kaggle that operate as a knowledge broker between companies aiming to outsource predictive modelling competitions and a network of over 31 000 data scientists that compete to produce the best solutions, have become an informal recruitment ground for companies looking for the best and brightest data analysts. Several studies have shown that participants also often expect financial compensation for working on open-innovation projects (Dahlander and Wallin, 2006). Crowdsourcing is after all a strategy for firms to outsource parts of their corporate R&D activities to a large group of individuals who might expect to be rewarded for their solutions. Lakhani *et al.* (2006) and Archak and Sundararajan (2009) identify a significant effect of extrinsic monetary motivation on the performance of contributors. Conversely, Frey *et al.* (2011) report the prevalence of intrinsic over extrinsic motivational factors in the case of innovation challenges posted by the online platform Atizo. This platform, however, invites solvers to participate in creative ideation tasks as opposed to submitting detailed solutions to scientific or technical problems or writing lines of codes or algorithms more typical of Kaggle or InnoCentive competitions. In addition, research has revealed the overall behavioural pattern of contestants' participation in crowdsourcing competitions. Archak (2010) reports that expert participants face tougher competition from their opponents in the competition phase of the contest. In order to reduce competition, they move first in the registration phase of the contest to deter entry of additional opponents in the same contest. Yang *et al.* (2008) find significant variation in the expertise and productivity of the participating users; a very small core of successful users contributes nearly 20% of the winning solutions on the online platform.

The second aspect is knowledge diversity and the need for disparate thinking (Howe, 2008). Having knowledge in diverse fields allows participants to blend disparate solution elements in novel ways (Hargadon, 2003; Majchrzak *et al.*, 2004). However, it would be difficult to get actionable solutions to technical problems or algorithms to predictive modelling problems from a crowd without the specialised knowledge. As with incentives, the type of competition would ultimately determine the target group and knowledge diversity required for the competition.

18.3.2 What is the right innovation problem to bring to the crowd

Crowdsourcing as a model to open up corporations and governments to fresh ideas and insights from online networks and communities is growing, with new competitions and initiatives being launched every week. The surge in the practical use of innovation contests among high-tech firms is striking, with more than 80% of those listed in the Standard & Poor's (S&P) 500 index having established virtual user communities to benefit from users' participation (Mahr and Lievens, 2012). However, little is known about when crowdsourcing contests should be used; what type of innovation problems should be solved by the crowd and which are better solved internally (Terwiesch and Xu, 2008). Similarly to open-innovation projects, companies need to determine whether the idea, knowledge or technology should be developed using internal resources or by adding complementary external resources to companies' in-place resource bases.

Crowdsourcing competitions have been applied to a variety of industries to solve both mundane and highly complex tasks. Crowdsourcing is not merely a Web 2.0 buzzword, but is instead a strategic model to attract an interested, motivated crowd of individuals capable of solutions superior in quality and quantity to those that even traditional forms of business can provide (Baldwin and von Hippel, 2009; Bogers *et al.*, 2010). From a data-analysis perspective, crowdsourcing is most suitable for dealing with experimental problems and developing radical innovations. However, this does not mean that only radical, non-critical business problems should be crowdsourced. The speed with which host companies such as Kaggle and Innocentive are able to develop, launch and evaluate competitions means that critical business decisions do not become bottlenecked by this process.

18.3.3 Security and privacy implications in data mining

Data mining allows us to discover information we do not expect to find in databases. Providing access to these databases via crowdsourcing raises a potential security/privacy risk to companies by making supposedly anonymous data open to the public to facilitate and stimulate the crowd's participation. As the shopper challenge competition illustrates, inquisitive players go to great lengths to unmask individual characteristics of the data and general trends as part of their self-motivation to challenging problems. Hence, simply removing or replacing individual details (e.g. names or sociodemographic data) would not be sufficient to ensure privacy owing to the 'inference problem' (i.e., when connections can be made with other data sets already in the public domain) (Clifton and Marks, 1996). This would violate data-protection laws by failing to protect personal information, and is therefore a key issue companies must consider when using this form of open-source research. Indeed, both America Online and Netflix disclosed data about their customers/users to help drive the crowd's ingenuity and in doing so risked violating data-security regulations (Barbaro and Zeller, 2006; Ohm, 2010). Unfortunately, developing effective methods of protecting consumer privacy remains an open-search problem (Narayanan and Shmatikov, 2008). Our research suggests that running private competitions, an invitation-only challenge in which players undergo background checks and compete under non-disclosure agreements could be a potential solution to overcome the risks associated with making commercially sensitive data open to the public.

18.4 Case study: Kaggle

Founded in 2010 by Australian entrepreneur and data analyst Anthony Goldbloom, Kaggle acts as a virtual knowledge broker between firms, governments and researchers seeking support to solve statistical/analytical problems and a network of skilled external contributors that compete to produce the best solution. At the end of a competition, the competition host pays prize money in exchange for the intellectual property behind the winning model. In little over a year of operation, Kaggle have become the leading online platform for hosting predictive modelling competitions. They have hosted high-profile competitions for Wikipedia, Delloitte and NASA as well as hosting the World Heritage Health Prize worth $US3 million that aims to predict and prevent unnecessary hospitalisations. In 2011, Kaggle received $US11.25 million Series A funding. The financing round was led by Index Ventures and Khosla Ventures with participation from PayPal co-founder Max Levchin, Google chief economist Hal Varian and Applied Semantics' (now Google Adsense) co-founder Gil Elbaz.

Kaggle has a network of over 31 000 data scientists, many of them at PhD level that crave real-world data to develop and refine their techniques and Kaggle harnesses what it describes as the 'cognitive surplus' of these data scientists to bridge the gap between data and knowledge for companies that are willing to fund and run suitable competitions. Kaggle therefore represents an extension of the 'connection-oriented' business model described by Wirtz *et al.*'s (2010, p. 275) 4C typology of internet business models in that they can be defined as providing a physical and/or virtual network infrastructure and their value proposition includes providing the prerequisites for exchange of information over the Internet.

Kaggle currently operates three types of competitions for their clients: private, in-class and open. A private competition is an invitation-only challenge in which players undergo background checks and compete under non-disclosure agreements. Participants in private competition are selected based on their past

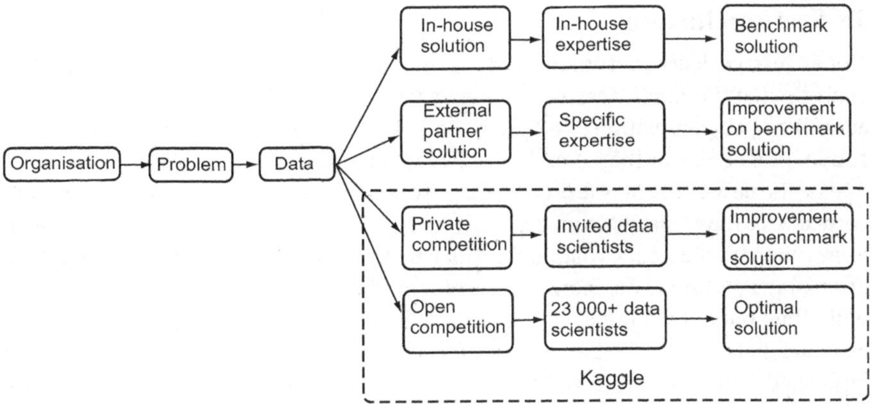

Fig. 18.1 Typology of data modelling contests.

performance in Kaggle competitions. Unlike open competitions, every competitor that accepts the invitation to compete wins some prize money, with larger prizes for those who produce the best results. Kaggle's in-class competition is a free-of-charge service that allows students to compete against their classmates in coursework-based challenges in order to learn statistical and data-mining skills.

In an open competition, the competition host prepares the data and a description of the problem. Kaggle offers a consulting service to help the host do this, as well as to frame the competition, anonymise the data, and integrate the winning model into their operations. The data is made available to all of Kaggle's data scientists, and participants submit their entries via an online portal. Competitive pressures drive participants to keep trying new ideas. Real-time feedback is given on a live leaderboard, so when somebody makes a breakthrough, others revise their own algorithms to outdo the leader's performance. This leapfrogging continues until participants reach the full extent of what is possible in a set time frame. At the end of the competition, the competition host pays the winner of the competition for the intellectual property of their winning solution.

An open competition allows an organisation to find an optimal solution to their problem because the live leaderboard encourages a large number of data scientists continually to update and revise their algorithms until they advance towards the full extent of what is possible with a given data set. The advantage of using an open competition to analyse an organisation's data over other methods is displayed in Fig. 18.1. In this diagram, the term 'benchmark solution' is used to describe the best possible solution available to a company using only their own in-house resources. This benchmark can be improved upon by engaging external partners such as universities or consultants or by opening up data to experts in other fields in the form of private and open crowdsourcing competitions. An optimal solution can only be found when the boundaries of what is possible with a data set have been explored. This is most likely in the context of an open competition where data scientists are encouraged continually to revise their algorithms/models in order to leapfrog other solutions.

18.5 Conclusions

Not all lessons learned from the early adoption of crowdsourcing as a tool for data analysis may be applicable to following firms. Huzingh (2011) suggests that in adopting open-innovation strategies; 'Followers may be more reluctant to organizational change, but they could also have sound reasons for delaying adoption: maybe the new concept is less attractive to them, making it inappropriate to merely copy the lessons learned from early adopters'(p. 7). However, firms that are open to new approaches may wish to consider that online competitions appear to be a promising solution to the increasing problems faced by companies in trying to deal with 'big data' (Manyika *et al.*, 2011). Market intelligence consultants IDC estimate that in 2011, the amount of information created in the digital universe and replicated will surpass 1.8 zettabytes (1.8 trillion gigabytes), growing by a factor of nine over a period of five years (Gantz and Reinsel, 2011). Recent discussions on the subject of 'big data' have been able to describe the problems faced by companies working with digital information on a massive scale: the inability to collate the data or to process the data. However, solutions are in short supply.

The design of effective crowdsourcing competitions that brings the right problem to the table, targets the right crowd and encourages the crowd's participation and motivation is central to the success of online communities as a vehicle for innovation. This chapter theorises that a truly open competition that utilises a live leaderboard to encourage activity is the best way to provide an optimal solution to a data-analysis problem, but that a private competition is the best way of improving a benchmark solution whilst retaining security with regards to data protection and commercial information. The process design of an effective crowdsourcing competition must also take into account the motivation of the crowd. Both intrinsic and extrinsic motivational factors are at play; however, traditional monetary rewards can act as a demotivating factor if not carefully planned.

18.6 References

ARCHAK, N. (2010) Money, glory and cheap talk: analyzing strategic behavior of contestants in simultaneous crowdsourcing contests on TopCoder.com. In *Proceedings of WWW'10: the 19th International World Wide Web Conference*, April 26–30, 2010, Raleigh, NC, USA. 21–30, http://pages.stern.nyu.edu/~narchak/wfp0004-archak.pdf.

ARCHAK, N. AND SUNDARARAJAN, A. (2009) Optimal design of crowdsourcing contests. *Proceedings of the Thirtieth International Conference on Information Systems, Association for Information Systems*, Atlanta. Association for Information Systems, http://pages.stern.nyu.edu/~narchak/optimalCrowdsourcingDesign.pdf.

BALDWIN, C. Y. AND VON HIPPEL, E. (2009) Modeling a paradigm shift: from producer innovation to user and open collaborative innovation. Harvard Business School, http://www.hbs.edu/research/pdf/10-038.pdf.

BARBARO, M. AND ZELLER, T. J. (2006) A face is exposed for AOL searcher no. 4417749. *NY Times*, http://query.nytimes.com/gst/fullpage.html?res=9E0CE3DD1F3FF93AA3575BC0A9609C8B63&pagewanted=all.

BENNETT, J. AND LANNING, S. (2007) The Netflix Prize. In: *Proceedings of KDD cup and workshop*, 12 August, San Jose, California, http://www.cs.uic.edu/~liub/KDD-cup-2007/NetflixPrize-description.pdf.

BOGERS, M., AFUAH, A. AND BASTIAN, B. (2010) Users as innovators: a review, critique and future research directions. *Journal of Management*, **36**(4), 857–875.

CLIFTON, C. AND MARKS, D. (1996) Security and privacy implication of data mining. In *Proceedings of the 1996 ACM SIGMOD workshop on data mining and knowledge discovery*, http://www.cs.purdue.edu/homes/clifton/document/dmkd.pdf.

DAHLANDER, L. AND WALLIN, M. W. (2006) A man on the inside: unlocking communities as complementary assets. *Research Policy*, **35**(8), 1243–1259.

DAVENPORT, T. H. AND HARRIS, J. G. (2006) *Competing on analytics: the new science of winning*. Harvard Business School Press, Boston (MA).

DAVENPORT, T. H., MULE, L. D. AND LUCKER, J. (2011) Know what your customers want before they do. *Harvard Business Review*, **89**(12), 84–92.

FRANKE, N. AND SHAH, S. (2003) How communities support innovative activities: an exploration of assistance and sharing among end-users. *Research Policy*, **32**(1), 157–178.

FREY, K., LUTHJE, C. AND HAAG, S. (2011) Whom should firms attract to open innovation platforms? The role of knowledge diversity and motivation. *Long Range Planning*, **44**, 397–420.

FULLER, J. (2006) Why consumers engage in virtual new product developments initiated by producers. *Advances in Consumer Research*, **33**(1), 639–646.

FULLER, J. (2010) Refining virtual co-creation from a consumer perspective. *California Management Review*, **52**(2), 98–122.

GANTZ, J. AND REINSEL, D. (2011) Extracting value from chaos. MA: IDC Go-to-Market Services, http://idcdocserv.com/1142.

HAIR, J. F. (2007) Knowledge creation in marketing: the role of predictive analytics. *European Business Review*, **19**(4), 303–315.

HARGADON, A. (2003) *How breakthroughs happen: the suprising truth about how companies innovate*. Harvard Business Press, Boston (MA).

HARRIS, J. G. AND LOWITT, E. M. (2007) How retailers compete on analytics to achieve high performance. Accenture, http://www.accenture.com/SiteCollectionDocuments/PDF/RetailersCompete.pdf.

HOWE, J. (2006) The rise of crowdsourcing. *Wired Magazine*, http://www.clickadvisor.com/downloads/Howe_The_Rise_of_Crowdsourcing.pdf.

HOWE, J. (2008) *Crowdsourcing: why the power of the crowd is driving the future of business*. Crown Business, New York.

HUIZINGH, E. K. R. E. (2011) Open innovation: state of the art and future perspectives. *Technovation*, **31**, 2–9.

HUMBY, C., HUNT, T. AND PHILLIPS, T. (2003) *Scoring points: how Tesco is winning customer loyalty*. Kogan Page Limited, London.

JEPPESEN, L. B. AND FREDERICKSEN, L. (2006) Why do users contribute to firm-hosted user communities? The case of computer-controlled music instruments. *Organization Science*, **17**(1), 45–63.

LAKHANI, K. R., JEPPESEN, L. B., LOHSE, P. A. AND PANETTA, J. A. (2006) The value of openness in scientific problem solving. Boston, MA: Harvard Business School, http://www.hbs.edu/research/pdf/07-050.pdf.

LEIMEISTER, J. M., HUBER, M., BRETSCHNEIDER, U. AND KRCMAR, H. (2009) Leveraging crowdsourcing: activation-supporting components for IT-based ideas competitions. *Journal of Management Information Systems*, **26**(1), 197–224.

MAHR, D. AND LIEVENS, A. (2012) Virtual lead user communities: drives of knowledge creation for innovation. *Research Policy*, **41**, 167–177.

MAJCHRZAK, A., COOPER, L. P. AND NEECE, O. E. (2004) Knowledge reuse for innovation. *Management Science*, **50**(2), 174–188.

MANYIKA, J., CHUI, M., BROWN, B., BUGHIN, J., DOBBS, R., ROXBURGH, C. AND BYERS, A. H. (2011) *Big data: the next frontier for innovation, competition, and productivity*. McKinsey Global Institute.

MORGAN, J. AND WANG, R. (2010) Tournaments for ideas. *California Management Review*, **52**(2), 77–97.

MORRISON, P. D., ROBERTS, J. H. AND MIDGLEY, D. F. (2004) The nature of lead users and measurement of leading edge status. *Research Policy*, **33**, 351–362.

NARAYANAN, A. AND SHMATIKOV, V. (2008) Robust de-anonymization of large sparse datasets. In *Proceedings of 29th IEEE Symposium on Security and Privacy*, May, Oakland, CA. 111–125, http://www.cs.utexas.edu/~shmat/shmat_oak08netflix.pdf.

OHM, P. (2010) Broken promises of privacy: responding to the surprising failure of anonymization. *UCLA Law Review*, **57**, 1701–1777.

POETZ, M. K. AND SCHREIER, M. (2009) The value of crowdsourcing: can users really compete with professionals in generating new product ideas? Paper presented at the DRUID Summer Conference 2009, June 17–19, Copenhagen, Denmark, http://www2.druid.dk/conferences/viewpaper.php?id=5682&cf=32.

PORTER, A. L. AND NEWMAN, N. C. (2011) Mining external R&D. *Technovation*, **31**, 171–176.

PRAHALAD, C. K. AND RAMASWAMY, V. (2004) Co-creation experiences: the next practice of value creation. *Journal of Interactive Marketing*, **18**(3), 5–14.

ROWLEY, J. (2007) Reconceptualising the strategic role of loyalty schemes. *Journal of Consumer Marketing*, **24**(6), 366–374.

SHAH, J. AND KRUGLANSKI, A. (2000) The structure and substance of intrinsic motivation, In: Sansone, C. and Harackiewicz, J. (Eds), *Intrinsic and extrinsic motivation: the search for optimal motivation and performance*. Academic Press, San Diego, CA, pp. 105–127.

SUROWIECKI, J. (2005) *The Wisdom of crowds – why the many are smarter than the few*. First Anchor Books Edition, New York.

TAPSCOTT, D. AND WILLIAMS, A. D. (2006) *Wikinomics: how mass collaboration changes everything*. Penguin Group, New York.

TERWIESCH, C. AND ULRICH, K. T. (2009) *Innovation tournaments*. Harvard Business School Press, Boston.

TERWIESCH, C. AND XU, Y. (2008) Innovation contests, open innovation and multiagent problem solving. *Management Science*, **54**(9), 1529–1543.

TSENG, M. M. AND PILLER, F. T. (2003) *The customer centric enterprise: advances in mass customization and personalization*. Springer-Verlag, Berlin.

VON HIPPEL, E. (2005) *Democratizing innovation*. MIT Press, Cambridge, MA.

WIRTZ, B. W., SCHILKE, O. AND ULLRICH, S. (2010) Strategic development of business models: implications of the Web 2.0 for creating value on the Internet. *Long Range Planning*, **43**, 272–290.

YANG, J., ADAMIC, L. A. AND ACKERMAN, M. S. (2008) Crowdsourcing and knowledge sharing: strategic user behavior on Taskcn. *Proceedings of the 9th ACM International Conference on Electronic Commerce*, Chicago, Illinois, USA. 246–255, http://www-personal.umich.edu/~ladamic/papers/taskcn/EC2008Witkey.pdf.

ZWASS, V. (2010) Co-creation: towards a taxonomy and an integrated research perspective. *International Journal of Electronic Commerce*, **15**(1), 11–48.

19

The role of information systems in innovative food and beverage organizations

J. G. Caudill, University of Tennessee, USA

DOI: 10.1533/9780857097248.4.343

Abstract: Concepts of information systems are introduced for use in innovative practices. Specifically, the role of technology in facilitating innovation and specific technologies that serve this end are explored. Supporting these roles are multiple technical tools for innovation, including information and communication technologies, computer-supported collaborative work environments, Web 2.0 contributions, technology supporting creativity, and the rising practice of crowdsourcing, all of which are explored. Some open source tools are introduced to support open innovation.

Key words: innovation, information systems, food and beverage, crowdsourcing.

19.1 Introduction

As the world's population continues to grow and agricultural activity is increasingly corporatized and centralized, competition in food production necessarily increases. Other industries of all kinds have found that in today's rapidly evolving markets innovation has moved from a source of competitive advantage to a competitive necessity. Wang *et al.* (2010; p. 447) state that '...high technology firm survival and competitive advantage rely upon R&D ability, and hence, innovation in extremely competitive environments'. This necessity is driven by several factors.

For many products the need for innovation is driven by an ever-shortening product life cycle. This can be seen by new product development teams focusing on reducing development cycle time as a primary goal (Mohammadjafari *et al.*,

2010). As consumers demand more frequent releases of new or different products, companies must create such products more quickly, thus shortening the time allowed for innovation and requiring more efficient innovative cycles. Although food and beverage products do not move as quickly as other industry sectors there is still a demand for new products and new versions of old products in order to stay visible and viable in the consumer market.

Adding to the complexity of shortened product life cycles is the globalization of the marketplace. Integrated supply chains increasingly push manufacturers of all types towards sole-sourcing decisions and lean-enterprise management encourages a culture of consolidation to streamline the supply chain. As opportunities to compete for both existing and emerging markets consolidate, there is increased pressure on competitors in the market to capture and retain relationships with others in the supply chain. Part of this pressure comes from the drive for continuous improvement, an effort requiring innovation in both products and processes.

Innovation alone is a substantial topic, and within a discussion of innovation, there are multiple subtopics, each worthy of dedicated study. Of particular interest are the mechanisms through which innovation occurs. Increasingly such mechanisms are based on utilizing the tools that can most efficiently promote innovative activities: technology. How technology is involved in innovative activity and examples of technology available for innovative processes are explored in this chapter.

19.2 The role of technology in innovation

One concept that is critical to any discussion of the role of information systems in innovation is that technology of any kind is simply a tool. Properly applied, tools can help skilled persons perform a job more efficiently, more quickly, or with greater complexity than a person working without the tool could accomplish. What the tool does not do is make the person operating it any more capable of performing the core of the work. In the case of innovation, technology does not make innovators more organized, more collaborative, or more creative. All of these things are certainly parts of an innovative process, and all of them can be enhanced by the proper application of information systems, but none can be present without the talents first residing in the individual who is performing the work of innovation. What tools such as the personal computer and Internet provide is more opportunity for more organizations to take an active role in innovative activities (Baldwin and Von Hippel, 2009).

As with any tool, using information systems requires that workers obtain applicable skills to capitalize on the advantages offered by the tool. Thus, the inclusion of new tools drives changes in the way people work, which connect to both skills and the policies which manage the workers. In a study on innovation and information technology Bartel *et al.* (2007; p. 1723) noted that, '…the adoption of new computer-based IT also increases the skill requirements of workers, notably technical skills, while also promoting the adoption of new human resource practices'. This is a theme that recurs; the pursuit of innovation in one area of an organization often drives innovative efforts in other areas. This is highlighted by the systems perspective of innovation; innovation is not a single idea, rather an innovation is built of many ideas, each leading to the next (Fagerberg, 2005). The organization shifts from an organization pursuing innovation to an innovative organization.

Information systems are excellent tools for many different types of work. Two, specifically, are communication and decision making. Communication has been a core function of computer technology for many years, most easily seen in the Internet's rise from the original DARPA project to today's 'always on' ubiquitous computing environment. Decision-making too has been a focus, and much of the academic discipline of management information systems is focused on this task, using information systems to optimize the outcomes of managerial decisions.

One of the more frequently seen components of information systems in innovation is the use of information and communication technology (ICT). Specifically, 'ICTs (information and communication technologies) foster a broad spectrum of innovation activities which involve the individual, organizational, industrial, and national levels of economic productivity' (Ho *et al.*, 2008, p 1). The important question to innovative organizations, however, is why ICTs do this.

At the core of ICT's importance to innovation is that innovative work is rarely a solitary activity. Because multiple people or multiple teams are involved in innovative projects their success depends on effectively and efficiently communicating with one another throughout the project life cycle. Robidoux and Andersen (2011) make the point that innovation is one of the benefits of a collaborative culture in an organization. Although many different technologies are discussed in relation to innovation the majority, at some level, contribute to communication and collaboration.

19.3 Innovative technologies in agriculture and food production

Some technologies benefit innovation across industries, but there are others that are uniquely suited to innovation in agriculture and food production. Other technologies are employed by multiple industries but have found unique uses for the food and beverage industry specifically. Some of these technologies include biosensors, simulation modeling, and networking and communications technology. Each of these tools plays a role individually, but they are also integrated as part of an overall system that benefits the innovative process.

A core component of any innovation process is a firm knowledge and understanding of both the current and desired state of the product or process in question. This particularly applies to the concept that a prime opportunity for innovation is often found in existing problems. Malins and Grant (2010; p. 182) offer that one definition of creativity is …'new ways of looking at problems' and that 'Innovation is the successful exploitation of new designs'. In agricultural and food production operations the leading technology in providing such information, in identifying possible problems, is the use of biosensors.

19.3.1 Biosensors

Luong *et al.* (2008; p.495) frame the scale and potential of the biosensor market as, 'The worldwide food production industry is worth about US$578 billion and the demand for biosensors to detect pathogens and pollutants in foodstuffs is expected to grow in the near future'. Biosensors can clearly provide safety data, but their value to innovative processes goes beyond this. Depending on the types of information the sensors deliver they may be critical to data generation for technology-integrated innovation.

There are two central themes of technology's role in innovation. One is communication, the other is information. Biosensors represent a significant potential to input information into the innovative process for agriculture and agricultural products. Since 1990, the EU–US Taskforce on Biotechnology Research has been pursuing multiple avenues of inquiry into the future of biotechnology, including the emerging field of animal and plant bioinformatics (Cichocka *et al.*, 2011). The term bioinformatics is in fact specifically defined as a process that utilizes computer science (Dictionary.com).

Biosensors are thus an early step in the development of innovative processes in that they generate data input to the decision system. Once such initial data has been collected it can then be analyzed, assessed, and utilized in the process of determining what might come next. Specifically, biosensors can create input for simulation models.

19.3.2 Simulation modeling

Simulation modeling is a popular tool in many disciplines, including both science and business applications. A simple but descriptive definition of a simulation is that it is, 'An imitation (on a computer) of a system as it progresses through time' (Robinson, 2004, p. 2). This imitation depends on timely and valid inputs of the system being modeled in order to create useful projections of what may happen to the system going forward. Taken a step further, adjusting certain inputs and observing the impact on the model outcome can be a highly useful tool in creating innovative approaches to systems.

Wang *et al.* (2007; p. 2) explain the value of high-quality inputs to a modeling system saying that, 'An integrated traceability system can improve process control and detect cause and effect when a product fails to conform to standards. It is also expected to improve operations planning so that raw material use is optimized'. Although this explanation may be applicable to almost any process it is of particular use in the food industry. Wang *et al.* (2007) specifically address the value of simulation to the food industry in that operational problems are often connected to raw material choice or batch size decisions. Such factors can be modeled and explored in simulations in order to optimize production processes.

19.3.3 Communication in innovation

As will be explored in more detail, innovation rarely occurs in a vacuum. Specifically, communication, and networks to support such communication, must be facilitated as part of the innovative process. Kuhne *et al.* (2010; p. 2) make the value of the network clear, explaining, 'According to the theory of "new economy", many scholars acknowledge the network in which a firm is embedded, as more important for the development and implementation of innovation than the firm itself'. They go even further, directly stating that the network is the location where innovation takes place.

This reliance on networks and communications is indicative of the increasingly complex nature of both science and business. The scale and scope of most operations today are simply beyond the ability of a single innovator to effectively manage. This complexity of innovation processes has driven the demand for networks of communication among many in the agrifood sector (Kuhne *et al.*, 2010).

Although these components of innovative processes are far from a comprehensive view of all the technology that goes into innovation they are indicative of the systematic process of innovation. The overall process can be expressed as three steps: input, processing, and output. Input, whether through biosensors or other technologies such as automated inventory tracking, point of sale systems, and customer feedback, all contributes to building a repository of data on the system. Data, however, are simply points of information. In order to utilize data they must first be transformed through processing into information.

The processing stage takes the collected data and applies analysis tools to generate information. In the example of simulation modeling, the data are used to build a picture of what the system looks like and what it is likely to look like in the future. Other simulations, such as weather patterns and market demand, may also be incorporated into a larger model that examines the influence of multiple forces on the product and the firm. Multidiscipline approaches, using inputs from multiple sources and specialties, have the potential to create a more realistic model of the firm's operating environment, which can lead to better informed, more accurate decision making processes.

Once data have been processed and information has been created then outputs must be shared for analysis and, ultimately, a decision. Although outputs may take many forms, the sharing of information, whatever its form, is an output process. The networking and communication technologies are most evident in this step.

Such activities are circular; decisions that are made as a part of the output step of the process ultimately serve as inputs to the next cycle of innovative activities. Having established a framework of innovation and defined some of the technologies for innovation, the next step is to understand some of the specifics for how technology serves the innovative process.

19.4 Technology's support of innovation

Technology is playing an ever-increasing role in almost every activity today. Far from being an exception, the work environment has, over time, actually led the adoption of technologies such as the personal computer, the Internet, mobile phones, and smart phones. The reason that business so often leads the way in the adoption of consumer electronic devices is that the technology is good for business. Efficiency, effectiveness and, ultimately, the bottom line for the firm are enhanced by the appropriate application of technology tools.

Competition goes through stages. As economies and industries mature, different efforts rise to be the most important competitive advantage. Over time these competitive advantages are adopted by almost everyone in the market and they shift into being competitive necessities. From the perspective of communications technology, society has been increasingly changed, and more and more people in more and more remote areas have gained access to information, as technology has moved from the printed word to radio and television to the Internet (Schneiderman, 2000). As just one example, in the mid-1990s it was a competitive advantage for a company to have a website for their product. Move forward just ten years and companies had to have a website because all of their competitors did and doing business without one put the firm at a competitive disadvantage.

Over time there have been many different areas of focus for competition. At the time of this writing one of the leading issues is innovation. Excelling in innovative products and services can provide an organization with a competitive advantage. Innovation alone, however, is for most firms a competitive necessity. Combining this necessity with the complexity of modern organizations demands that technology tools are applied to create innovation.

Kuhne *et al.* (2010; p. 2) define innovation as, 'the ongoing process of learning, searching and exploring which results in new products, new processes, new markets and new forms of organization'. To make this happen a range of tools can be applied. ICT, computer supported collaborative work (CSCW), Web 2.0 tools, technology as a creative medium, and most recently the practice of crowdsourcing are all parts of the modern innovative organization. Each of these different concepts are described and their relation to innovative processes is explored in the following subsections.

19.4.1 Information and communication technologies (ICT)

As introduced earlier, ICT is, at its core, the sharing of information among innovators to enhance the innovative process. What ICTs actually do in an organization fall into five different categories: understanding idea sources, documenting ideas and sources, distribution and sharing of ideas for cross-application, idea design, testing, and refinement, and idea commercialization (Awazu *et al.*, 2009). Each of these five functions supports the three stages of the innovative cycle.

Input is the first step of the innovation cycle. Understanding idea sources and documenting ideas and sources are both components of this first step. In order to be useful to an organization, an idea must first be captured and understood. Knowing where the idea came from is an essential component of this process. The original source of the idea can then be used to validate the concept or to return to the source for clarification or additional discussion. Documenting the details of the idea and its source ensures that the knowledge, once created, is not lost.

Processing is the second step of the innovation cycle. Three of the components of ICT fall into this step. Idea design, testing, and refinement are all part of processing an innovative concept. Gathering ideas is a process of collecting initial concepts. Taking those concepts and transforming them into a workable new product or process is designing an idea. Once the design has been completed then it must be tested to determine its viability and suitability. Based on the results of this testing the idea can be further refined in order to optimize its value to the firm. Often the process of refinement generates a new idea that resets the cycle, and processing of the new idea occurs. Only when an idea has been processed sufficiently to reach a point where additional refinement is either impossible or impractical does the cycle move to the final stage.

The final stage of the innovation cycle is output. For an organization engaged in commerce the commercialization of an idea is the ultimate output of the innovative cycle. This involves taking the fully developed, tested, and refined idea and moving it into the marketspace for consumer use. Again, this may not be the end of the cycle as consumers will ideally offer feedback that can be accepted as new idea inputs thus leading to a new cycle of innovation.

Clearly the use of ICTs involves taking advantage of technology to communicate. Each of the roles described involves different people in the innovation cycle creating, storing, or accessing information about the idea. This is a relatively abstract definition, but technology's role in communication during innovation can be better seen through the concept of CSCW.

19.4.2 Computer-supported collaborative work (CSCW)

CSCW is founded on the concept that technology is flexible enough to not only support collaboration during the innovation cycle but also to support different individuals in the way that best suits their needs (Herrmann, 2008). As with ICT, CSCW also has five different components, each a different form of a technology tool: supporting the large picture – visualization of rich material, malleability of shared material and stimulation of variations, support of convergence within evolutionary documentation, smooth transitions between modes of creative collaboration, and integration of communication with work on shared material (Herrmann, 2008). Each of these five roles support the innovative process and more clearly focus on the collaborative aspect of innovation.

Although CSCW is most clearly applicable to the processing stage of the innovation cycle, it applies across the cycle as a whole. Visualization of rich material, although applicable at any level, can certainly apply to the input stage when ideas are first being generated. Moving forward to processing all of CSCW is directly applicable. Visualizing rich material is a key component of idea design; designing new ideas demands considering the idea's interaction with the surrounding environment.

Shared material and stimulation of variations along with the support of convergence with evolutionary documentation are part of idea design but also of testing and refinement. Testing and refinement are evolutionary processes that rely on being able to vary ideas from their original states. Transitioning between modes of creative collaboration also supports the cycle as a whole as each stage demands a different intellectual approach to satisfy the tasks at hand. With the work changing in each stage creative collaboration must shift to address each responsibility. Finally, integrating communication with work on shared material also supports every stage of the cycle, as everyone involved in the innovative effort must communicate with and support one another throughout the effort.

19.4.3 Web 2.0 and the creative process

Although communication is critical, it is not the only contribution to the innovative process. One of the key purposes of communication among innovators is to facilitate the creative process. This process is enhanced through the proper application of Web 2.0 technologies and can also be clearly seen in the use of technology as a creative medium.

In discussing communication as a part of innovation what is really happening is that users are creating the content that is being used. Wirtz *et al.* (2010) define four factors that compose Web 2.0 interactivity: social networking, interaction orientation, personalization/customization, and user-added value. Comparing these Web 2.0 characteristics to what occurs in ICT and CSCW, there is a clear fit among the different approaches. Through Web 2.0 innovators have the opportunity to document their ideas and their findings, add value to the discussion, and tailor the environment to their own work styles, all while effectively documenting the work in which they are participating.

Shifting perspectives from inside the firm to outside Web 2.0 also offers promising opportunities for consumers to participate in the innovative process. This will play a key role in crowdsourcing, to be explored later. Culturally, consumers have shifted from consuming information to creating information (Fischer, 2009). With consumers contributing their own thoughts to the innovative process they expand the capacity of the firm to take part in innovation. This can be explained as, 'End-user development is an essential component of this transformation, but its impact is much broader: this transformation represents a change and new opportunity for social production, for mass collaboration, for civic and political life, and for education' (Fischer, 2009, p. 4).

Furthermore, the focus of this technology is on facilitating communication. The difference with Web 2.0 is that outside players are included in the discussion. If innovation is seen as a creative process then the inclusion of more ideas is a helpful addition. In discussing creativity as a component of innovation Leonard and Swap (1999; p. 4) defined innovation as, '...the embodiment, combination, and/or synthesis of knowledge in novel, relevant, valued new products, processes, or services'. If creativity is a critical component of innovation, how then does technology support creativity?

Just as technology does not make decisions, but helps to support decisio-making by providing tools that enhance the process, technology also does not generate creativity. Technology does, however, offer helpful tools that enhance the creative process. Shneiderman (2007; p. 2) explains that, 'Creativity support tools extend users' capability to make discoveries or inventions from early stages of gathering information, hypothesis generation, and initial production, through the later stages of refinement, validation, and dissemination'. As with many tools, technology that supports creativity has more than one role. The two things that technology can do to support creative work are specific tasks that support discovery and the capacity to generate multiple alternatives (Shneiderman, 2007).

This again ties into the basic innovation cycle. Discovery can include understanding idea sources and idea design. The process of generating multiple alternatives can also play a role in idea design as well as being a factor in idea refinement. The process as a whole is entirely creative, communication of creative ideas is necessary for success, and enhanced creativity provides additional opportunity for product and process innovation.

19.4.4 Crowdsourcing

For many organizations an opportunity to be more creative comes from bringing more people into the creative process. Recently, there has been an emerging practice called crowdsourcing. Multiple sources cite Howe (2006) as coining the term and first defining it:

> Simply defined, crowdsourcing represents the act of a company or institution taking a function once performed by employees and outsourcing it to an undefined (and generally large) network of people in the form of an open call. This can take the form of peer-production (when the job is performed collaboratively), but is also often undertaken by sole individuals. The crucial prerequisite is the use of the open call format and the large network of potential laborers.

Schenk and Guittard (2010; p. 2) provide a simpler but no less accurate definition, as, 'Crowdsourcing means outsourcing to the crowd'. Crowdsourcing as a practice is focused on generating real solutions for real problems and successful contributors are paid real money. 'Crowdsourcing is mainly seen as a business solution and an alternative form of outsourcing' (Geiger *et al.*, 2011, p. 4).

In crowdsourcing a firm, or a designated agent for the firm, puts a request online for a community of experts or consumers to answer a question. Sometimes this is just an opinion on a design or an image, and sometimes it is a complex industrial problem. Regardless of what the problem is, the details are shared and individuals or groups from geographically diverse areas provide input or work to develop a solution. Feedback is delivered to the firm and the best information is used to benefit the process.

To date, crowdsourcing may be best recognized as a marketing tool, a way for consumers to tell sellers about their preferences. Threadless.com allows members to join online, submit shirt designs, and rate designs on a scale of zero to 5. The highest scoring designs are selected for production and provide frequent new products to the customers (Brabham, 2008). Although this may be what many think about in relation to the term crowdsourcing there is potential for it to be applied to complex industrial and scientific problems just as effectively. As just one example InnoCentive.com utilizes crowdsourcing to develop solutions to scientific problems by posting challenges to the crowd and accepts suggested solutions. Winning solutions receive cash prizes from the company (Brabham, 2008).

In crowdsourcing, firms are using the same focus on creativity and communication that is important to all innovative efforts, but applying those concepts on a larger scale. Instead of just co-ordinating multiple individuals or multiple groups within the firm, or across partner firms, crowdsourcing leverages the power of the Internet to reach a potentially worldwide group of interested contributors to maximize the number of people engaged in the problem solving exercise. Often this advice is not free, but only winning or successful contributions are compensated, which provides exceptional value to the firms.

19.5 Free tools for innovation

Technology for innovative processes does not have to be expensive. Many free and open source solutions are available to enhance the innovative process for companies. Virtually all of the popular Web 2.0 tools are provided online at no cost to the user. If information security is a concern then there is a cost associated with hosting the tools internally, but the programs themselves, or highly similar substitutes, are often available at no cost.

At a higher level of complexity there are many open source solutions available for simulation software. One such program, Xholon, available at www.primordion.com/Xholon/, is described as an 'open source tool for multi-paradigm modeling, simulation, design, execution, and transformation'. There are other open-source simulators such as OpenSim (www.opensimulator.org) and Open CASCADE (www.opencascade.org) that provide three-dimensional modeling and virtual environment simulators.

There are also some applications designed specifically to model and manage agricultural operations. CuteFarm is an open source application designed to manage farm and agriculture resources. IRRISTAT is focused on data management and analysis of experiment data. These are just some examples of the resources that are available and new ones are constantly being developed. The important point here is that these are technical tools to aid in innovation that do not require a financial investment to acquire. This means that small and medium enterprises can have access to the technology that they need to effectively function as innovative organizations.

19.6 Future trends

Future trends in technology as a component of innovation are likely to be centered on technology trends as a whole. Moore's law, the concept that computing power at a given price doubles every eighteen months, is often being surpassed today in many different measures of computer performance. One of the most visible markets in which this is occurring is the miniaturization and mobility of electronic devices.

As more powerful technology becomes more portable and better connected over faster networks true ubiquitous computing will become a reality. This deep penetration of connectivity into individuals' lives will undoubtedly impact work and, by extension, innovation. Communication will continue to be easier and faster and information will be more accessible. This may serve to further increase the speed of innovation cycles with a proportional decrease of product life cycles. One of the major trends is likely to be more speed.

The price of this speed and connectivity is the additional risk to security. Already information security is one of the fastest-growing disciplines in information systems and there are no signs of that changing in the near future. Decreasing the length of product life cycles necessarily increases competitive pressures and that, in turn, increases the value of knowing what competitors are doing. With much of the work being performed through digital mediums, security becomes a vital component of successful innovative projects.

Both procedures and technologies must evolve as future security concerns threaten operations. Hardware, software, and network security measures will all become more stringent and must become more effective, particularly in relation to wireless communications and access to mobile devices. Increasing security traditionally has meant decreasing ease of use, but future devices will probably incorporate fully developed and easy-to-use biometric security features that will both provide excellent security and also be nearly invisible to the user.

19.7 Conclusion

Innovation is a highly complex and cross-functional discipline. This complexity, however, is more than equaled by its value to an organization. In order to be successful in the modern marketspace organizations must aggressively pursue innovative practices.

There are many different perspectives on innovative practice, each as important to the success of the overall system as the others. The goal of this chapter has been to focus on building a helpful toolkit of technology resources that enhances and improves innovative efforts. The available tools will change quickly, and it is likely that, from the time of this writing to the publication of the book, new software applications or different versions of existing applications will be released. The tools themselves, and their versions, are of little importance.

What is important is that technology is adopted with a focus on the desired outcome of the project. The strategy of the organization should dictate what kinds of technology are utilized; the availability of technology should not drive strategy. Communication and creativity are critical components for the success of the innovation cycle. Technology provides excellent tools to make these processes work better. Innovators should use what works and allow the tools to provide the help they need to excel.

19.8 Sources of further information and advice

Innovation is a growing topic in many fields and much of the discussion centers on the use of information systems as a component of innovative practices. There are many excellent books available on the topic and readers are encouraged to explore specific topics of interest. A recent work that focuses on innovation as a whole with chapters dedicated to information systems and innovation is *Technological, managerial and organizational core competencies: dynamic innovation and sustainable advantage* by Nobre, Walker, and Harris from IGI Global.

D'Atri, Ferrara, and George released a full book on the interaction of technology and innovation through Springer titled *Information technology and innovation trends in organizations*. For researchers interested in focusing on smaller organizations, *Information systems – creativity and innovation in small and medium-sized enterprises* by Dhillon, Stahl, and Baskerville, also through Springer, is of interest.

In addition to books on the subject, there are also journals dedicated to innovation. Some peer-reviewed journal options include *The Innovation Journal*, *Economics of Innovation and New Technology*, *International Journal of Innovation Science*, and the *International Journal of Innovative Technology and Creative Engineering*. Resources from a wide variety of sources are numerous and growing. In a field as dynamic as innovation new sources and new information quickly becomes available and readers are encouraged to use the references provided here as a starting point, but to explore broadly to discover what is most recent and most helpful.

19.9 References

AWAZU, Y., BALOH, P., DESOUZA, K., WECHT, C., KIM, J. AND JHA, S. (2009). Information-communication technologies open up innovation. *Research Technology Management*, **52**(1), 51–58.

BALDWIN, C. AND VON HIPPEL, E. (2009). Modeling a paradigm shift: from producer innovation to user and open collaborative innovation. *Harvard Business School Finance Working Paper No. 10–038*.

BARTEL, A., ICHNIOWSKI, C. AND SHAW, K. (2007). How does information technology really affect productivity? Plant-level comparisons of product innovation process improvement and worker skills. *Quarterly Journal of Economics*, **122**(4), 1721–1758

BRABHAM, D. (2008). Crowdsourcing as a model for problem solving: an introduction and cases. *International Journal of Research Into New Media Technologies*, **14**(1), 75–90.

CICHOCKA, D., CLAXTON, J., ECONOMIDIS, I., HOGEL, J., VENTURI, P. AND AGUILAR, A. (2011). European Union research and innovation perspectives on biotechnology. *Journal of Biotechnology*, doi:10.1016/j.jbiotec.2011.06.032.

FAGERBERG, J. (2005). Innovation: a guide to the literature. In: Fagerberg, J., D. Mowery and R. Nelson (Eds), *The Oxford handbook of innovation*, Oxford University Press, Oxford, pp. 1–26.

FISCHER, G. (2009). End-user development and meta-design: foundations for cultures of participation. In *Proceedings of the second international symposium on end user development*, Springer, Berlin, pp. 3–14.

GEIGER, D., SEEDORF, S., SCHULZE, T., NICKERSON, R. AND SCHADER, M. (2011). Managing the crowd: towards a taxonomy of crowdsourcing processes. *Proceedings of the Seventeenth Americas Conference on Information Systems,* (AMCIS), 4–7 August, Detroit, MI.

HERRMANN, T. (2008). Design issues for supporting collaborative creativity. *Proceedings of the 8th International conference on the design of cooperative systems*, pp. 179–192.

HO, S., KAUFFMAN, R. AND LIANG, T. (2008). A growth-theoretic empirical analysis of simultaneity in cross-national ecommerce development. *Proceedings of the 41st annual Hawaii international conference on system sciences* (HICSS 2008), Washington, DC, IEEE Computer Society.

HOWE, J. (2006). The rise of crowdsourcing. *Wired*, 14.6.2006.

KUHNE, B., GELLYNCK, X. AND WEAVER, R. (2010). Network connections and innovation capacity in traditional agrifood chains. *Proceedings for the 116th EAAE seminar 'spatial dynamics in agri-food systems: implications for sustainability and consumer welfare'*, October 27–30, Parma, Italy, European Association of Agrcultural Economists.

LEONARD, F. AND SWAP, W. (1999). *When sparks fly: igniting creativity in groups*. Boston, MA: Harvard Business School Press.

LUONG, J., MALE, K. AND GLENNON, J. (2008). Biosensor technology: technology push versus market pull. *Biotechnology Advances*, **26**(5), 492–500.

MALINS, J. AND GRANT, E. (2010). Designing creativity tools to support business innovation. In: B.T. Christensen, S. Boztype, and T. Kristensen, Eds. *Proceedings of the DESIRE '10 conference: creativity and innovation in design*, 181–188.

MOHAMMADJAFARI, M., AHMED, S., DAWAL, S. AND ZAYANDEHROODI, H. (2010). The effects of electronic collaboration in reducing production time: product design process in SMEs. *Proceedings of the international multiconference of engineers and computer scientists*, Vol. 3.

ROBIDOUX, C. AND ANDERSEN, R. (2011). Cultivating a culture of collaboration. *Society for Technical Communication, 58th Annual Conference*, pp. 90–98.

ROBINSON, S. (2004). *Simulation: the practice of model development and use*. John Wiley & Sons, Ltd. West Sussex, England.

SCHENK, E. AND GUITTARD, C. (2010). Towards a characterization of crowdsourcing practices. *Journal of Innovation Economics*, **1**(7), 93–107.

SCHNEIDERMAN, B. (2000). Creating creativity: user interfaces for supporting innovation. *ACM Transactions on Computer-Human Interaction*. 7(1). 114–138.

SHNEIDERMAN, B. (2007). Creativity support tools: accelerating discovery and innovation. *Communications of the ACM*, **50**(12), 20–32.

WANG, X., LI, D. AND O'BRIEN, C. (2007). Optimisation of traceability and operations planning: an integrated model for perishable food production. *International Journal of Production Research*, **47**(11), 2865–2886.

WANG, C., CHIN, Y. AND TZENG, G. (2010). Mining the R&D innovation performance processes for high-tech firms based on rough set theory. *Technovation*, **30**, 447–458.

WIRTZ, B., SCHILKE, O. AND ULLRICH, S. (2010). Strategic development of business models: implications of the Web 2.0 for creating value on the Internet. *Long Range Planning*, **43**(2/3), 272–290, doi: 10.1016/j.lrp.2010.01.005.

20

Effective organizational and managerial company frameworks to support open innovation: overview and the case of Heinz

V. Lazzarotti and R. Manzini, Cattaneo University – LIUC, Italy

DOI: 10.1533/9780857097248.4.356

Abstract: Empirical evidence is presented on how openness in an organization can be managed and supported, by analysing the experience of Heinz, a company that embraced the philosophy of open innovation about three years ago and started a contextual process of organizational change. The human and organizational factors and managerial tools are discussed along with other elements that define an enabling environment for open innovation.

Key words: open innovation, collaboration, organizational change, business management.

20.1 Introduction

Studies on innovation and technology management promote the view that the successful implementation of an open approach to innovation is challenging, whatever are its prevailing goals. Recent studies (Bérézai, 2010; Chiaroni *et al.*, 2010; Goers, 2009; Jiménez-Barrionuevo *et al.*, 2011) give evidence that a favourable social-organizational context is needed to allow the satisfactory deploying of open innovation: the company's capabilities, human and organizational factors, and several managerial tools are mentioned among the elements that define such an enabling environment.

This chapter aims to enrich the empirical evidence on how this environment can be built and how openness can be managed and supported. The chapter is organized in three sections as follows. Section 20.2 reviews relevant published studies on the topic to find into evidence for the need for organizational and management tools to support open innovation, with particular attention to the best known experiences in the food and drink industry. Section 20.3 describes the specific experience of Heinz that embraced the philosophy of open innovation about three years ago and started a contextual process of organizational change. In particular, a specific unit dedicated to open-innovation organization and management was created. The unit tasks are described in detail in order to understand how Heinz has been able to successfully support the opening of its innovation process. Section 20.4 provides some recommendations for managers that have to face similar situations of change and transition towards open innovation.

20.2 The need for organizational and management tools to support open innovation

There is the fundamental assumption that open approaches cannot be successfully implemented without an enabling organizational context supported by appropriate management tools designed for the purpose. Indeed, this assumption is widely held within the innovation and technology management literature, as seen in various interesting studies, some of them significant for the food industry (e.g. Bilgram *et al.*, 2008; Birkinshaw and Gibson, 2004; Cassiman and Veugelers, 2006; Chesbrough *et al.*, 2006; Dahlander and Gann, 2010; Dodgson *et al.*, 2006; Huston and Sakkab, 2006; Pisano and Verganti, 2008; Raasch *et al.*, 2008; Sakkab, 2002). In other words, these studies stress the importance of the 'right conditions' (in terms of collaboration capabilities, organizational factors and managerial tools) to make any open approach successful. In this sense the experience of Procter & Gamble (P&G) is paradigmatic. Sakkab (2002) and Huston and Sakkab (2006) describe the strategic planning process that characterizes Procter & Gamble's open-innovation approach (i.e. the connect and develop program). This highly structured program allowed P&G's innovation approach to be completely changed. On one hand, an extraordinary effort was conducted to build a global network by building technological platforms to harness ideas and innovations (i.e. P&G's proprietary networks of: 'technology entrepreneurs', senior employees working worldwide in order to identify promising product ideas and technologies; 'supplier' IT platforms to combine ideas from internal R&D and the company's top 15 suppliers; or open networks such as: NineSigma and InnoCentive to connect P&G to a wide audience of companies, universities, government and private laboratories and consultants). On the other hand, managerial actions 'to push culture' were introduced through a system of incentives to make sure that the best ideas, wherever they come from (inside or outside), came to the surface; the employees were not afraid that the new approach to innovation might eliminate jobs or that P&G would lose capabilities. In other words, P&G aimed to overcome the 'not invented here' syndrome. Instead, effectively managing externally acquired technologies seems to require just the development of complementary internal capabilities as it is largely recognized in the innovation and technology management literature: internal research is a prerequisite for being perceived as an attractive partner and for absorbing external knowledge (i.e. the concept of absorptive capacity by Cohen and Levinthal, 1990, more recently reviewed by Zahra and George, 2002).

Thus, the main message from the P&G experience is that 'open innovation' is far more complicated than 'the more openness, the better' (Dahlander and Gann, 2010). The change of innovation approach, from closed to open, requires a structured approach that involves not only tools to identify technological trends, but also organizational mechanisms to manage people and culture (Bérézai, 2010). The 'want find get manage' (WFGM) approach, widely applied also in the food industry (Garcia Martinez, 2011) clearly proves this fact. It is not within our scope to analyze this approach here in detail, even though it is important to note that, by observing the experience of companies that applied such framework, organizational interventions (i.e. through the creation of specific units and roles) emerge as key factors to successfully implement open innovation. Already highlighted in the general contribution by Chesbrough and Crowther (2006), the importance of specific organizational roles for enabling the implementation of open innovation in the food and drink industry is demonstrated by the experiences of General Mills, Kraft, PepsiCo and Mars (Goers, 2009; Bérézai, 2010; Garcia Martinez and Briz, 2000; Garcia Martinez, 2011). In particular, the creation of cross-functional teams seems to strongly contribute to the establishment of effective relationships with the outside world by allowing also the overcoming of 'the not invented here syndrome'. In practice, interfunctional teams are characterized by an extensive communication for brainstorming, project-related information and feedback between the different participants. Therefore, interorganizational projects can directly build the network by establishing interdependencies, frequent interactions, and information flows among participants. Moreover, through the establishment of collaborative projects, employees have an incentive to share their knowledge and ideas, because it contributes to collective performance, and managers also signal that collaboration is important. Therefore, the presence of interorganizational project teams promotes an organizational climate more inclined to collaborations with third parties (Jolink, 2009). A clear example of such organizational interventions is still provided by the P&G experience with the creation of the Global Technology Council, made up of its business technology directors, corporate heads and key geographical R&D leaders, to represent all the of company's competencies. The main task of the Global Technology Council is to explore how to leverage technologies: it serves as an incubator for exploratory research and early stage product development. It is a global management tool, because it can see almost every new-to-the-world product P&G has in its pipeline and encourages collaboration within and outside the company.

Very interesting is also the General Mills's experience, which uses a variety of activities to support open innovation and whose goal is the establishment of an effective linkage between the internal teams and the external actors:

- the external partner development group, whose purpose is the creation of creative business models and partnering relationships, so that both General Mills and its partners achieve a certain value from the collaboration;
- the external supported team, which is a cross-functional team that, every two weeks, reviews the projects openly, shares insights and ensures that appropriate communication takes place between the different actors involved;
- a connected innovation team, which develops new tools and methodologies and shares a substantial amount of this information through publications and speaking engagements. The team also acts as a catalyst to inspire divisional resources.

These programs undoubtedly render the identification and realization of an innovative project more effective, and the presence of cross-functional teams is a key point, enabling high levels of communication and sharing the heterogeneous competences of its members, coming from different business functions within and outside the company. Moreover, the presence of cross-functional teams helps to break down the barriers between the company and the outside world, increasing the likelihood of collaborations with third parties (Bérézai, 2010).

Undoubtedly, the problems of the 'not invented here syndrome' and the connected fear of losing intellectual property (IP) are highly publicized management challenges regarding the implementation of an open-innovation model (Bérézai, 2010). The root of the problem lies in the organizational culture, which can be hard to overcome if the top management's support for open innovation is not integral and not translated into well-designed managerial tools. Further good evidence of an effective model to manage collaborations in an open context is provided by the SiW (sharing is winning) model, adopted by Nestlè (Traitler *et al.*, 2011). Driven by the goal of value creation, Nestlè takes care to clearly define the mutual benefits among the partners and to establish specific performance targets for the starting collaborations. To achieve this, the managerial tool used is the master joint development agreement (MJDA). The MJDA outlines the scope of the expected collaboration, firstly in general terms, and, in a second more detailed part, it delineates the projects, the expectations, the employed resources, the timelines and the IP issues. The MJDA seems to be a poweful tool for speeding up the opening process, allowing partners to readily discover the opportunities of the collaboration and, at the same time, protecting themselves from risks (Traitler *et al.*, 2011).

In summary, this section is intended to highlight the importance of organizational and managerial support of open innovation. It is clear, however, that the experiences quoted here do not allow an in-depth grasp of the scope of the interventions. Moreover, authoritative authors state that there is a need for further enrichment of this topic, especially in terms of empirical research (Lichtenthaler and Lichtenthaler, 2009). The next section describes in detail the experience of Heinz towards the adoption of an open-innovation approach. After some information about strategy, the innovation process and its openness in Heinz, the focus is on the tools developed by the company to organize and manage open innovation.

20.3 Case study: Heinz's strategy, business and organization

Heinz's mission is to become the premier company in the food sector and offer delicious food throughout the world. To establish itself as a leading food company, Heinz must offer the best in terms of value for customers, employees and skill-s. Heinz's mission is supported by significant values that are included in the model code PREMIER:

- passion: the passion for victory and for their brands, products and people, which ensures a higher value for its shareholders.
- risk tolerance: to create a philosophy designed to encourage and reward initiative and willingness to take risks with prudence.
- excellence: to always provide the best quality and to keep the leader position at any time.
- motivation: to celebrate success and recognize and reward the achievements of individuals.
- innovation: to be innovative in everything the company does, from products to processes.
- empowerment: to enable talented employees to take the initiative to do the right thing.
- respect: to act with the utmost integrity and respect for all.

20.3.1 Food quality and safety

Quality standards and safety are core to the mission of the company. They have driven Heinz's growth since 1869, when the founder, Henry John Heinz, began selling his first product, horseradish, in clear glass bottles, so consumers could see Heinz's advantage over the competition through the transparency of the pack. Today, the robust quality assurance principles, procedures, standards and technology help maintain Heinz's reputation as one of the most trusted brands in the world.

In 2000, Heinz instituted an online global database that allows every raw material and ingredient that are used in food products to be tracked and traced. The database, which is continuously updated and operates 24 hours a day, seven days a week, enables the employees to maintain critical information on the qualifications of suppliers and share the guidelines on ethical trading with them.

The quality system tracks food safety requirements, records required data reported by Heinz suppliers, and enables them to monitor the quality and safety performance of factories and ingredients.

20.3.2 Relationship with suppliers

The suppliers are carefully screened and selected by Heinz through the vendor management process. The company expects suppliers to uphold its supplier guiding principles and global operating principles. If it is discovered a supplier is found not to be following the company's guidelines, the supplier must take immediate corrective action, or Heinz will discontinue the relationship. Another component of the vendor management program is continuous improvement. The global vendor teams, along with research and development (R&D), closely work with suppliers to monitor and improve their compliance, process control and delivery systems.

20.3.3 Research and development (R&D)

The research and development teams around the globe are renowned for their food and packaging innovation. There is a widespread belief that the greatest insight into research and development comes from people who use Heinz foods every day. By listening to what the consumer requests, and responding with fresh, new ideas that deliver the traditional Heinz great taste, healthy goodness and packaging convenience, Heinz is able to maintain a leadership position.

The company has a R&D centre worldwide, the Global Innovation and Quality Center, which provides technical direction, assistance and advice to business units around the world. Inside, there are more than 200 chefs, food technologists, researchers and packaging designers, experts in nutrition and quality control. The center covers 100 000 square metres with:

- research kitchens, where they develop new varieties of all the brands of Heinz;
- a feedback centre, where they connect with consumers in person and get their valuable input;
- state-of-the-art test kitchens, where their latest innovations top the menu for product sampling and taste tests;
- a simulated retail environment that helps them enhance their marketing prowess and observe consumer behaviour; and
- a corporate library that houses the extensive history and expertise of its 140-year-old company.

The structure makes it possible to take advantage of Heinz's most innovative thinking from all over the world to continually develop innovative products with associated packaging innovations.

> The Innovation and Quality Center you see today is testament to our resolve to utilize innovation to spur growth. Our global innovation teams are energized, like never before, with a strong focus on delivering new products, new packages, better nutrition, better taste and consumer value. Mr. R. Johnson, CEO (Business Wire, 19 September 2005).

20.3.4 Innovation

> Constant attention to quality and excellence still guides our mission and

> our focus on innovation. Heinz aims to do what we do best: a mixture of product excellence and modern innovation (Johnson, 2005).

You may notice, as evidenced by the official website of Heinz (www.heinz.com/our-foodinnovation), three types of innovation:

- nutrition innovation;
- agricultural innovation;
- packaging innovation.

The first type of innovation concerns the implementation of the product in the strict sense. Most product innovations in recent years have developed thanks to the opening of the huge Heinz research center worldwide, the 'global innovation and quality center'. A commitment to nutrition innovation also includes new ways to connect, listen and learn from consumers. Engaging with the people who use the products every day Heinz has always been at the heart of the definition of business innovation. The 'agricultural innovation' aims to maintain high standards of purity and nutrition, continuing to seek solutions for more sustainable farming. This type of innovation is developed especially by HeinzSeed, owner of the tomato seed that is used throughout the range of products. Today HeinzSeed provides more than six billion seeds every year and produces all natural hybrid tomato varieties with high performance. The tomato product is called a 'pure innovation' as the result of traditional farming techniques, without genetic modification, but with high performances, excellent durability, appreciable taste, resistance to diseases and reduction of pesticide use. Heinz uses more than 2.5 million tons of tomatoes per year, making its tomatoes a primary ingredient in the world.

The program HeinzSeed is a great example for the company's commitment in this area, because it allows the company to control product quality, to improve the farmers' quality of life in developing countries and to minimize the impact on land as well as to improve the quality of food.

With regard to the 'packaging innovation', we can consider that each package is the result of intensive research. Ensuring that the products retain their characteristics when consumed is only one of the objectives of Heinz: it is equally important to make them easy to use, easy to store and easy to recycle. An example of this is the product 'Heinz Dip & Squeeze', which marks the first remake of ketchup package for the foodservice industry in 42 years. Other examples of innovative products in this respect are the bottles of ketchup Top-Down and Frigo Door Fit and baked beans that can be inserted directly in the microwave, Snap Pots.

A recent packaging innovation is given by the easy opening of Heinz soups, packaging innovation for both the consumer and the environment. The cans of reduced weight also minimize the overall weight for their transport, with fuel economy and improvement of the overall efficiency. As a global company, Heinz is very aware of the impact it has on the environment.

20.3.5 Heinz's open-innovation model

Innovation is clearly part of the Heinz vision and values:

> Innovation: We spot consumer and customer needs and meet them with simple, creative solutions (Johnson, 2005).

Innovation at Heinz is pursued through technology platforms, which in Heinz are intended as 'technological fields of activity', in which tools, instruments and solutions are studied and developed with the aim of achieving the goals of the company's innovation strategy. In order to be a technological platform, the scope and objectives should not be too limited (for example, not referred to a single product) nor too large (for example, not concerning too many scientific disciplines at the same time). For example, a technology platform has been created for reducing the content of salt and sugar of Heinz food products.

Heinz embraced the philosophy of open innovation five years ago, because in all technology platforms some external actor is involved. However, this does not mean that the problems and risks of an open-innovation approach are underestimated. In particular, the risk of a high 'dispersion' of innovation activities in too many fields of activity may lead to a loss of focus on the company's innovation strategic objectives. In recognition of this awareness, the Heinz approach to open innovation involves a clear and explicit definition of the aim and scope of each partnership to ensure agreement with the company's strategy. In other words, Heinz defines the technological fields and areas in which the partners should work as well as the expected value of their contribution and proposals. In order to actually implement this kind of approach, Heinz involves external partners only in relation to specific technology platforms that are defined to be consistent with the company's innovation strategy. The Heinz board is well aware that this may limit the potential for open innovation, in terms of innovative contribution and creativity, however, this approach would ensure the required compliance with the company's strategy.

For each technology platform, the gaps to be filled, in terms of tools, instruments and solutions, are identified. Then, a scouting activity is started to search for partners able to give a contribution to fill that gap, i.e. partners that possess the appropriate competences and technologies to work on the identified technology need. Once the potential partners are identified, a process begins aimed at better understanding the partners and to achieve reciprocal trust. This process goes through sharing with them the company's mission and vision, explaining the innovation strategy, and clarifying the innovation goals. Only if a 'good feeling' is finally found with the partners, if they are explicitly interested in contributing to the company's needs, if trust is actually established and if mutual benefits are clear for all the actors involved, then can the collaboration actually begin. At this point, with the guarantee of mutual trust, the partners can work without strict control, with a management-by-objective approach. Tasks are assigned to the partners, but with a quite large scope and without definition of strictly specified methodologies. This would maximize the innovative contribution and the potential of the partnership, and would allow Heinz to exploit all inputs, suggestions and triggers coming from the partners. In other words, within the aim and scope of the partnership defined by Heinz, partners are left free to choose the organization and research method most suitable to achieve the goals and to maximize the benefits of the collaboration.

Collaborations described so far, i.e. concerning the research and development of innovative solutions and tools for technology platforms, mainly involve universities and research centres as partners. Thus, open innovation mainly concerns idea generation, exploration and experimentation of innovative solutions. In some cases, however, even partners coming from other sectors of activities are involved, such as, in particular, chemical and pharmaceutical companies. These first types of collaborations, called 'intellectual partnerships', concerning the early phases of the innovation funnel, are probably the most relevant for Heinz, because they are strictly connected to the strategic innovation goals.

However, other collaborations concern the following part of the innovation funnel, i.e. full development and industrialization of innovative ideas. In these cases, partnerships mainly involve suppliers, especially of machinery, tasks are much more precise, and the management of the partnerships is aimed at going straight to the market, at the right time and with the specified quality and cost profile.

Finally, the role of competitors in the Heinz open-innovation model should be mentioned. Competitors are kept far from the innovation process because the risk of spill over and imitation is perceived to be very high. However, there is some

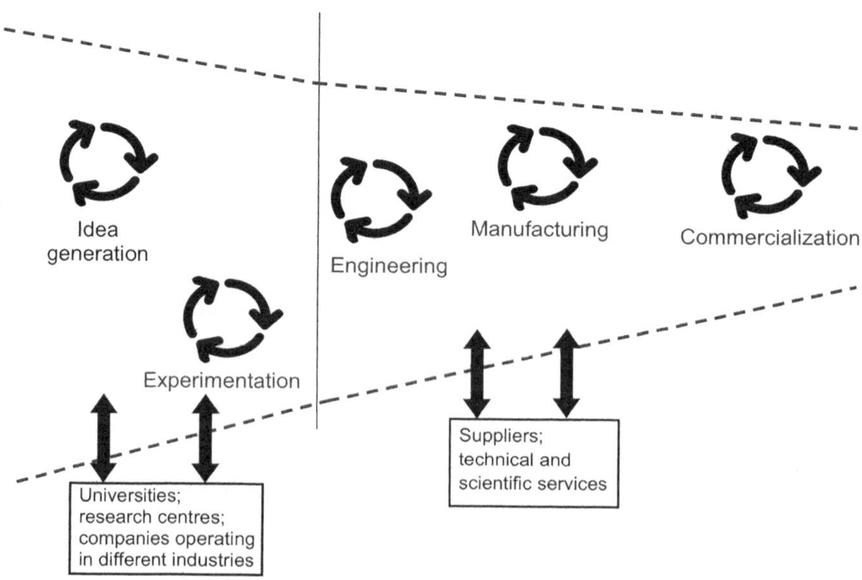

Fig. 20.1 Heinz's open-innovation model.

interaction with them mainly concerning initiatives in the design and organization of training programs on topics of common interest, such as, for example, the industrialization of innovation or the problem of managing IP in the food and drink industry.

The open-innovation model (Fig. 20.1) adopted in Heinz is now described.

20.3.6 Organizing and managing open innovation

About three years ago, Heinz introduced a specific unit dedicated to the organization and management, the global external innovation team (GEIT), which acts as a sort of 'hub' between the external world and the internal units involved in innovation activities. The GEIT has two main tasks. First, and most relevant, the GEIT has to scan the external environment (technology scouting) in search of potential partners, i.e. in search of the technologies and competences that may help Heinz in achieving its innovation strategy goals. The GEIT works on each technology platform generated by the company's strategy, with the aim of bringing internally those technologies able to satisfy the needs related to the platform. In this way, the technology platform is characterized and enlarged by the work of the GEIT. This scouting activity generates a type of ranking of potential partners, characterized and distinguished by their competences and technologies. Second, the GEIT has the task to make the potential partnership active and to make it work. In this phase the GEIT:

- defines the organizational form and the governance of the collaboration;
- writes the contract to be signed with the partners (with the support of the legal staff, in particular for IP rights);
- involves internal people from the various internal units (R&D, marketing and manufacturing), defining the team of each open-innovation project;
- defines specific performance targets, timelines and resources devoted to collaboration.

In order to accomplish all these tasks, the GEIT is organized on the basis of the geographical areas (Europe, Asia Pacific, North America, Latin America and Africa) with one expert for each business area dedicated full time. In two of the three business areas: ketchup and sauces; ready meals and snacks, the GEIT work is limited to that described above, i.e. it ends with the definition of the partnership and it is not involved in the management of the ongoing partnership. On the contrary, in the business area infant and nutrition, the GEIT is also involved in the operative activities of the partnership. For this reason, the GEIT group in this case also involves a project manager for each collaboration project, who follows the whole open innovation project, until commercialization.

With this organization, open innovation does not create many problems: dispersion of activities is avoided by focusing the work of partners on specific technology platform needs. Opportunistic behaviour of partners, spill over and technical problems are limited by the careful work of the GEIT in selecting the right partners, establishing mutual trust with them, and defining a robust contract.

However, some difficulties may emerge in open-innovation projects. The most relevant are related to the fact that some partners, in particular universities and research centres, do not have a project management culture, do not know the project management techniques for supporting activities, and lack the necessary time–outcome–market orientation. This lack of competence and management culture leads in some instances to mistakes in planning by universities and research centres and, as a consequence, the planned activities, time and costs can not be respected.

Another problem of open innovation concerns the difficulty of having a clear and precise track of all initiatives, and this prevents the company from capturing the benefits deriving from experience and learning. Hence, it is difficult to improve the culture of open innovation and the managerial and organizational skills used in technological partnership.

Finally, open innovation is dramatically affected by the huge amount of administrative and bureaucratic work that has to be done when innovation is conducted in partnership, especially with universities and public entities.

In summary, the management of open innovation turns out to be particularly effective and efficient in the Heinz company: the sharing of strategic objectives within the technology platform, and the presence of an organizational unit that manages and co-ordinates the innovation, and that carefully selects the external roles involved, are certainly factors that determine a lower failure rate of the outcomes of open innovation.

20.4 Conclusions and managerial implications

As shown in previous studies, Heinz's experience makes clear that the opening up of the innovation process can not be something makeshift. Indeed, the way to the opening is very sensitive and time-consuming, and it requires appropriate changes in attitude that must be supported by appropriate organizational and managerial initiatives. The creation of specific organizational units and roles together with the use of project management techniques seem to be essential requirements for a successful implementation of open innovation. As the Heinz case study shows, these tools provide consistency of collaborations with the company strategy, discovery of opportunities, good communication, sharing of the relevant expertise and facing of IP issues. Moreover, once started, the collaboration needs to be managed in order to be directed steadily towards the desired objectives of value creation, thus the definition of performance targets is another important lever. However, as noted in the Heinz case study, obstacles and risks are always there, particularly with certain types of partners (i.e. universities) who are accustomed to completely different working practices.

The strong message that emerges from such situations is, therefore, that, in general, it is necessary to change the company culture to implement successful initiatives of openness which are able to involve the skills needed to create value. Managerial and organizational interventions are certainly the first step in this direction and make it achievable in the long run.

20.5 References

BÉRÉZAI P (2010) Future R&D strategies in food & drinks: evolution from orthodox approaches to open-innovation models, *Business Insights*, March. 74–82.

BILGRAM V, BREM A AND VOIGT K (2008) User-centric innovations in new product development: systematic identification of lead users harnessing interactive and collaborative online-tools, *International Journal of Innovation Management*, **12**(3), 419–458.

BIRKINSHAW J AND GIBSON C (2004) Building ambidexterity into a organization, *MIT Sloan Management Review*, Summer, 47–55.

CASSIMAN B AND VEUGELERS R (2006), In search of complementarity in innovation strategy: internal R&D and external knowledge acquisition, *Management Science*, **52**(1), 68–82.

CHESBROUGH H AND CROWTHER AK (2006) Beyond high tech: early adopters of open innovation in other industries, *R&D Management*, **36**(3), 229–236.

CHESBROUGH H, VANHAVERBEKE W AND WEST J (2006) *Open innovation: researching a new paradigm*, Oxford: Oxford University Press.

COHEN WM AND LEVINTHAL DA (1990) Absorptive capacity: a new perspective on learning and innovation, *Administrative Science Quarterly*, **35**(1), 128–152.

CHIARONI D, CHIESA V AND FRATTINI F (2010) Unravelling the process from closed to open innovation: evidence from mature, asset-intensive industries, *R&D Management*, **40**(3), 222–245.

DAHLANDER L AND GANN D (2010) How open is innovation?, *Research Policy*, **39**(6), 699–709.

DODGSON M, GANN D AND SALTER A (2006) The role of technology in the shift towards open innovation: the case of Procter & Gamble, *R&D Management*, **36**(3), 333–345.

GARCIA MARTINEZ M (2011) Future R&D strategies in food & drinks: evolution from orthodox approaches to open innovation, *IFAMA, Annual World Symposium*, 20–21 June, Frankfurt, Germany.

GARCIA MARTINEZ M AND BRIZ J (2000) Innovation in the Spanish Food & Drink Industry, *International Food and Agribusiness Management Review*, **3**, 155–176.

GOERS SK (2009) Open innovation: driving growth by applying principles of open innovation, Kraft Foods, open access at http://openinnovation.berkeley.edu/speaker_series/Goers_Berkeley_02_23_09.pdf.

HUSTON L AND SAKKAB N (2006) Connect and develop: inside Procter and Gamble's new model for innovation, *Harvard Business Review*, **84**(3), 58–66.

JIMÉNEZ-BARRIONUEVO, MM, GARCÍA-MORALES, VJ AND MOLINA LM (2011) Validation of an instrument to measure absorptive capacity, *Technovation*, **31**, 190–202.

JOHNSON WR (2005) H. J. Heinz Company celebrates opening of Global Innovation and quality Centre, www.businesswire.com/news/heinz/20050919005905/en/H.J.-Heinz-Company-Celebrates-Global-Innovation.

JOLINK, M (2009) People management to stimulate networking: a study in the food industry, Books on Demand.

LICHTENTHALER U AND LICHTENTHALER E (2009) A capability-based framework for open innovation: complementing absorptive capacity, *Journal of Management Studies*, **46**(8), 1315–1338.

PISANO GP AND VERGANTI R (2008) Which kind of collaboration is right for you, *Harvard Business Review*, **86**(12), 78–86.

RAASCH C, HERSTATT C AND LOCK P (2008) The dynamics of user innovation: drivers and impediments of innovation activities, *International Journal of Innovation Management*, **12**(3), 377–398.

SAKKAB N (2002) Connect & develop complements research & develop at P&G, *Research-Technology Management*, **45**(2), 38–45.

TRAITLER H, WATZKE HJ AND SAGUY IS (2011) Reinventing R&D in an open innovation ecosystem, *Journal of Food Science*, **76**(2), 62–68.

ZAHRA SA AND GEORGE G (2002) Absorptive capacity: a review, re-conceptualization and extension, *Academic Management Review*, **27**(2), 185–203.

21

Innovating with brains: the psychology of open innovation

B. A. Sabel, University of Magdeburg Medical School, Germany and Institute of Automation, China, K. Sabel, Consulting-Berlin, Germany and J. T. Preston, TEM Capital and Continuum Energy Technologies, USA

DOI: 10.1533/9780857097248.4.369

Abstract: The psychological aspects, so-called 'soft factors' necessary for open innovation are investigated in order to show why the majority of innovation failures are caused by the management problems not by lack of time or technology. Steps of innovation identified include a 'creative phase' and an 'ability to prevail' phase, to reach full realization despite hurdles. Favorable psychological factors for innovation identified are passion and intrinsic motivation, communication skills, charisma, ambition, creative urge goal orientation and the willingness to partner with others with complementary skills. Cognitive aspects identified are curiosity, novelty seeking behavior, divergent thinking, creativity, frustration tolerance and intelligence. Unfavorable factors identified are greed (an unwillingness to share with those who can help), pressure, anxiety and fear as well as ego-issues; and lack of know-how.

Key words: innovation, psychology, investing, inventor, cognition behavior.

21.1 Introduction

On the most fundamental level, innovation is done by people and not by structures or organizations. This chapter focuses on the human 'soft' factor; however, it is not a review of the available academic literature, but a condensation of the authors' personal experience in innovation in various roles as founders in executive, scientific and finance and technology transfer managers in both Germany and the USA. We believe that the accounting of personal experiences with an analytical view may provide the reader with a greater practical insight into the psychology of innovation than the consideration of empirical studies and academic arguments.

From this experience it becomes clear that innovation is not only a matter of technical know-how but most of all a matter of human behavior, created by our brains and describable by concepts of psychology. Indeed, innovation is shaped in many ways by psychological aspects, called 'soft factors'. Much has been said about how to optimize organizations and innovation support structures, but to understand the real challenges of innovation, and positively handle them, we need to better understand how innovation players tick and which soft factors drive them to successfully create something that was not there before, i.e. bring ideas to life. It is the personality of the innovators, using perhaps their brains' 'novelty seeking centers' (Kumaran and Maguire, 2007) that has a significant, yet unappreciated, impact on innovation. Though people have different technical skill sets, it is not only these skills of 'formal intelligence' that matter in innovation, but perhaps even more so 'soft' factors such as certain personalities, motivations and 'emotional intelligence'.

This is particularly apparent in the struggle between the academic and commercial world as both function differently. To make innovation successful, we need to acknowledge such differences and find ways to help both worlds to work together well. This chapter describes from the view of practical experience several personal (psychological) 'soft' factors that may be considered as key elements of innovation.

Many significant innovations have come out of academic institutions, particularly in the life science field. The innovator must cover a wide range of different perspectives to succeed. He/she must function within the rules of academia: openness, publications, conflict avoidance, etc. At the same time the innovator must create financial incentives to attract investors, customers, employees, and partners to join in the venture. The financial incentives, in turn, promote secrecy, intellectual property protection, participation in the economic value and other non-academic issues. Additionally, the innovator must have sufficient knowledge of the industry to recognize a problem for which he/she has a solution. The innovator then must determine whether that solution offers value to customers which is greater than the current practice or competitive options. This is just the first part of the process discussed in the following section.

21.2 Innovation is all about psychology

Psychology is the science of behavior, cognition (thinking) and emotions. When comparing the relevance of psychology ('soft factors') compared with the hard facts consider how to commercialize an innovative idea. Which scenario would you pick? A little money but experienced and smart people or a lot of money and incompetent people. Or, to put it in other terms: a 'B' product developed by an 'A' team is more likely to succeed than an 'A' product developed by a 'B' team. In fact, as a recent analysis shows, reasons for write-offs of VC-funded start-ups are as follows: 64% of innovation failures are caused by bad management; only 29% results from insufficient time/funding and only 7% is attributable to technology (Earlybird Venture Capital, personal communication, 2011; www.earlybird.com).

Clearly, it is not money or structure that innovates, it is the innovation team. Think of those 'garage start-ups' that started with not much more than a brilliant idea and little cash (few thousand dollars) but developed into outstanding success stories such as Apple or Google.[1] The innovators' psychological skills mattered at the time of birth of the innovation; open innovation is primarily a 'people issue'.

Psychological factors in innovation relate to issues such as emotions, motivations, novelty-seeking behavior, creativity, and the ability to anticipate the future, but it also requires frustration tolerance and perseverance. These are collectively referred to as 'social competence' or 'emotional intelligence'. It is the ability to handle such soft factors that makes the difference between a 'cheap idea' and a true innovation. To just have a cool idea is worth nothing unless it is translated into something real.

Innovation is usually not an activity of single inventors but a group effort of various players with complementary skills. Irrespective of whether the innovation is small (e.g. an optimization solution for a small production process) or big (such as the invention of the automobile), it always involves different kinds of players.

To understand the role of the different partners in innovation, let us consider the most fundamental evolution of the innovation: recognition of the problem (the creative part), finding the solution (which is often mistaken as the 'creative step'), developing a product and bringing it to market. This requires technologists (scientists and engineers), management and investors. Unless these different players work together well, failure is probable. They all have different roles and responsibilities in all phases of the innovative process.

21.3 Phases of innovation

There are different phases in the evolution of an innovation. Psychology is important in each of them. Steps 1 and 2 are usually considered to be the most important in innovation. But for an idea to be called an innovation, it needs to reach full realization, i.e. enter the market place. So unless the project reaches step 11, it is not an innovation that moves the world. The steps are:

1. recognition of a problem or conceiving an opportunity;
2. the invention of the solution;
3. a feasibility check of the solution (economics or value proposition for the customer, benefit, market need, realization potential, technical and commercial goals etc.);
4. feasibility study (technical feasibility, legal and competition analysis);

[1] Although Apple started with very little money, Steve Jobs was able to utilize significant resources from other sources such as the graphical interface designed by Xerox PARC. Many first-rate managers find ways to leverage third party assets. Google replicated this strategy by utilizing many iPhone features in its Android phone.

5. determination and appropriation of funding;
6. establishing a project management system and team that includes potential customers;
7. design and development of a prototype (including simulation and tests);
8. market tests of prototypes and improvements/adjustments working with a lead customer;
9. mass production and the associated distribution logistics;
10. market entry with advertisement and PR; and
11. distribution.

The next question is which innovation player contributes to each phase of the innovative step. Ideally, the founding innovator is involved in all phases and, given the complexity of each task, psychological know-how is useful in all of them. Success in today's business climate requires continuous innovation to stay ahead of competitors; new product features need to be added, new designs conceived, technical variations invented, etc. This requires new innovations and continuous nurturing of know-how. The iPhone and other Apple products provide good examples with major new releases every two years. Innovation cannot stop.

Psychologists studying innovation have mostly focused on the first and second step, the 'creative phase'; suffice it to say that it takes a creative and imaginative personality to come up with new ideas and inventions, but to bring a project to realization also needs an 'ability to prevail', i.e. making progress towards full realization despite problems and hurdles. Often, the critical step of innovation is step 3, where the idea has to survive many attacks by reality checks (value proposition, legal and patents issues, market size etc.), power games and ego-issues ('I want to control the project and get the recognition for it'), professional competition within the team ('the funding is better spent in my own department'), personal conflicts ('I help my friend, no matter what'), etc. Step 3 is the key phase of innovation and perhaps the most susceptible to psychological influences. It is a delicate phase that makes or breaks an innovation. Step 3 also contains a critical success factor: that a product can be made at a price low enough to provide real value to customers. Creating a product that has a quantifiable positive value proposition for customers is the single most important step toward building a business. This factor turns out to be very difficult for inventors because it requires interactions with customers and knowledge of the customer's business. Picking up the phone to make a cold call to a potential customer is often a frightening experience.

The role of psychology in the phases after this step 3 is not innovation-specific but applies to business at large. This concerns general issues of project realization such as personal professionalism, team building, project management, marketing expertise and financial planning. It also includes the usual 'micropolitics', questions of who is in power, who is on top and who is on the bottom. Needless to say, that many aspects of psychology are involved here as well, but they are not specific to innovation. These are rather general issues taught in management and business administration courses. However, there are many 'soft factors' worth considering in the process of innovation (Sabel, 2008).

21.4 The influence of soft factors on the success of innovation

The soft factors are now identified and their effects on the success of innovation are investigated.

21.4.1 Personality factors

Following the theory of Sternberg and Lubart (1991) innovative people are able to adopt and initiate innovative ideas with a high growth potential at a time when such ideas are largely unknown or unpopular. Innovators stick with an idea against all odds or road blocks and carry them all the way through to realization. They also face opposition from within and from outside. To quote Albert Einstein: 'Great spirits have always encountered violent opposition from mediocre minds'.

To handle all this requires specific personality characteristics such as intelligence (including the ability to anticipate the future), profound knowledge and a high level of motivation or passion. But innovative behavior also requires a conducive environment ('innovation climate'). The statement by the former German Minister of Economics, Ludwig Erhard, that 'fifty percent of the economy is psychology' also applies to innovation. Innovators are typically open to new ideas and experiences ('early movers'), have a high degree of self-confidence, are highly effective, have great persistence power and a lot of ambition. However, they also tend to be non-conforming, impulsive, dominant and typically master the art of non-convergent thinking and reasoning, (Patterson, 2002). They also love to experiment and accept calculated risks. However, the right personality or innovation talent is not enough, if other soft factors are ignored.

21.4.2 Anxiety and fear vs. 'lightness of being'

Anxiety and fear (angst) may, under certain circumstances, facilitate creativity and innovative drive, leading to new ideas or solutions during situations of existential relevance, such as the loss of a job or immediate danger to one's life. However, in the normal working world, angst is poison to innovation and probably the most damaging soft factor of them all. To be afraid to fail takes away the necessary 'childlike lightness of being' and ruins the playful attitude that is a great source of new associations and unexpected ideas in the early phase of innovation. In the working world, anxiety may be produced by constant performance pressure and negative stress, lack of appreciation or an aggressive and competitive working environment. Threat does not work well for innovation either. Though threat or punishment might work in some situations ('you are not achieving enough, try harder'), one cannot be specifically punished for a non-existent idea, i.e. being blamed for not having a new idea that nobody had before.

A much better way to enhance innovation is to reward the desired behavior. Reward can be achieved by bonus payments for small or large innovations, and encouragement followed by recognition of the inventor's achievements. Rewarding is a stronger and more stable measure to modify behavior than punishment.

One major advantage of the American innovation engine is the tendency to have shared ownership through stock options or grants. Owning a piece of the company is a very strong incentive and aligns interests. However, this is different in other cultures such as in Germany, where the average employee is not much interested in sharing ownership, but rather in a fixed salary and pension plan contributions (Sabel, 1993).

21.4.3 Ego issues

There are good and bad sides of the 'ego-coin'. The good side is that a person with a strong and healthy ego is able to prevail despite road blocks and adverse events and can, in a positive way, dominate his/her peers, thus advancing new ideas and achieve innovation milestone against the odds. This is linked to the 'persistence' topic discussed in 21.4.9. On the other hand, a 'selfish ego' is typically a sign of weakness: the result of insecurity and lack of selfconfidence or simply lack of competence and knowledge. Such persons tend to compensate their insecurity by being overly pushy, blame others, unreasonably dominating, loud, emotionally eruptive, hypersensitive etc. They cannot listen to, nor accept, good arguments and they are extremely susceptible to whatever feeds their ego and hypersensitive to challenges of their ego. When one talks about 'ego issues' it is the latter, toxic or unhealthy type of ego problem that is meant. Here, the person makes decisions based on the principle 'is it good for me', i.e. a personal, emotional benefit such as standing above others, rather than deciding for the benefit of the group, the project or the company. To put it in a graphic way: people with an 'ego issue' can talk, but they fail to hear or see.

Ego issues of management

Because most innovation happens in start-ups (Section 21.7), the ego issue of the management is worth considering. The key person is the CEO as he/she makes all major decisions in the company. They usually have a strong ego, otherwise they would not be fit for the leadership position. A healthy ego is not afraid to be challenged, can change its mind in view of convincing arguments, is not afraid to delegate, and involves others in the decision process. A person with an unhealthy ego, in contrast, makes single-handed decisions, does not listen to advice ('resistant to being consulted'), and puts himself first over the goals of the project and company. He/she supports suggestions and projects that feed their ego and ignores objective facts, irrespective of whether the decision is the right one for the company or the project. They also do not pass credit to others because they feel more important (elevated) when they belittle the efforts of others. This kind of attitude kills the innovative drive of the company and guarantees failure. Such individuals are innovation killers and generally fail.

Ego issues of investors

The same issues apply to investors. If an investor has an ego issue, this will lead to a conflict with management, in particular with the CEO. A CEO cannot work freely and effectively if they are master-minded (controlled) or micro-managed by a dominant investor. Investors should generally have the goal of value creation for the company in mind, i.e. making the company successful, but investors with ego issues constantly seek opportunities to make the company or project look like 'their achievement', brightening the glow of their own significance ('now it's my baby'). It is as if they carry out the CEO's job without accepting the responsibility of it. If you have such an investor in your company, the likelihood for failure is greatly increased.

Ego issues of founders

Founders are the fathers or mothers of the project. They are heavily invested in terms of time, money and passion in the project and have brought it to a level of success so that investors or investment committees believe in the project and fund it. They are usually players with a motivation and passion much greater than commercial success; they are the champion of the idea as such. It is, after all, their brain-child. Like management, founders usually have a strong ego. The first step in building a talented team often involves sharing ownership. Here, an ego issue can arise if the founders are asked to give up control (or shares) of the company and put it into the hands of management. What founders need to realize is that having a small share of a valuable pie is worth more than being the sole owner of a worthless pie. They sometimes do not see that 10% of a lot is more than 100% of nothing.

In a typical start-up company, the founder is a technologist or scientist and the investors help recruit a CEO to work with the founder. The probability that founders and the CEO end up with problems is high. So this is an issue that needs to be carefully considered if the innovation is to succeed. Unless the founders are appointed as the chief technology officer (CTO) and are thus under the control of the CEO, they often end up in a conflict with the CEO. A technology-based founder typically wishes to be the only one in charge of technical and scientific matters and views himself/herself as the only one that can understand technical work. This often creates problems if someone else is appointed CTO.

In the ideal scenario, all three players, founder, CEO/CTO and investors should be a harmonious team, but if this does not work, the CEO and investors will team up, and the founder will lose. Unfortunately, the probability is small that all three, founder, CEO/CTO and investor have no ego problems or no competition anxieties. This is why most company failures are caused by management problems and not by lack of time (funding) or technical problems.

One may wonder why investors typically align with the CEO. The reason is simple: both wish to generate money (for themselves or the fund) and that is the language they understand. Many investors were trained in finance and not in science or technology. A positive cash flow is the key to survival of any company. It is important that all three players embrace this simple measure of success. Getting quickly to positive cash flow enhances options and builds freedom for the business.

21.4.4 Passion and other motivators

Motivation is what determines the energy and direction of human behavior. A motivated individual shows much energy and pushes forward in a certain direction, trying to achieve some specific goal. Energy without direction or direction without energy is, in essence, a lack of motivation. Motivation in the work place depends, in general, on many soft factors, such as appreciation by coworkers and superiors, and mutual respect and tolerance to different views. A good test of motivation is to ask a person how much they enjoy going to the office in the morning. However, to appreciate the role of motivation in the innovation process, main characteristics to consider are: passion, charisma and ambition.

Passion is the most energetic and goal-directed kind of motivation. It is a strong desire to succeed and the goals are usually very specific (like being in love). The source of the passion is usually from within a person, some strong inner urge, higher-goal plan ('I want to make the world a better place') or the desire to prove some point to oneself or others ('The world is not flat, it is round', 'the sun is the center of the planetary system and the earth circles around it'). This motivation from within is called 'intrinsic motivation'. It is the most powerful, most stable and longest-lasting type of motivation. It is something that all parents wish for their children. It is typically induced by role models and appropriate reward combined with freedom of choice, but it is resistant to and not easily modifiable by punishment.

In contrast, 'extrinsic motivation' is induced by outside factors, such as money, benefits, perks or recognition by peers, or punishment. It may be a powerful motivator in the short term, but like a bush fire it is unstable and short-lived. A key management challenge is to continually find methods to stimulate extrinsic motivation by aligning interests. If, for example, positive cash flow is the most important near-term goal, the staff should have bonuses tied to achieving this goal.

Because innovation involves many risks, failures and errors from trial, passion and intrinsic motivation are important for the success of the innovation project. In the early phase, in particular, of an innovation, there are typically few resources for attractive extrinsic motivators. Passion is therefore essential in the early stage. However, passion is important not only in the beginning phase but during the entire evolution. Generators of passion are when management passes credit to subordinates, shares blame for mistakes, and focuses on strength, not on weaknesses; killers of passion are greed, conflict policies, committees, big egos, fights over budgets and responsibility, politics (when power takes priority over the right decision), and lawyers who are trained to find problems and risks rather than solutions and opportunities.

Inventors and founders typically have a lot of passion for their 'brain child' and they are the most long-lasting and stable psychological contributors to innovation, with a long-term 'vision' (years or even decades). But just as when in love, the downside of passion can be that it blinds you to reality (such as the market and the risks). In contrast, management and investors tend to be more extrinsically motivated as their first and foremost goal is financial success, and this is not typically very long-term (3–5 years maximum). This difference in time horizon can often be at odds with views of the founder scientists/technologists on the team. This is one possible source of conflict between management on the one hand and founders/inventors on the other. Managing this tension between these two worlds of thinking may be the key to innovation success.

21.4.5 Communication and charisma

The ability to communicate well and to convince others of the new idea is a central element in bringing the innovation to life (i.e. into the market). It is also essential to be able to fight adversaries successfully, as there are plenty of them around. Therefore, excellent communication skills and charisma, the ability to fascinate others on an emotional level for the new idea, is an important element in the psychology of innovation. When combined with passion, charisma of the innovator can be a great facilitator to convince others to invest in the project and support its realization, bearing in mind that others usually have a partial understanding of the details of the project so their judgment relies on emotional decisions in that they trust the inventor. If the inventor does not have the passion, why should others support a risky and radically new idea to make the project fly?

21.4.6 Ambition

Innovation is not something for people that are fully satisfied and who think of safety and comfort first. The innovator needs the drive to achieve something special, to conquer the world.

21.4.7 Creative urge and innovative power

The desire to be creative and innovative is usually not only motivated materially, but it is intrinsic. The desire to create, find new solutions, (such as reducing a work load or increasing efficiency, the removal of road blocks or saving cost) can all motivate innovation. Motivations that block innovation are lack of care, resistance against the employer or the company or the fear of being punished for suggesting change. Intrinsic motivation also means having fun being creative, mastering the challenge, the enthusiasm for a topic or self-realization are important elements of innovative behavior (Amabile, 1988). Thus, monetary incentives have little effect on innovation drive, but non-monetary rewards do: personal recognition and public acknowledgement.

21.4.8 Team spirit

An innovation team is made up of many people with different backgrounds. It goes without saying that life is colorful, many people with different education, nationality, skills, expectations and motivations, and a mixture of such backgrounds is a plus. There should also be a healthy mix of males and females in the team and, everything else being equal, they will outperform unisex teams. When a team spirit is achieved, a project can be completed on time and on budget. To be a good team player is a particularly important personality/psychological factor relevant for innovation because innovation environments are usually interdisciplinary, involving different tasks. The greater goal of success must dominate personal goals in a cohesive team. It is also critical to have aligned interests where all members of the team succeed or fail together.

As an oversimplification, what we can accomplish in life is a function of:

1. how hard we work,
2. how smart we work,
3. how much courage we have, and
4. how much leverage we have.

The first three attributes are easy to understand, but the concept of leverage is poorly understood by most people. As an example, if you were starting a software company and Bill Gates offered to call some venture capitalists on your behalf, his five-minute phone call could save you six months of hard, smart work. Teams have diverse leverage that can be utilized effectively if the right spirit exists. One key aspect of building a team is to always try to attract people that are smarter than you or better connected. This behavior runs counter to a strong controlling ego, but it is one of the greatest keys to success.

Leverage is one of the primary reasons for needing a Board of Directors and Scientific Advisory Board. In addition to challenging concepts, boards open doors to outsiders who can add value.

21.4.9 Cognition

Cognition is the way the brain thinks, the assumptions it makes, the way it interprets the world and the way in which information is selected and stored. The following innovation-related cognitive factors are discussed in detail.

Curiosity and novelty-seeking behavior

Curiosity is the force at the very beginning of the innovation process. To seek novel experiences and discover novel things is a basic instinct of human behavior and, indeed, it is the driving force behind evolution. Novel approaches are needed in a changing world to survive in individual life and to assure survival of the species in evolution; new niches have to be found in the everlasting competition for scarce resources. In that sense, innovation is an essential element of evolution. It is also the beginning of scientific and technical progress, and, down the road, the basis for economic progress.

The brain has certain centers dedicated to novelty that are particularly activated when new, unexpected visual scenes are observed (Kumaran and Maguire, 2007).

When there is a continuous lack of novelty, as in a prison cell, this produces boredom, depression and frustration. Entrepreneurs and inventors have features in common with playful children that seek new games to play and challenges to master new things. That is why innovators sometimes appear to be a bit weird, unconventional and perhaps unpredictable. This is both a strength and a weakness in the innovation context. It is important to be tolerant of the creative and innovative individuals. Their unconventional thinking is the basis for the creation of new ideas.

Divergent thinking and creativity

Divergent thinking is a cognitive ability to conceive of as many possible solutions to a problem as possible, which is the opposite of 'convergent' thinking, where the goal is to find a single solution. Divergent thinking is typically assessed by subsections of intelligence tests, where one task could be this: 'Tell me all the things you can do with a brick'. The innovative mind has to be able to handle both divergent and convergent thinking: their mind first scans many divergent options and then comes up with a convergent and practical solution. Divergent thinking is often associated with creativity, the novel assembly of thoughts that were already there, embedded in other contexts.

Most creative acts are not completely new ideas, but they are rather a kind of reassembly of existing or known elements, like connecting many random pixels to a new image. Creative people typically have many ideas (divergent thinking), conceiving such ideas all the time, but also dropping many again to come to the final solution. With their experience and knowledge (or the reality check by family, friends or colleagues), only the most valuable ideas survive and converge to realization. Of course, it may happen that one could have a single idea that turns out to be an innovation hit. However, the more ideas created and checked against reality, the greater the probability that one succeeds.

Inventors are revolutionaries who want to change the world, challenging long-held beliefs, overcoming accepted barriers and creating something new. They are usually not afraid to make mistakes and they are typically not perfectionists. To quote Albert Einstein: 'Anyone who has never made a mistake has never tried anything new'. Creative people are the human capital in the innovation process and their way of thinking (and sometimes their way of behaving) has unexpected elements. The creative person is sometimes also a little crazy because, to quote Einstein again: 'a really good idea is one that seemed to be impossible to realize'.

Creative people do not always like to follow the rules of the expected or seek peer acceptance. New ideas, by definition, have not been there before, yet they rely on memories, associations and new combinations of thoughts. The famous story of Kekulés discovery of the nature of the benzene ring is an example of it. Kekulé, a German chemist, was working on the structure of aromatic bonds and was an expert on carbon–carbon bonds and how they assemble. One day he had a dream of carbon atoms bonding to pearl-string-like long rows which then moved like a snake around in space. The snake then bit into its own tail to form a ring-like structure. From this association Kekulé conceived the discovery of the benzene ring which formed the basis of aromatic chemistry.

Creativity is the ability to leave established paths of thinking and travel into uncharted territory. Creative thoughts arise from within the brain mostly during the relaxation state when connections are less defined, modifiable and able to reassemble in new ways. Memory traces (such as the string of atoms and the shape of a snake) can then be combined in new ways. Such relaxation states of the brain are the time just before falling asleep, during dreams, under the shower or during meditation. Here, the 'old thoughts' can be assembled to 'new combinations'. Tension, stress, or fear and anxiety, in contrast, are creativity killers (unless we face a situation of threat or danger).

Just as the reassembly of existing thought elements is the basis of creativity, true innovations often happen when different academic/technical fields interface (merge); such as informatics meeting graphics, psychology meeting brain physiology, optics meeting space technology. Therefore, the creative person needs sound knowledge of the subject so there are more little 'puzzle pieces' inside his/her brain that can be recombined to create more useful associations and new connections. However, the generation of any odd idea that is devoid of technical know-how is not considered to be creative; an idea has to be practicable and of functional use to qualify as creative.

Frustration tolerance and persistence

A lot of energy is needed to get a new project started, internally accepted and developed to completion. This requires patience, dedication and complete conviction that the idea is valuable. Innovations tend to come with lots of problems and challenges, set-backs and failures. This needs persistence and frustration tolerance. It is why an 'innovative personality' requires the ability to stick to the new idea and not give up too easily. It is almost contradictory to the divergent thinking of creativity; so the successful innovator has to somehow unify these opposites.

Intelligence

Intelligence is a more general psychological term that covers many different mental aptitudes such as attention, remembering, learning, planning, orientation, imagination power, argumentation/logic reasoning, introspection ability and the ability to anticipate. However, it also includes creativity and divergent thinking, as already discussed.

21.5 The psychology of the innovation team

We now examine the psychology of the innovation team, how the gap between academia and commerce affects open innovation, and the effect investor relations has on the success of innovation.

21.5.1 Team psychology

Because it is the brains (psychology) of a group of people working together, and not machines, that create innovations, the psychology of the team needs to be examined. The principle is this: human factors before technical factors. Therefore, there should be no compromises on people, ever! Value the team of entrepreneurs, pick the best and brightest and give them what they need; hire 'A' people and they will hire other A people; if you hire 'B's, they try staying important (or on top) by hiring 'C's; For B people, it is ego over progress ('me first'); only their own ideas are brilliant (the NIH syndrome: 'not invented here'). Furthermore, the leadership of the team should reward passionate behavior by talking about the dream of creating the future, while avoiding governing by pressure or fear. Pressure and fear are poison for innovation. Inventing is not only hard work, it is a playful thing that works best with a certain 'lightness of being' and with a specific purpose in mind.

To use an analogy of Antoine de Saint-Exupéry (the novelist of the famous book 'The Little Prince'): 'If you want people to build a boat, don't tell them to work hard chopping trees, cut them into boards, assemble them on the beach and knit the sails. Rather, tell them that one day they can travel to the unknown, far away land that lies beyond the sea'. It is dreams that kindle extraordinary powers in people, setting free unexpected desires to reach for the unknown. Young scientists need these dreams and learn ways to get there by following experienced innovation role models, i.e. their teachers; therefore, engaging innovators who have done it successfully before is a valuable asset for any innovation system.

However, it takes much more than a creative inventor to make for a good innovation team. Management is needed and investors (institutional or personal) to bring a project to the stage of realization. But there is a fundamental gap between these two 'camps' which needs to be overcome for innovation to work. That is perhaps the biggest potential challenge of the innovation team.

21.5.2 The gap between academia and the commercial/financial world

There are two types of players in the innovation space: the inventors (usually academics) on the one hand, and business management/investors on the other. The academic and the business worlds operate very differently and follow different rules and principles. This is not only a source of a rich variety of new solutions, but it is also a potential source of conflict. It is perhaps the greatest danger for a start-up company if both 'camps' (technical and business) do not get along. The reason is that they follow different principles in life and have a different view of what is important to succeed. Table 21.1 summarizes in short some of the differences. However, Table 21.1 is not applicable to everyone, it is just an exaggerated illustration of principles and should be interpreted cautiously. Moreover, most people incorporate aspects of both worlds.

When business people and academics talk, they may think they understand each other when often they have no idea what the other is really saying. Therefore, knowing how the other thinks helps reduce conflict and, in practice, can be a rich resource of ideas and ways of solving issues.

21.5.3 Investor relations

It is not just the technical and the management team that determines the success of an innovation project (e.g. a start-up), but it also depends on the quality and

Table 21.1 Features of the academic and the business world

Feature	Academics (inventors)	Business (managers, investors)
Motivation	Knowledge 'as such'; recognition	Responsibility (delivering results)
Passion (motivation)	Discover, change the world, and get famous	Get rich (maybe change the world, get famous)
Principles of thinking	Creative, non-committing, 'romantic', divergent with the capability to focus deeply once a topic is selected	Results/profit-oriented, more focused
Financial goals	Typically secondary	Salary (bonus) driven
Risk taking attitude	Tenure creates a relatively low-risk environment	Calculated risks are normal business activities
Time horizons	Long term	Quarterly, yearly

characteristics of the investor. Investors improve the chances of success if they open doors (provide leverage), have deep pockets, contribute hands-on in a qualified manner and focus on value creation. However, they decrease the chances of success if they focus on quick wins by hyping the business, failing to support the innovation drivers ('founders-out' attitude), have ego issues ('we know it all') and play control freak. Often investors focus only on the CEO with the correct view that management is the most important determinant of success. However, if the CEO is the problem, it often takes too long to recognize the problem unless the investor is engaged throughout the organization. Less CEO-centric attitudes might be a bit more complicated to handle, but it is not only the CEO that brings the innovation to market. It is the team as a whole, and proper governance needs the competition of the best and most reasonable thoughts, not a dictatorship.

21.6 The innovative environment of academia

We now examine why academia can be a useful source of innovative ideas but can also fail to live up to its potential for innovation.

21.6.1 Academia as the fountain of innovation

Academia is a great fountain of innovative ideas. Many start-up companies have their origins in some academic project, or were started by students and scientists working in academia. Academic research and development is usually funded or cosponsored by the government, it provides an environment where any avenue that fantasy suggests can be explored, free from the typical economic pressures of the commercial world. 'Everything that is really great and inspiring is created by the individual who can labor in freedom.' (Albert Einstein). Academia is also an unlimited pool of young talent.

As identified by Michael Porter (1990), economic development in a region typically depends on a cluster of strengths:

1. innovation drivers such as strong academic environment and research facilities,
2. robust industrial activities,
3. wealthy individuals or investment firms with a long term perspective, and
4. demanding customers.

Innovation and economic progress are tightly linked. As Werner von Siemens, the founder of Siemens, stated: 'research in the natural sciences is always the secure foundation for technical progress, and a country's industry will never be able to gain and sustain a leading position, if the country is not, at the same time, a leader in scientific progress'. Many countries have realized this, including the recently emerging market BRIC countries (Brazil, Russia, India, and China). Together with Japan and the major Western nations in Europe and the United States, there has never been so much research effort. This is, in fact, an important time in history because 'ninety percent of the scientists and engineers that have ever lived in the history of mankind are alive today.' (D. B. Merrifield).

Examples of such innovative, science-based environments in the USA are Silicon Valley (near Stanford University) and Route 128 in Boston, USA (near Harvard and MIT). Innovation environments are also found in the vicinity of universities in many other places in the world. In Europe they include Cambridge, ETH Zurich, Fraunhofer Institutes, and Munich. In the USA the university-driven innovative regions are Silicon Valley, Boston, Southern California, Houston, North Carolina and Seattle. In the above examples, the entrepreneurial engines of Silicon Valley and the Boston area dominate the others. Clearly, there are aspects of those regions that are highly conducive for innovation. Silicon Valley, for example, has created an innovation engine that has significant spin-off activity from existing firms as well as strongly entrepreneurial academic institutions such as Stanford. What separates MIT and Stanford from many other universities is a culture that includes:

1. encouraging entrepreneurs,
2. encouraging interactions with industry (consulting),
3. problem solving or 'can-do' attitude,
4. sharing risk with innovators – taking equity in lieu of royalties, and
5. regional advantages – investment money, existing industries and demanding customers.

MIT has spun off hundreds of biotech and medical-device companies and Harvard Medical School oversees some of the country's top research hospitals. This is a regional synergy that feeds growth.

Large corporations tend to be very good at incremental innovation, but they lack the ability to create more radical innovations. Such corporations are not willing to embrace totally new paradigms easily. However, such paradigm shifts are fundamental to both scientific evolution (Kuhn, 1962) and economic growth. In some industries such as pharmaceuticals, such radical innovation is vital to fill a pipeline of new high-margin products as old products come off-patent. Thus, more and more large firms have had to outsource some R&D to start-up spin-offs or simply purchase small (VC-funded) companies with new technologies. In addition, many large corporations have relocated research labs to innovative environments, thus explaining the economic prosperity of certain regions (such as California, Cambridge MA, and Munich Germany). However, this innovation potential is only partially realized because not all academics are interested in collaboration with industry beyond their own lab bench.

21.6.2 Why academics avoid bringing innovation to market

Many academic environments do not live up to their innovation potential because many professors, scientists, post-docs and their students have concerns about impairing their impartiality through interactions with industry and others simply are not familiar enough with the innovation process. Many academics worry that they would lose their cherished freedom if they work with industry. Such concerns include being bound by contractual obligations, non-compete clauses, secrecy arrangements and financial attachments that may create the impression that their judgment is biased and not pure and clean. Jealousy is another issue: what do my colleagues think about me being involved with industry? Do they think that I am conflicted or can be bribed?

Yet another concern is loss of control. If an academic has ventured out to innovate, they may worry that they lose control over their baby ('...my precious...') when an investor comes on board and that they have to give up too much ownership of the innovation. It is difficult for academics to take the risk that they could be better off by sharing ownership with value-added people. If everything works well the academic would have a smaller ownership of business with great potential rather than total ownership of an invention that they are incapable of taking to the market: 10% of something very valuable is of greater benefit to them than owning 100% of something that is worthless. In addition, university environments are not always conducive to innovation; most universities make too little effort, if any effort at all, to teach entrepreneurship. There is also peer pressure to avoid commercial activities, especially for young faculty members (though they are the most valuable for start-ups). There are also few role models of successful entrepreneurs and there tends to be an ivory-tower attitude that connections to industry are 'demeaning' or 'reducing credibility' or they fear 'a pact with the devil'. Most academics have the mission to publish and they get rewarded for it. However, in innovation, the mission is not to publish but to make a difference in the world. It is needless to stress that innovation is not accomplished without a business structure in place to market the new product. However, as discussed above, the place of radical innovation is not in the large corporations, but in small start-ups.

21.7 Start-ups and small to medium-sized enterprises (SMEs): open innovation by default

With his statement 'Radical innovation never originates with the market leader' Utterback (1994) stresses that radical innovation does not typically come from the market leaders (elephants) who are the representatives of the 'old technology'. Radical innovation is disruptive by definition, it challenges established concepts and non-entrepreneurial firms respond poorly to disruptive ideas. Around 70% of new jobs in the USA come, not from elephants, but from 'gazelles', small start-ups. That is why many nations in Europe, Asia and the USA have elaborate start-up funding programs and grants: seed research is funded in academia and start-up activities by private investors (e.g. 'angels' or venture capitalists) and leveraged by state support.

Our findings based on involvement in hundreds of start-up companies are:

- The best and brightest young people drive innovation. It is important to have a culture (like Silicon Valley) where such people are willing to take risks.
- The fruits of success should be shared with those who enabled success. The US culture generally enables broader equity participation than does Europe.
- Innovation requires hard work and passion and innovators encounter many barriers. The passionate find ways to overcome the barriers.
- Patient capital is needed for innovation. However, smart capitalists (those who extend the innovator's capabilities by providing leverage) are invaluable.
- Sources of quality ideas such as academic research institutions can become instrumental in the innovation process, if the institutions support an innovation culture.
- Collaboration of academics with industry is very important, because the industrial customer provides the definition of the problems that need to be solved. Recognizing a problem is the first step toward finding a solution to the problem. Universities that encourage academics to consult with industry enhance innovation.
- Government policies that support business creation are helpful.
- Do not rely on experts of the 'old technology'; they are the most threatened by something new (NIH syndrome). Moreover, new inventions threaten the value of the company's assets that are based on the old technology.
- Avoid committees: committees are rewarded for avoiding mistakes rather than facilitating something new. Therefore, they normally have a negative psychology predisposition.

21.8 Predicting innovation success: the 'Preston' equation

When one of the authors, John Preston, was Director of Technology Development at MIT, he developed a formula (half as a joke) that describes the probability of success for a new business as:

$$Ps = (Pat*Qt) * (Pam*Qm) * (Pai*Qi)$$

where Ps = probability of success, Pa = passion, Q = quality, t = technologists, m = managers, i = investors and each parameter is ranked between zero and one, zero meaning it is terrible and one meaning it is perfect. When these parameters are multiplied together, all ones guarantee success, any zeros guarantee failure. Although this formula is overly harsh, it points out the key parameters that innovators need to consider. It also highlights that most successes today are a team effort of people with diverse experiences: technology, management and financial.

21.9 Future trends

The people who can best predict the future are the people shaping it today. The opportunities for innovation have never been better. These opportunities are created not only by the desire for a better quality of life for an expanding global population, but also by the stresses that a larger more affluent population creates on the environment. These stresses cover every material we need: drugs and medical devices, water, oil, gas, metals, food and air to name a few. Innovations that allow us to better utilize our natural resources become a necessity for survival of mankind. Innovations in the medical field, in turn, are key to our quality of life as they improve our health and mental wellbeing. The revolution in media technology is opening up vast unexplored opportunities to deliver new messages ranging from the movement that led to changes of a half dozen Middle East governments, to new ways to launch a new popcorn product or try on luxury gowns, watches and similar products in one's own home. In short, new communication patterns change nearly every aspect of life from government to commerce to socializing with friends. Innovators see this as an opportunity.

In order to realize this potential, we must foster the factors that promote innovation. These factors include functions of the brain that support objective thinking: recognizing problems, connecting diverse dots to find solutions and being passionately motivated to succeed. Innovation factors include regional clusters: pools of technology talent (perhaps from academia), capital sources that want to create long term value, and customers that want the best products at the best price.

Given the role of academic institutions promoting innovation in Silicon Valley and Boston, it is clear that research universities in most other parts of the world are generally underutilized as a driver for the economy, even when there is a chance of fundamental system change as in Germany after the reunification (Sabel, 1993). Many of the barriers are internal to the university, such as an ivory tower perspective and the view that industry has a corrupting influence. Such attitudes are widespread but outdated and destroy an enormous potential for doing 'good' by rationalizing the prevention of a small amount of 'bad'. The bad events that can come from university interactions with industry are numerous, but they happen very rarely. They normally center around a conflict of interest in which a researcher changes research results to match an outcome that is financially rewarding personally. These types of conflicts are easier to spot in today's transparent environment. Potentially good outcomes from interactions between academia and industry are also numerous, but they are harder to identify.

The spark of invention is usually at the core of these positive outcomes and if driven by entrepreneurs with intuitive minds this often leads to innovation. To quote Einstein again: 'The intuitive mind is a sacred gift and the rational mind is a faithful servant. We have created a society that honors the servant and has forgotten the gift'.

Innovation is our best pathway to shape a better future.

21.10 References

AMABILE, T.M. (1988) A model of creativity and innovation in organizations, in: Staw, B.M. and Cummings, L.L. (eds.), *Research in organizational behaviour*, Vol. 10, JAI Press, Greenwich (CT), pp. 123–167.

EARLYBIRD VENTURE CAPITAL (2011), personal communication.

KUHN, T.S. (1962) *The structure of scientific revolutions*, University of Chicago Press, Chicago.

KUMARAN, D. AND MAGUIRE, E.A. (2007) Match mismatch processes underlie human hippocampal responses to associative novelty. *Journal of Neuroscience*, **27**, 8517–8524.

PATTERSON, F. (2002) Great minds don't think alike? Person-level predictors of innovations at work, in: Cooper, C.L. and Robertson, I.T. (eds.), *International Review of Industrial and Organizational Psychology*, Chichester, UK, pp. 115–144.

PORTER, M.E. (1990) *The competitive advantage of nations*, Simon and Schuster, New York.

SABEL, B.A. (1993) Science reunification in Germany: 'a crash program'. *Science*, **260**, 1753–1758.

SABEL, B.A. (2008) The softer side of science. *Nature*, **455**, 694.

STERNBERG, R.J. AND LUBART, T.I. (1991) An investment theory of creativity and its development. *Human Development*, **34**, 1– 31.

UTTERBACK, J. (1994) *Mastering the dynamics of innovation*, Harvard Business School Press, Cambridge, MA.

Index

academia
innovative environments, 382–4
avoidance of bringing innovation to market, 384
source of innovation, 382–4
agri-food sector, 74–7
agricultural innovation, 362
alliance framework, 239
appropriability factor, 27–8
articulation, 191
asymmetric partnership
case studies of collaborations between companies, 287–9
Ben and Jerry's supporting socially responsible start-ups, 289
(INP) and (SiW) model, 287–8
Tate & Lyle the novel salt-reduction technology, 288
Unilever open-innovation infrastructure, 288–9
open innovation in multinational companies and SMEs, 276–91
culture, complexity and communication problems, 281–3
difficulties, 278–81
risk, reward and balance importance, 283–4
overcoming obstacles to achieve successful company partnerships, 284–7
company organisation and culture, 285–6
ongoing relationship management, 286–7
process technology readiness and shared roadmaps, 284–5
setting up the deal, 286
shared vision and agreed business model strategy, 284

balanced scorecard, 220
Before-Noccior, 32–4
biosensors, 346
Bivio Vela, 101

central location testing (CLT), 124
closed innovation, 29–32
innovation process at Lindt, 29–30
organisational units that deal with innovation and their competencies, 30
cluster organisation (CO)
Dutch agri-food sector and Food Valley Organisation, 177–83
effectiveness in open innovation in regional innovation system, 174–88
future trends, 184
theoretical background, 175–7
role in open innovation, 177
co-creation, 127–32
consumer as innovation strategy in food and beverage industry, 139–50
Molson Coors Brewing Company UK, 142–4
open innovation style, 145–6
research design, 142
co-innovation
openness and tradition, 25–36

Lindt case results, 34–6
Lindt case study method, 28–9
Noccior innovation project, 32–4
open and closed innovation at Lindt, 29–32
open innovation studies, 26–8
phases, 256–60
evaluation, 259–60
formalisation, 258–9
implementation, 259
initiation, 256–7
partner selection, 257–8
Unilever and flavour suppliers, 254–72
Flavour Operating Framework implementation and development, 262–7
future trends, 272
co-operation agreement, 82–4
cognition, 378–80
curiosity and novelty-seeking behaviour, 378–9
divergent thinking and creativity, 379–80
frustration tolerance and persistence, 380
intelligence, 380
collaboration agreement, 89
collaborative product innovation (CPI), 157–8
food service industry, 154–71
future trends, 170–1
open innovation in the food industry, 156–60
drivers, 156–7
initiatives, 158–60
main characteristics, 159
sous vide technology in the US, 160–70
combination, 191
communication, 282–3
innovation, 67, 347
community
selecting, engaging and managing partners for open innovation, 235–52
building and structuring relationships, 239–43
community building, 249–51
ecosystems, 243–6
human factors, 246–9
identifying and securing partners, 236–9
competence gaps, 53–6
competence sourcing, 54–5
complementary industry convergence, 43
complexity, 281–2
computer supported collaborative work (CSCW), 349–50
computerisation, 134
confidential disclosure agreement (CDA), 238
Connect & Develop open-innovation program, 246
consumer
co-creation value as innovation strategy in food and beverage industry, 139–50
key factors to successful open innovation, 150
Molson Coors Brewing Company UK, 142–4
open innovation style, 145–6
research design, 142
food and beverage industry innovation, 109–34
co-creators, 127–32
consumer-centric company culture, 115–17
future trends, 133–4
objective, 111–15
project process, 117–27
consumer behaviour, 111
consumer-centric corporate culture, 115–16
incorporating consumer mindset into company culture, 116
consumer-conscious employees, 116–18
consumer-driven innovation
co-creation value in food and beverage industry, 139–50
consumer-centric company culture, 115–17
consumers as co-creators, 127–32
food and beverage industry, 109–34
future trends, 133–4
Molson Coors Brewing Company UK, 142–4
organisational structure, 144
objective, 111–15
consumer-related variables influencing food and beverage choice and consumption, 112–13
environmental variables influencing food and beverage choice and consumption, 114
product-related variables influencing food and beverage choice and consumption, 113

open innovation style, 145–6
project process, 117–27
concept classification based on liking and purchase intent scores, 123
consumer-driven tolerance for a quality attribute, 126
consumer input, 118–20
overview, 117–18
preference mapping of savoury snack bars, 125
sample, 122–7
selected consumer research techniques application, 121
test design considerations, 120
tools, 120–2
consumer input, 118–20
business case, 118
concept, 118
development, 119
launch, 119
opportunity identification, 118
post launch evaluation, 119–20
validation before launch, 119
consumer segments, 113
controlled open innovation
Lindt case results, 34–6
opposing forces of the openness degree, 35
openness and tradition, 25–36
Noccior innovation project, 32–4
open and closed innovation at Lindt, 29–32
research method for Lindt case study, 28–9
review of open innovation studies, 26–8
creative process, 350–1
crowdsourcing, 249, 351–2
potential tool for data analysis, 332–40
design and management of predictive modelling competitions, 335–8
Kaggle case study, 338–9
predictive modelling competitions, 333–5
Cuisine Solutions, 168–9
Culinary Brands, 168
culture, 281
CuteFarm, 352

data analysis
crowdsourcing and online communities, 332–40
design and management of predictive modelling competitions, 335–8
Kaggle case study, 338–9
predictive modelling competitions, 333–5
deep ties, 194–5
demand articulation support, 182, 183
DEWmocracy, 131–2
DSM, 56
Dutch agri-food sector, 177–8
demand uncertainty obstructing innovation, 178

ecosystem network, 297
ego issues, 374–5
founders, 375
investors, 374–5
management, 374
environment, 115
European Patent Office (EPO), 8
evaluation phase, 259–60
exploitative ties *see* deep ties
explorative ties *see* wide ties
external innovation expenditure, 84–5
R&D department of Spanish firms and agri-food firms, 85
external knowledge acquisition
countries distribution, 19
data sources and types, 8–9
first European patent applications in agrifood patent families (1998–2006), 9
external knowledge sources, 19
firm differences, 19–20
open innovation trends in food industry, 3–20
technological knowledge types, 19
External Partner Development (XPD), 243
extrinsic motivation, 376

Flanders' FOOD, 198, 206–7
Flavour Operating Framework Partnership
implementation and development, 262–7
implementation phase, 265–7
initiation phase, 262–3
partner-selection phase, 263–4
partnership-evaluation phase, 267
partnership-formalisation phase, 264–5
food and beverage industry
co-creation value with consumer

innovation strategy, 139–50
key factors to successful open innovation, 150
Molson Coors Brewing Company UK, 142–4
open innovation style, 145–6
research design, 142
consumer-driven innovation, 109–34
consumer-centric company culture, 115–17
consumers as co-creators, 127–32
future trends, 133–4
objective, 111–15
project process, 117–27
knowledge exchange, innovation and knowledge exchange, 190–3
role of information systems in innovative organisations, 343–54
agriculture and food production technologies, 345–7
future trends, 353
innovation free tools, 352
innovation support technology, 347–52
technology in innovation, 344–5
food industry
collaborative product innovation (CPI), 154–71
future trends, 170–1
open innovation, 156–60
sous vide technology in the US, 160–70
external knowledge acquisition trends for open innovation, 3–20
countries distribution, 19
data sources and types, 8–9
external knowledge sources, 19
firm differences, 19–20
measuring open innovation, 6–8
open innovation study results, 9–18
reasons for open innovation, 4–6
technological knowledge types, 19
networks for knowledge exchange and innovation, 189–208
Flemish case studies, 195–207
food and beverage industry, 190–3
future trends, 207–8
role, 193–5
open innovation role in industry convergence, 39–58
challenges and role, 43–7
coping in the NFF sector, 51–6
evidence, 47–51
future trends, 58
studies, 40–3
types of open innovation in industry convergence, 58
food product, 114
food technology, 4
mostly incorporated technological knowledge types, 10–11
Food Valley Organisation, 174–88, 179
data and methods, 180–1
response rates for various types of companies, 180
functions, 179–80
results, 181–3
functions assessed by different types of companies, 182
functions assessed by large companies and SMEs, 181
formal interaction, 194
formal relationship, 194
formalisation phase, 258–9
'free revealing' process, 3

General Mills, 358–9
global external innovation team (GEIT), 365
Global Innovation and Quality Centre, 361
global reach, 303
globalisation, 27

Heinz
case study of strategy, business and organisation, 360–6
food quality and safety, 360
innovation, 362
open-innovation model, 363–5
organising and managing open innovation, 365–6
relationship with suppliers, 360–1
research and development (R & D), 361
effective organisational and managerial framework for open innovation, 356–66
managerial implications, 366
need for support tools, 357–9
home use testing (HUT), 124
hub and spokes network, 296–7
hybrid products, 51
Hypermodern Cuisine, 163–4
idea-driven innovation, 68
inbound open innovation, 6
industry convergence

challenges and open innovation
role, 43–7
coping, 45–7
'industry recipes' and converging sectors, 45
missing competence and path-dependency, 44
standards and absorptive capacity, 44–5
evidence in food and pharmaceuticals, 47–51
regulations, 50–1
future trends, 58
open innovation role in food and pharmaceutical industries, 39–58
studies, 40–3
input-side and output-side convergence, 41
patterns, 40–1
regulation, 42–3
special situation for innovation, 43
informal interaction, 194
informal relationship, 194
information and communication technology (ICT), 345, 348–9
information systems
innovation support technology, 347–52
computer supported collaborative work (CSCW), 349–50
information and communication technology (ICT), 348–9
role in innovative food and beverage organisations, 343–54
agriculture and food production technologies, 345–7
future trends, 353
innovation free tools, 352
technology in innovation, 344–5
innovation
information systems in food and beverage organisations, 343–54
agriculture and food production technologies, 345–7
free tools, 352
future trends, 353
support technology, 347–52
technology in innovation, 344–5
innovation acceleration, 67
innovation co-operation, 82–4
breadth, 84
degree in Spanish firms and agri-food firms, 83
innovation ecosystem, 67
innovation funnel, 30
innovation partnerships (INP), 287
innovation performance
agri-food sector, 74–7
impact of open innovation, 74–92
measurement, 219
open innovation in Spanish agri-food firms, 80–5
external information search, 82
idea sources, 81–2
openness effect, 85–91
firm innovative performance for agri-food firms, 90
firm innovative performance for Spanish firms in general, 88
variables correlation analysis on firm behaviour, 87
Spanish agri-food firms, 77–80
innovation process support, 182
innovation system (IS), 175, 176
institutional barrier, 100
intellectual property, 28, 70, 99, 106, 150, 242, 320, 359
management, 278
intellectual property rights (IPR), 264
intermediation
company innovation research, 64–5
future trends, 71–2
collaborative sustainability, 72
innovation portals, 72
value chain open innovation, 72
idea-driven to need-driven innovation, 68
Kraft melt-chocolate bars, 68–70
open innovation capabilities, 65–8
company growth, 65–6
open innovation cycle acceleration, 63–72
internal innovation expenditure, 84–5
R&D department of Spanish firms and agri-food firms, 85
internalisation, 191
International Flavours and Fragrances (IFF), 158
International Patent Classification (IPC), 8
internationalisation, 101–2
interrupted catering system, 162–3
intrinsic motivation, 376
IRRISTAT, 352

Kaggle, 338–9
data modelling contests typology, 339
knowledge accumulation, 191
Knowledge Discovery and Data Mining (KDD) Challenge Cup, 334

knowledge diversity, 337
knowledge exchange
  Flemish case studies, 195–207
    Belgian networks studied, 197
    network role in food industry innovation, 201–3
    networks process, 198–201
    results, 203–7
    sections of semistructured interview guide and main respondent categories, 196
    similarities and differences highlights, 204–5
  future trends, 207–8
  knowledge exchange and innovation in food and beverage industry, 190–3
  networks for innovation in the food industry, 189–208
  role, 193–5
knowledge exploitation, 192
knowledge intensive business service (KIBS), 97–9
knowledge transformation, 192
Kraft, 68–70

laddering, 123
large company perspective
  difficulties of open innovation, 278–9
    brand protection, 278
    culture, 279
    financial stability, 279
    managing intellectual property, 278
    technology readiness, 279
lead users, 224
learning, 190
Likert Scale, 180
Lindt
  controlled open innovation l approach, 25–36
    case results, 34–6
    case study method, 28–9
    Noccior innovation project, 32–4
    open and closed innovation, 29–32
    open innovation studies, 26–8

Magic corner, 321
maitre chocolatiers, 31
market-driven output-side convergence, 42
  NFF sector, 49–50
    food and drug market trends, 49
    product functions of nutraceuticals and functional foods *vs.* drugs, 50
Mars, Incorporated
  want find get manage (WFGM) framework for open-innovation management
    open-innovation framework, 322–28
    open innovation history, 317
    open-innovation model, 317–21
    want find get manage (WFGM) framework for open-innovation management, 315–30
master joint development agreement (MJDA), 359
means–end chain-theory, 123
medium-sized enterprise (SME)
  knowledge exchange and innovation, 192–3
melt proof chocolate bar, 68–70
mergers and acquisitions (M&A), 54, 55
Molecular Gastronomy *see* Hypermodern Cuisine
Molson Coors Brewing Company
  UK, 139–50, 142–4
  organisational structure, 144
Mountain Dew, 131–2
multinational companies
  difficulties, 278–81
    large company perspective, 278–9
    small-company perspective, 279–81
  food and beverage, challenges of strategic external networks for open innovation, 294–310
    discussion, 304–5
    future trends, 305
    recommendations, 305–10
    research methodology, 298–9
    results, 299–304
  SMEs and, asymmetric partnership in open innovation, 276–91
    collaborations between companies, 287–9
    culture, complexity and communication problems, 281–3
    overcoming obstacles to achieve successful company partnerships, 284–7
    risk, reward and balance importance, 283–4

national innovation system, 175–6
need-driven innovation, 65, 66–8, 68
Nestlé Health Sciences, 55
networks
  Flemish case studies, 195–207
    Belgian networks studied, 197
    process, 198–201

results, 203–7
role in food industry innovation, 201–3
sections of semistructured interview guide and main respondent categories, 196
similarities and differences highlights, 204–5
future trends, 207–8
knowledge exchange and innovation in food and beverage industry, 190–3
knowledge exchange and innovation in food industry, 189–208
role in knowledge exchange and innovation, 193–5
New Cookery *see* Hypermodern Cuisine
New Food Economy, 5
new product development (NPD), 46–7, 53, 110, 117
Noccior, 32–4
phases and involved partners, 34
non-patent citation (NPC), 7–8, 9–10, 10–11, 11–12, 16
non-pecuniary open innovation, 6
nondisclosure agreement (NDA), 278, 321
Novartis, 56
nutraceuticals, 47–8
emerging inter-industry segment of NFF, 48
food and pharmaceuticals convergence, 48
nutraceuticals and functional food (NFF) sector, 47, 57
market-driven output-side convergence, 49–50
open innovation to cope with convergence, 51–6
technology-driven input-side convergence, 48–9
nutrigenomics, 49
nutrigenomics organisation (NuGO), 54
nutrition innovation, 362

online communities
crowdsourcing tool for data analysis, 332–40
design and management of predictive modelling competitions, 335–8
Kaggle case study, 338–9
predictive modelling competitions, 333–5
open business models, 241
Open CASCADE, 352
open communication, 266
open innovation, 29–32, 155, 217
asymmetric partnership in multinational companies and SMEs, 276–91
collaborations between companies, 287–9
culture, complexity and communication problems, 281–3
difficulties, 278–81
importance of partnership, 276–8
overcoming obstacles to achieve successful company partnerships, 284–7
risk, reward and balance importance, 283–4
cluster organisation effectiveness in regional innovation system, 174–88
Dutch agri-food sector and Food Valley Organisation, 177–83
future trends, 184
theoretical background, 175–7
convergence-related challenges and role, 43–7
coping, 45–7
'industry recipes' and converging sectors, 45
missing competence and path-dependency, 44
standards and absorptive capacity, 44–5
coping with convergence in the NFF sector, 51–6
close competence gaps, 53–6
competence differences of actors, 52
competence gaps arising in NPD, 53
missing competence, 52–3
countries distribution, 19
cycle acceleration through intermediation, 63–72
capabilities, 65–8
company research, 64–5
creating good environment, 70
future trends, 71–2
idea-driven to need-driven innovation, 68
Kraft melt-chocolate bars, 68–70
ownership of the problem, 70
resource allocation, 71
span of control, 70–1
timelines and budget, 71
data sources and types, 8–9
first European patent applications in

agri-food patent families (1998–2006), 9
discover style, 145–6
effective organisational and managerial framework, 356–66
case study of Heinz's strategy, business and organisation, 360–6
managerial implications, 366
need for support tools, 357–9
experimentation, 31
external collaboration challenges in multinational food and beverage corporation, 294–310
discussion, 304–5
future trends, 305
recommendations, 305–10
research methodology, 298–9
results, 299–304
external knowledge acquisition trends in food industry, 3–20
firm differences, 19–20
Flemish case studies, 195–207
Belgian networks studied, 197
network role in food industry, 201–3
process of knowledge exchange in networks, 198–201
results, 203–7
sections of semistructured interview guide and main respondent categories, 196
similarities and differences highlights, 204–5
future trends, 58, 207–8
impact on innovation performance, 74–92
agri-food sector, 74–7
measurement in Spanish agri-food firms, 80–5
openness effect, 85–91
Spanish agri-food firms, 77–80
influence of soft factors, 373–80
anxiety and fear or lightness of being, 373–4
personality factors, 373
innovation process at Lindt, 29–30
organisational units that deal with innovation and their competencies, 30
knowledge exchange and innovation in food and beverage industry, 190–3
manufacturing, 31–2
Lindt open innovation funnel, 32
measurement, 6–8
networks for knowledge exchange in the food industry, 189–208
new concept generation, 31
new developments, 31
packaging, 31
phases, 371–2
practices in the food industry, 156–60
CPI initiatives in the food industry, 158–60
drivers, 156–7
main characteristics, 159
psychology, 369–87
environments of academia, 382–4
future trends, 386–7
start-ups and SME's, 385
success prediction and Preston equation, 385–6
team, 380–2
public research centres, private technical and scientific service providers, 97–107
actors in the technology market, 98
managerial implications, 105–7
reasons, 4–6
role, 193–5
role in food and pharmaceutical industries convergence, 39–58
selecting, engaging and managing partners from transactions to communities, 235–52
building and structuring relationships, 239–43
community building, 249–51
ecosystems, 243–6
human factors, 246–9
identifying and securing partners, 236–9
soft factors influence on innovation success
ambition, 377
cognition, 378–80
communication and charisma, 377
creative urge and innovative power, 377
ego issues, 374–5
passion and other motivators, 376–7
team spirit, 377–8
strategic external networks, 294–310
studies review, 26–8
external factors, 26–7
firm-specific factors, 27–8
study results, 9–18
differences across countries, 11–13
differences across firms, 13–15

patents and applicant citations by country, 12
top patenting companies and citations by sector (1998–2006), 14–15
want find get manage (WFGM) framework of Mars, Incorporated, 315–30
framework, 322–28
history at Mars, 317
Mars model, 317–21
open-innovation academy, 247
open-innovation community
cause and effect linkages between performance parameters, 227–8
exemplary strategy map, 227
strategy map relationships, 227–8
financial perspective, 225
goals and performance measures, 225
innovation-performance measurement, 219
learning perspective, 224–5
customer's goals and performance measures, 225
member perspective, 223–4
goals and performance measures, 223
process perspective, 220, 222–3
goals and performance measures, 222
scorecard development, 215–34
future trends, 229–30
implementation, 228–9
overview, 217–19
schematic diagram, 221
shortcomings of existing approach, 220
virtual communities, 217–19
building and management model, 218
open-innovation model
Heinz, 363–5
schematic diagram, 364
openness degree, 26
OpenSim, 352
organisation framework
effective managerial framework to support open innovation, 356–66
case study of Heinz's strategy, business and organisation, 360–6
managerial implications, 366
need for support tools, 357–9
outbound open innovation, 6–7
packaging, 125
packaging innovation, 362
partnership
building and structuring relationships, 239–43
defining boundaries, 240
intellectual property (IP), 242
organisation structuring, 242–3
small company perspective, 243
spoils sharing, 241
strategic alliances, 239–40
ecosystems, 243–6
building, 244
physical, 245–6
supplier networks, 244–5
human factors, 246–9
choosing and training people, 247
communication and conflict, 247–9
I win vs we win, 246
maintaining relationship, 249
identifying and securing partners, 236–9
cultural fit assessment, 238
identification and selection criteria, 237–8
partner definition, 237
readiness to build partnerships, 237
strategic context, 236
managing co-innovation partnership, 254–72
co-innovation between flavour suppliers, 260–2
Flavour Operating Framework implementation and development, 262–7
future trends, 272
recommendations, 271
success factors, 269–71
selecting, engaging and managing for open innovation, 235–52
community building, 249–51
patent citation (PC), 6, 7–8, 9–10, 10–11, 11–12, 13, 16
Patent Statistical Database, 8–9
patented food technology, 5
knowledge source, 9–10
origin of citations in agrifood patents by sectors (1998–2006), 10
open innovation trend, 16–18
external knowledge incorporation in agri-food sectors, 16, 17
external knowledge incorporation in sector 14, 17

external knowledge incorporation in sector 25, 18
external knowledge incorporation in sector 14-25, 18
path dependence, 238–9
pecuniary open innovation, 6
pharmaceutical industry
open innovation role in industry convergence, 39–58
challenges and role, 43–7
coping in the NFF sector, 51–6
evidence, 47–51
future trends, 58
studies, 40–3
types of open innovation in industry convergence, 58
PLATO, 198, 199, 202, 206–7
predictive modelling competitions, 333–5
design and management, 335–8
identification of right crowds, 335–7
right innovation problem, 337
security and privacy implications in data mining, 337–8
distinguishing features and ideas competitions, 335
preference mapping, 124
Preston equation, 385–6
priori, 222
private technical and scientific service providers
open innovation, 97–107
role, 99–100
psychology
innovation team, 380–2
academia and commercial/financial world gap, 381
features of academic and business world, 382
investor relations, 381–2
team psychology, 380–1
open innovation, 369–87
environments of academia, 382–4
future trends, 386–7
influence of soft factors, 373–80
phases, 371–2
start-ups and SME's, 385
success prediction and Preston equation, 385–6
public research centres, 98
open innovation, 97–107
role, 100–1

qualitative consumer testing, 123
quality assurance (QA), 119
quality control (QC), 119
quantitative concept testing, 123

regional innovation system (RIS)
cluster organisation effectiveness in open innovation facilitation, 174–88
Dutch agri-food sector and Food Valley Organisation, 177–83
future trends, 184
theoretical background, 175–7
relational alignment, 257–8
research and development (R&D), 26–8, 64, 78, 84–5, 102, 103, 110–11, 117
collaboration in the NFF sector, 55–6
number of patents to industrial background of filling companies, 55
networks for developing new technology platforms for NFF sector, 54
research methodology, 298–9
resource-based view (RBV), 45
Riso Scotti, 97–107, 101–5
innovation, 102
managerial implications, 105–7
open innovation approach, 103–5
illustration, 105
process innovation, 103
product innovation, 102–3

SAPiens research project, 228
science–technology relationship, 6
scorecard
implementation, 228–9
data sources of success measures, 232–3
member survey, 234
open-innovation community management, 215–34
development, 219–28
future trends, 229–30
overview, 217–19
senior buy-in, 300
sharing is winning (SiW) model, 287, 359
simulated test market (STM), 126
simulation modelling, 346
small-company perspective
difficulties of open innovation, 279–81
absorptive capacity, 280–1
changing points of contact, 280
company roles, 280
how to engage, 279
lack of start-up knowledge, 281

power imbalance, 280
slow decision cycles, 280
small to medium enterprise (SME)
difficulties, 278–81
large company perspective, 278–9
small-company perspective, 279–81
multinational companies and, asymmetric partnership in open innovation, 276–91
collaborations between companies, 287–9
culture, complexity and communication problems, 281–3
overcoming obstacles to achieve successful company partnerships, 284–7
risk, reward and balance importance, 283–4
small to medium-sized enterprise (SME), 74–6, 175, 181–3, 183–4
socialisation, 191
SODA-LO, 288
soft skills, 247
sous vide, 160–70
case study, 165, 168–70
R&D effort and net sales of Cuisine Solutions, 169
methodology, 164–5
case information, 166–7
overview, 160–4
cooking process, 161
food supply chain, 162
Spanish agri-food firms, 74–92
innovation, 77–80
input, 78
output, 80
percentage of firms, products and process, 79
open innovation measurement, 80–5
external information search, 82
idea sources, 81–2
specialisation, 99–100
spillovers, 3
staff turnover, 300
strategic alliances, 239–40
partnership governance factors, 240
strategic external networks, 294–310
benefits and drawbacks, 297–8
definition, 295
open innovation of multinational food and beverage corporations, 294–310
discussion, 304–5
future trends, 305
hub and spokes network and ecosystem network, 297
recommendations, 305–10
research methodology, 298–9
purpose, 296
results, 299–304
challenges, 300–4
current status for open innovation in food and beverage industry, 299–300
types, 296–7
strategic open-innovation networks, 295
strategical alignment, 257–8
substitutive industry convergence, 43

T profile, 247
tactical open-innovation networks, 295
talking can, 139–50
co-creation of value with consumers, 141–2, 146–9
value creation innovation by open innovation, 147
open innovation style, 145–6
research design, 142
target consumer, 113
technological alignment, 257–8
technological convergence, 27
Technological Innovation Panel (PITEC), 77, 79
technological intensity, 26–7
technological knowledge, 3, 4, 5
types mostly incorporated in food technology, 10–11
applicant citations by sector and cited technological fields (1998–2006), 11
technological turbulence, 27
technology-driven input-side convergence, 41–2
NFF sector, 48–9
technology ladder, 148
ten-step guide, 306–10
candidates selection, 308
definition and candidate selection, 309
define the need, 306
develop clear selection criteria, 306
partner selection criteria attributes, 307
engagement process, 308
evaluate and rank partners, 308
external audit, 307–8
internal audit and gap analysis, 306–7

network review, 309–10
nurture the network, 308–9
refresh network periodically, 310
running network steps, 310
thermochromic device, 148

Unilever
co-innovation phases, 256–60
evaluation, 259–60
formalisation, 258–9
implementation, 259
initiation, 256–7
partner selection, 257–8
Flavour Operating Framework
Partnership implementation and development, 262–7
future trends, 272
managing co-innovation partnerships, 254–72
co-innovation between flavour suppliers, 260–2
university, 98, 100–1
university–industry collaboration, 100, 105

VLAZ, 198, 201, 203, 206–7

W. R. Grace, 165, 168
WANT, 143, 146, 151
want find get manage (WFGM) framework, 358
open-innovation framework, 322–8
decision making process, 324
Find and selecting the right partner, 325
Get and setting up win-won solutions, 326–8
getting access to right competencies, 327
Manage and making the most out of relationships, 328
Mars ecosystem mapping, 325
want find get manage model, 322
Want and creating consumer value, 322–5
open-innovation management of Mars, Incorporated, 315–30
history, 317
open-innovation framework, 322–8
open-innovation model, 318–22
Mars' global open-innovation structure, 319
organisational structure, 318
reward and incentive systems, 321–2
team leading cultural change, 319, 320–1
Web 2.0, 350–1
wide ties, 195
win–win strategy, 149

Xpholon, 352